ANCIENT INVERTEBRATES
A N D THEIR LIVING RELATIVES

Harold L. Levin
Washington University

PRENTICE HALL, Upper Saddle River, New Jersey 07458

Library of Congress Cataloging-in-Publication Data

Levin, Harold L. (Harold Leonard),
    Ancient invertebrates nad their living relatives / Harold L. Levin. —
1st ed.

      p.   cm.
    Includes bibliographical references and index.
    ISBN 0-13-748955-2
    1. Invertebrates, Fossil.   2. Invertebrates. I. Title
QE770.L485 1999
562 — dc21   98-2690
                                    CIP

Executive Editor: Robert A. McConnin
Executive Managing Editor: Kathleen Schiaparelli
Production Editor: Kim Dellas
Page Layout: Richard Foster, Kim Dellas
Manufacturing Manager: Trudy Pisciotti
Cover Designer: Bruce Kenselaar
Cover Photo: Photograph of *Isoteles* courtesy of Wards Natural Science Establishment

The author holds the copyright to all photos/illustrations that do not have a credit line.

© 1999 by Prentice-Hall, Inc.
Simon & Schuster/A Viacom Company
Upper Saddle River, NJ 07458

Printed in the United States of America

10 9 8 7 6 5 4 3 2 1

ISBN 0-13-748955-2

Prentice-Hall International (UK) Limited, *London*
Prentice-Hall of Australia Pty. Limited, *Sydney*
Prentice-Hall Canada Inc., *Toronto*
Prentice-Hall Hispanoamericana, S.A., *Mexico City*
Prentice-Hall of India Private Limited, *New Delhi*
Prentice-Hall of Japan, Inc., *Tokyo*
Simon & Schuster Asia Pte. Ltd., *Singapore*
Editora Prentice-Hall do Brasil, Ltda., *Rio de Janeiro*

Contents

Preface

Students in introductory geology courses are taught that William Smith used assemblages of fossil invertebrates to identify, correlate, and map geologic formations. By 1811, he had published his basic concepts in a work titled *Strata Identified by Organized Fossils*. Since that time, courses in invertebrate paleontology have been a standard component of the curriculum required for an academic major in geology. This book is intended as an introduction to the basics of invertebrate paleontology so that students will be able to recognize the more important and abundant fossils in the field and to understand how fossils can be used in stratigraphic correlation and interpretation of sedimentary environments. The text is written for the undergraduate student who has never had a course in paleontology, and who may have little background in biology. For this reason, an attempt is made to keep the language simple and to eliminate all but essential terminology.

Early chapters in *Ancient Invertebrates and Their Living Relatives* provide background about fossils and about the processes responsible for past and present changes in the diversity of life. The value of fossils for determining the geologic age of strata, their value for investigating ancient environments, and for delineating the paleogeography of the past are examined. Mass extinctions and their causes, and changes in the total aspect of life through time, are also discussed in these early chapters. Chapter 3 surveys the oldest fossil record of life and describes the beginnings from which the rich array of subsequent invertebrates sprang with apparent abruptness. The text then proceeds with a phylum-by-phylum discussion of those invertebrate groups that have left a significant fossil record.

As suggested by the title of the book, the examination of major groups of fossil invertebrates is accompanied by information about their living relatives. Because students relate to invertebrates in today's world, this correlation of past to present animals enhances interest and provides relevance. The information is also useful to biology students seeking an introduction to invertebrate zoology. The text is copiously illustrated with over 250 figures. For uniformity of style and conformity to text, most of the illustrations have been redrawn by the author from published sources.

Acknowledgments

*F*ossil Invertebrates and Their Living Relatives could not have been written without extensive recourse to the important contributions of those at the forefront of paleontological research. Their published works on particular taxonomic groups were indispensible. I am also grateful for the instruction I received from the volumes of *The Treatise on Invertebrate Paleontology*, and The Paleontological Society's *Special Publications* and *Short Courses in Paleontology*. The text benefited greatly from an insightful and diligent group of reviewers who forced me to rethink explanations and discover errors. I am also deeply grateful for the assistance and direction provided by my friends at Prentice Hall, and for their professionalism and commitment to quality. Executive editor Robert A. McConnin provided the stimulus and encouragement that moved the project forward to the production stage, and production editor Kim Dellas guided the work from manuscript to galleys, to page proofs, and to finished product.

My thanks conclude with a special note of gratitude to my wife, Kay, who has cheerfully endured my lengthy preoccupation with writing this text.

CHAPTER 1

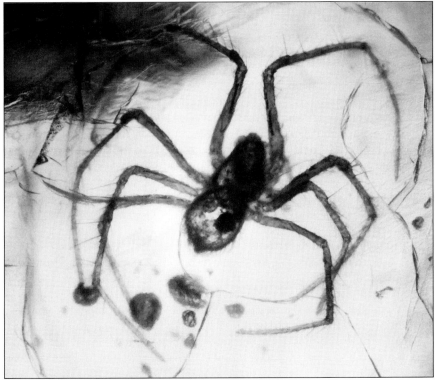

Fossil spider preserved for approximately 47 million years in amber. The specimen was taken from the Claiborne Group in Arkansas. (Courtesy of W. B. Saunders.)

FOSSILS AND EVOLUTION

There is nothing constant in the universe, all ebb and flow, and every shape that's born bears in its womb the seeds of change.

Ovid (43 B.C.–17 A.D.), *Metamorphosis, XV*

Fossils are the remains or tangible indications of life of the geologic past. They may consist of the variously altered remains of parts of plants or animals that one can hold in one's hand, or consist of indirect evidence of the former presence of organisms such as tracks, burrows, or merely imprints. Fossils have extraordinary value to Earth scientists, for without them little would be known about the history of life on our planet.

Fossils are the objects that form the basis for the interesting science of ancient life known as **paleontology**. That people find paleontology intriguing is clearly evident from the frequency with which fossil discoveries and interpretations grab the headlines in newspapers and magazines. These articles, however, may not convey a sense of the impressive breadth of paleontology. Paleontologists examine the spatial and temporal distribution of organisms to discover the intricate pathways by which the organisms evolved, how groups of organisms are related to one another, how they interact with neighboring life forms, and how they respond and adapt to their physical environment. Some paleontologists study only animals (paleozoologists, vertebrate paleontologists, invertebrate paleontologists), and some only plants (paleobotanists). Others examine postmortem changes in organisms (taphonomists), some use fossils to date and correlate strata (biostratigraphers), and some use them to decipher environmental conditions that existed long ago. As with any science, paleontology does not exist in isolation. It is closely allied to biology, profiting greatly from knowledge of the anatomy and habits of existing species, and providing the biologist with evidence of evolution. In certain areas of study, the fields of biology and paleontology merge, so that many paleontologists prefer to designate themselves as paleobiologists. Paleontology is also a partner to geology. It provides a way to find the relative geologic ages of strata, correlate strata from place to place, and determine the conditions under which the strata were laid down. For this reason, paleontology has historically been considered a specialized field within geology.

Because fossils are the indispensible ingredients of paleontology, it is appropriate to begin this discourse about ancient invertebrates and their living relatives with a discussion of how fossils are formed, how interpretations based on fossils are affected by uncertainties of preservation, how fossils are classified, and how fossilized organisms evolved.

FOSSILS AND PRESERVATION

The study of all the processes that occur between the death and fossilization of an organism, including the cause and manner of death, is called **taphonomy**. In this regard, paleontologists use the term **thanatocoenose** to define an assemblage of fossils that formerly lived together and were preserved where they lived. For a community of living organisms, the term **biocoenose** is employed. Determining the cause of death is an often difficult task. In most cases, fossil invertebrates do not provide evidence of why they died. Was it old age, toxins, disease, changes in temperature, salinity, light retardation because of turbidity, suffocation beneath a heavy influx of sediment, or the attack of one or more predators (see Box 1.1)? Whatever the cause of death, it is certain that many things may happen to an organism after it dies. It may be immediately

BOX 1.1   SCARS OF PREDATION

The preservation of evidence of predator and prey relations is not commonly found in fossils. When viewing an *in situ* assemblage of fossil marine invertebrates in their rock matrix, for example, it is often difficult to determine if death came as a natural consequence of aging, or whether toxins, disease, changes in the temperature or salinity, suffocation beneath a heavy influx of sediment, or some other factor or event was the cause. Occasionally, direct evidence of death by predation can be found. In 1995, for example, R. H. Mapes, M. S. Sims, and D. R. Boardman reported finding evidence of shark attacks on cephalopods that lived in a seaway covering part of the central United States about 280 million years ago. As will be described later in this text, cephalopods are marine mollusks with a tentacled head and shell divided into chambers by partitions called *septa*. The cephalopod prey, individuals of the species *Gonioloboceras goniolobum*, had their shells punctured by nearly circular wounds. The size and location of the punctures correspond exactly to the sharp, daggerlike central cusps of a contemporary ancient shark known as *Symmorium reniforme*. The brittle shell material around the perimeter of the punctures was inwardly crushed and surrounded by both radial and concentric fractures. It was apparent that the shark's mode of attack was to break away part of the victim's living chamber in order to get at the soft body parts inside. As is the habit of some present day sharks, the cephalopod providing the meal was probably vigorously shaken so as to dislodge the soft tissue. Once abandoned, the damaged shell chambers would fill with water and sink to the seafloor to become part of the thanatocoenose.

Predators other than sharks also must have preyed on cephalopods. In 1960, E.G. Kauffman and R.V. Kesling described the shell of the large Cretaceous cephalopod *Placenticeras* bearing puncture wounds that corresponded in size, shape, and pattern to the teeth in the jaws of the giant marine lizard *Mosasaurus*.

consumed and recycled into other organisms that benefit from the store of nutrients in its body. It may suffer oxidation or other forms of chemical attack, or it may be destroyed by abrasion, crushing, disarticulation, and fragmentation. Fortunately, some conditions inhibit such destructive processes and promote preservation. One of these is burial beneath sediment that falls onto the ocean floor or that is deposited by streams or wind. More rarely, burial may result when organisms are trapped in tar, resins of trees, or engulfed in falls of volcanic ash.

In addition to burial, the possession of a hard mineralized skeleton improves the probability of preservation. After death, flesh and other soft tissues are likely to be destroyed by bacteria, scavengers, and chemical decay. Bone and shell, however, are often left behind. If sufficiently durable and resistant to dissolution and abrasion, skeletal parts may remain at least partially intact for millions of years. More often, however, chemical changes occur that either destroy the fossil or alter its composition. Certain marine invertebrates, for example, secrete calcium carbonate in the form of the mineral aragonite. Aragonite is relatively unstable and tends to dissolve or recrystallize to form the more durable calcium carbonate mineral calcite. Although calcite is more stable, the process of recrystallization may destroy original structures in the shell, many of which are needed for precise identification of the organism.

A significant number of fossils of animals and plants from the geologic past have been preserved by dissolution of original shell substance with essentially simultaneous deposition of new and different mineral matter. The process is called **replacement**, because as each particle of the original material was removed, it was replaced by silica or calcium carbonate (or less frequently by sulfides, phosphates, or oxides of iron). Replacement is sometimes remarkably precise, preserving growth lines, relics of internal organs, and other delicate structures.

Permineralization is another process that can improve the probability of preservation. Water containing dissolved silica or calcium carbonate may have circulated through the sediment in which the organism or organic structure was entombed and precipitate mineral matter in pore spaces and cavities formerly occupied by soft parts. This in-filling process serves to strengthen bone, shell, and wood and improve its resistance to crushing caused by the weight of overlying rock or because of crustal compression associated with mountain building.

The expression "soft parts" refers to those components of an organism that are not mineralized. Soft parts are sometimes preserved as thin films of carbon that result from a process termed **carbonization**. Leaves and tissues of soft-bodied organisms such as worms and jellyfish tend to lose their volatiles (readily vaporized compounds) when compressed between layers of fine sediment. The carbon, however, may remain behind as a blackened silhouette that reveals something of the form and nature of the once living organism. In some cases, phosphate, carbonate, hematite, or pyrite may be precipitated as a film over the soft tissue, faithfully copying its texture and features. The Cambrian Burgess Shale fauna, described in Chapter 3, contains a wealth of carbonized or partially carbonized fossils that have provided a bonanza of information about life of this earliest period of the Paleozoic. The carbon residue of many of the fossils is coated with a shiny film of aluminum silicate that appears to be the residue of original clay minerals that covered the animals after death. Preservation of such soft-bodied or lightly skeletonized animals requires exceptional physical and chemical conditions, and probably involves decay-induced mineralization of tissues. Strata containing such extraordinary fossils in abundance are called **Lagerstätten**.

Fossils can also occur as **molds** and **casts**. When pressed into fine soft sediment such as clay, nearly any organic structure might leave an imprint of itself. Among many invertebrates, the shell that is enclosed in sediment may be dissolved, leaving a cavity bearing the surface features of the shell. The cavity is termed a *mold*. If the external features of the shell are impressed into the enclosing rock, the mold of the exterior of the shell can be more precisedly termed an **external mold** (Fig. 1.1). An **internal mold**, on the other hand, is a mold of the interior of the shell, and shows traces of muscle scars and various features associated with lost soft parts. Many shells enclose a hollow space. Familiar examples include the spiral cavity within the shell of a snail, or the space between the two halves (valves) of a clam shell. Internal molds develop when those cavities become filled with sediment (Fig. 1.2). The internal mold is also called a **steinkern**, meaning "stone core." Less commonly, the space between the internal and external mold may be filled with mineral matter, forming a **cast** that shows the original form of the shell (Fig. 1.3).

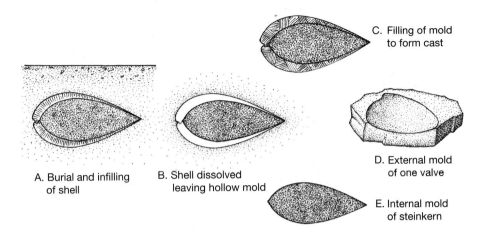

Figure 1.1   Diagram to explain molds and casts. (A) Cross section through the valves of a clam that has been buried and filled in with sediment. In (B), the original shell matter is dissolved, leaving a cavity or mold in the sediment or rock. In (C), mineral matter has been precipitated within the mold to form a cast that resembles the original shell. (D), the impression of the exterior of one valve of the clam in the enclosing sediment forms an external mold (E), the lithified sediment that once filled the original shell or the cast constitutes an internal mold.

It is not surprising that most abundant fossils consist of durable remains of hard parts. Soft tissue, however, may on rare occasions escape destruction and provide fossil remains. For such extraordinary preservation, there is little that is superior to the organic substance known as *amber* (Fig. 1.4). Amber is the hardened resin of ancient conifer trees. When the resin was exuded by the living plant, it served as a sticky trap for small organisms. Insects are the most common organisms preserved in amber, although spiders, crustaceans, snails, hair, feathers, plant remains, bacterial spores, and even small lizards and frogs have also been found. In 1992, preserved strands of DNA were obtained from an insect trapped in amber over 40 million years ago; more recently, DNA from a 125-million-year-old insect was reported.

Amber deposits are known in rocks ranging in age from Early Cretaceous to Holocene. The most famous are those found along the coast of the Baltic Sea southwest of the Gulf of Riga. Ancient stands of conifers in this area were inundated during a marine transgression in early Oligocene time. Marine sediments buried the trees along with the associated amber. Subsequently, chunks of amber eroded from the soft marine clays were distributed across adjacent areas by streams, wave action, and glaciers.

Bacterial spores preserved in amber are not the only microbes from the geologic past to have been preserved. Unicellular organisms enclosed in the siliceous rock chert are known from several places around the Earth (Fig. 1.5). Those from the Precambrian eras provide some of the earliest evidence about the nature of primordial life on our planet.

Figure 1.2 An internal mold (steinkern) of a gastropod shell. After death and decay of the animal, sediment filled the interior of the conch and hardened. Subsequently, the original calcium carbonate was dissolved, leaving behind the filling that formed the internal mold.

Figure. 1.3 Natural cast of a Devonian bivalve.

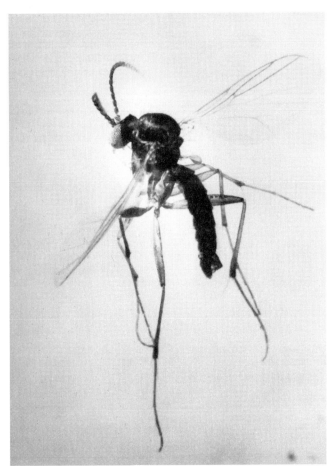

Figure 1.4   The body of a fly that has survived over 45 million years after becoming trapped in the sticky resin of an Eocene conifer. The resin hardened to amber, well known as a gem material as well as an entrapping medium for insects, spiders, and even small vertebrates. (Courtesy of W. Bruce Saunders.)

Evidence of ancient life does not always consist of the replacements, permineralizations, molds, and casts that we call **body fossils**. Much can be learned about how an animal lived or what it looked like by examining its tracks, borings, trails, burrows (Fig. 1.6), and even preserved excrement. Such markings are termed **trace fossils**, and the study of trace fossils is termed **ichnology**. Trace fossils (Fig. 1.7) of ancient invertebrates may provide an indication of the animal's size, shape, living habits, how it obtained its food, how it moved, how it rested, and how it may have escaped burial during periods of rapid deposition. In rocks lacking body fossils, trace fossils may provide the only indication that life was present at a particular time and place. For example, our oldest indications of animal life on land consist of trace fossils found in ancient soils.

Together, trace fossils and body fossils provide the documentation needed for writing the history of life. That history is not complete, for there are annoying gaps in the fossil record. The gaps are present wherever remains have been destroyed by predators, scavengers, agents of decay, or where they have been lost to erosion or destroyed by other geologic processes. Also, the fossil record does not always provide a

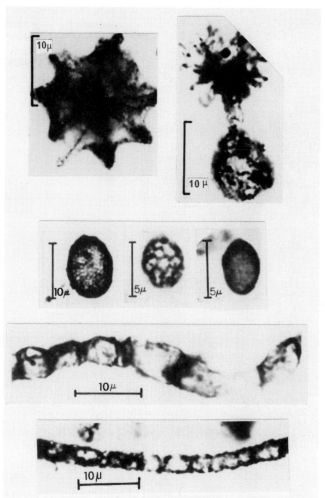

Figure 1.5   Preservation of microorganisms in chert. The fossils occur in the 1.9-billion-year-old Gunflint Chert. The two specimens at the top with umbrella-like crowns are *Kakabekia umbellata*. All three of the subspherical fossils are species of *Huroniospora*. At the bottom are two septate, filamentous species of *Gunflintia*. (Courtesy of J. Wm. Schopf.)

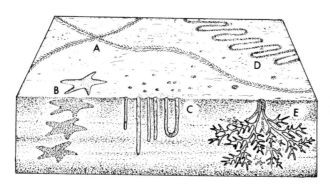

Figure 1.6   Trace fossils that reflect animal behavior. (A) crawling traces, (B) resting traces, (C) dwelling traces, (D) grazing traces, (E) feeding traces. (From H. L. Levin, *The Earth Through Time*, 5th ed. Philadelphia: Saunders College Publishing, 1966.)

Figure 1.7   Trace fossils. The fingerlike structures are the sediment-filled tubes of a worm that produced multiple burrows in order to recover particles of food efficiently in soft bottom sediment. These trace fossils occur in Lower Pennsylvanian strata near Jasper, Arkansas.

balanced view of life, for it is more complete for marine organisms than for those living on land. It is better for animals with mineralized skeletons than for creatures that lacked shell or bone. The missing parts of the fossil record, however, provide some interesting puzzles, offering many opportunities to use logic and inference in re-creating the form, function, and habits of once-living things.

THE RANK AND ORDER OF LIFE

THE LINNAEAN SYSTEM OF CLASSIFICATION

Because of the large number of living and fossil organisms, naming them in random fashion would be chaotic and inefficient. Realizing this, the Swedish naturalist Carl von Linné (1707–1778), who is also known by his latinized name Carolus Linnaeus, formulated a systematic method for naming animals and plants. As originally formulated, the Linnaean system uses morphology (the form and structure of organisms) as the basis for classification, and employs what is known as **binomial nomenclature** at the species level. In this scheme, the first name is that of the genus, and it designates a group of animals or plants that appear to be related because of their general similarity. For example, certain mollusks that lack a shell and have eight sucker-bearing arms or tentacles are placed in the genus *Octopus*. There are many kinds of octopods, and therefore the second or trivial name denotes a morphologically distinct and restricted group belonging to the genus *Octopus*. Thus, *Octopus joubini* is a dwarf species that lives in the empty shell of marine snails, whereas *Octopus vulgaris* differs in its larger size, with tentacles of some adults exceeding 2 meters in length. Linné made no attempt to indicate evolutionary relationships among the animals he classified. His purpose was to stabilize biological nomenclature by grouping together those organisms that possessed certain distinctive morphological features.

CONCEPTS INVOLVED IN CLASSIFICATION

The Species   The *species* is the fundamental category in the classification of living things. A **biological species** consists of a group of individual organisms that are morphologically similar to one another and are actually or potentially able to interbreed and produce fertile offspring. The ability of species to reproduce offspring of the same kind indicates that individuals within the species are closely related genetically. Although individuals of species are generally similar, they are not identical. They exhibit variation, and because of variation the description of a single individual cannot include the range of variations present in a species. Many kinds of individual variation may exist, including differences between the sexes or those associated with the development from juvenile to adult stages. If such differences are not recognized, one might err by applying two or more names to different members of the same species. Because the lives of extinct creatures cannot be directly observed, the problem is of particular concern to the paleontologist engaged in the naming and classification of fossil species.

Observable difference between males and females of the same species is termed **sexual dimorphism**. An example of sexual dimorphism from the fossil record of invertebrates is found among some extinct ammonoid cephalopods (Fig. 1.8 A,B) In such ammonoids, the female possesses the larger shell (conch). Perhaps a larger shell is needed to hold eggs. The larger shell is termed a **macroconch**. The male possesses a smaller, often more tightly coiled shell called a **microconch**. In some species, the microconch has lateral extensions of the aperture called **lappets**, which may have been used for clasping the female during mating.

All of us are aware of morphological differences that occur during the development of an individual from the time it is a fertilized egg to the time it has become an adult. One of many examples from the fossil record is seen among the tiny bivalved crustaceans knowns as ostracodes (Fig. 1.8 C-E). Young ostracodes grow in discontinuous stages called **instars**. As the body of the juvenile grows, the shell is periodically dis-

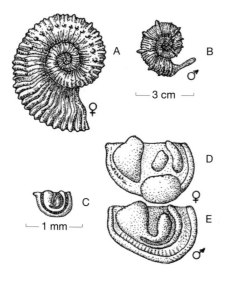

Figure 1.8   Examples of morphologic variations within species. (A,B) An illustration of sexual dimorphism of an ammonoid cephalopod (*Cosmoceras spinosum*). In their original description, the dimorphs were considered separate species. The larger shell (macroconch) is interpreted as that of the female. The lower three drawings are right valves of the ostracode *Craspedobolbina clavata*. The valve of the fifth instar (C) differs from both the adult male (D) and the adult female (E), which themselves exhibit marked sexual dimorphism.

carded and replaced by a new one. The replacement occurs about eight times before the fully adult stage is attained. The shells of the instars differ from those of the adult, not only in size and shape but often in ornamentation, hingement, and internal markings. For this reason, only the adult can be used as a basis for taxonomy (classification). Even with adult specimens, however, care is needed to recognize the sexual dimorphism that also is characteristic of ostracodes.

Yet another cause of observable morphologic differences among members of a given species involves alternation of sexual and asexual generations. An example familiar to micropaleontologists is the **alternation of generations** seen in the single-celled microorganisms called foraminifera (see Fig. 4.20). In the generation formed by the union of sex cells or gametes, the initial chamber (proloculus) is small, whereas the shell itself is large. This generation asexually produces daughter cells that grow a small shell having a large initial chamber. Here again, there is the danger that morphological dfferences between the dimorphs may be erroneously considered different species or even different genera.

One might surmise that biologists have an advantage over paleontologists in deciding on the validity of a proposal to provide a species name to a groups of similar animals. Biologists are able to work with living organisms and may be able to observe whether or not the organisms can successfully interbreed. Unfortunately, however, the breeding habits of over 80 percent of all living animals are not known. Thus, the recognition of species for both the biologist and paleontologist relies heavily on morphologic traits that are constant within the group. As a result, the definition of a new species entails some degree of subjectivity, and there will be differences of opinion among scientists as to the nature and amount of morphologic difference deemed necessary to distinguish one species from a related one. One must determine the range of variation within the proposed species and reach a well-reasoned decision about the kinds and levels of variation that can exist. For fossil species, it is often useful to compare the range of variation among individual fossils with the range of variation among members of the most closely analogous living (extant) species. When no suitable comparisons with living groups can be made, one should ascertain that the fossil species being defined possesses variations no greater than would be expected in a living population. For most groups, there are discontinuites in the range of variation that permit recognition of discrete species.

TAXONOMY

Taxonomy is concerned not only with the classification of organisms but also with the formulation of criteria for the identification of particular groups. It is used whenever organisms are named or identified. **Systematics** is sometimes regarded as equivalent to taxonomy. Systematics, however, is broader in scope in that it attempts to indicate evolutionary relationships, genetics, and speciation, as well as classification. In taxonomy, various categories of organisms are arranged in a hierarchy of groups. For example, a **genus** (plural, **genera)** is a group of species. Similarly, a **family** is a group of genera; an **order** is a group of families; a **class** is a group of orders; a **phylum** (plural **phyla**) is a group of classes; and a **kingdom** is a group of phyla. Members of each taxonomic category share a range of features by which they can be recognized. The term **domain** has recently been added to the heirarchy to encompass two or more kingdoms of organisms. To again use the octopus as an example, it is a member of the Domain Eukarya, Kingdom Animalia, Phylum Mollusca, Class Cephalopoda, Subclass Coleoidea, Order Octopoda, Family Octopodidae, Genus *Octopus* of which one of several species is *Octopus vulgaris*. The second term, *vulgaris*, is referred to as the **trivial name**. Additional categories can be formed by adding prefixes such as "super-," infra-," or "sub-." The term **taxon** (plural **taxa**) can be applied to any group in the classification. It is interesting that Linne´ did not establish such higher categories as phyla, classes, and orders. These terms were adopted after his death.

NAMING AND CLASSIFYING ORGANISMS

Because of the incredible diversity of organisms, scientists at the turn of the twentieth century recognized the need for rules to govern biological nomenclature. At international meetings they established "commissions on nomenclature," and these commissions formulated the rules. For example, *The International Code of Zoological Nomenclature* (ICZN) is the governing document for animal taxonomy. Specialists in the biology of plants refer to the rules of the *International Code of Botanical Nomenclature* (ICBN). *The International Code of Nomenclature for Bacteria* (ICNB) provides rules for naming bacteria. There are, in addition, formal codes developed for fungi, and even viruses. The codes state that no two genera in the same kingdom (or subject to the same code) may have the same name. It stipulates that scientific names must be expressed in Latin or be Latinized, and that names are printed with a distinctive typeface, such as the italics used in *Octopus vulgaris* above. Further, the name of the genus must be a single word (nominative singular) and must be capitalized. The actual name of a species is the combination of the name of the genus and the trivial name. Thus, it is a *binomen*. It begins with a lowercase letter and must agree grammatically with the generic name (the genus). To prevent duplication of names, the codes contain a *law of priority*, which stipulates that the first published name for the genus and species is the only correct name. In the event that a species in inadvertently described a second time and given a different name, the second name is "suppressed" as a junior synonym. Conversely, when the same name is used for two dissimilar organisms, only the first designation is valid. Many additional rules in the code assure uniformity in biological nomenclature.

When Linné formulated his binomial system of nomenclature, he assumed that all life could be divided into only two components, namely the Kingdom Animalia and the Kingdom Plantae. At the time, two kingdoms seemed to logically separate photosynthetic plants from mostly motile, food-ingesting animals. As biologists learned more about unicellular organisms, however, it became evident that many of these creatures were simultaneously photosynthetic, motile, and able to ingest food. The pond-dwelling flagellate *Euglena* (Fig. 1.9) is an example. To deal with organisms having such a combination of plant and animal characteristics, the biologist Ernst Haeckel proposed a third kingdom, the Protista. The term **Protoctista** is now preferred for this

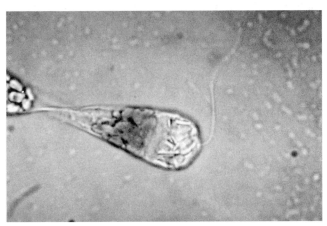

Figure 1.9 *Euglena*, a unicellular green flagellate, contains chloroplasts that enable it to carry on photosynthesis. Thus, it can live like a green plant when light is available. It can also, however, absorb organic nutrients directly, and possesses a mouth, gullet, and the ability to propel itself through water (Courtesy of H. W. Nichols.)

kingdom. The Protoctista include a vast array of yellow-green and golden-brown algae, euglenoids, dinoflagellates, and organisms once simply called protozoans.

Eventually, the three-kingdom system also proved inadequate. There exist unicellular organisms that differ from most protoctistans in that they lack a cell nucleus, lack certain cell organelles (membrane-bounded bodies in the cell that have specific metabolic functions), and reproduce only asexually. Such organisms are termed **prokaryotes**. These fundamentally important differences required the establishment of a fourth kingdom designated the **Kingdom Monera**. It was next recognized that fungi deserve separate placement. Like animals, fungi depend on a supply of organic molecules in their environment, yet they absorb nutrients through cell membranes as do plants. The **Kingdom Fungi** became the fifth division in the commonly used classification (Fig. 1.10).

Although the five-kingdom system is widely used in biology textbooks, it too has its shortcomings. It is based largely on **phenotypic traits**, which are *observable* traits that arise from genetic processes. As we have seen, phenotypic traits have been essential to biological classification and the study of evolutionary relationships. An even more convincing indicator of such relationships, however, is the structure of large molecules in the cell and in the sequence of amino acid components in such molecules as ribonucleic acid, or RNA. This molecular basis for phylogeny (see discussion later in chapter) has been particularly instructive in determining affiliations among microbial organisms, although it has provided a better understanding of the evolution of larger organisms as well. Molecular phylogeny indicates that such superficially dissimilar groups as plants, animals, and fungi are actually closely related, and that the Kingdom Monera of the five kingdom system actually contains two distinctive kinds of prokaryotic organisms. To show more validly the evolutionary relationships as revealed by molecular sequencing studies, it has been proposed that all life be divided into three domains, namely the **Bacteria**, **Archaea**, and **Eukarya**. Among the many organisms within the Domain Bacteria are cyanobacteria, along with purple sulfur bacteria, and various nonphotosynthetic groups of microbes. The Domain Archaea includes methane-producing bacteria and an interesting group of heat-loving microbes called thermophiles that populate hydrothermal vent systems. Fungi, plants, and animals with their molecular similarities are placed within the Domain Eukarya.

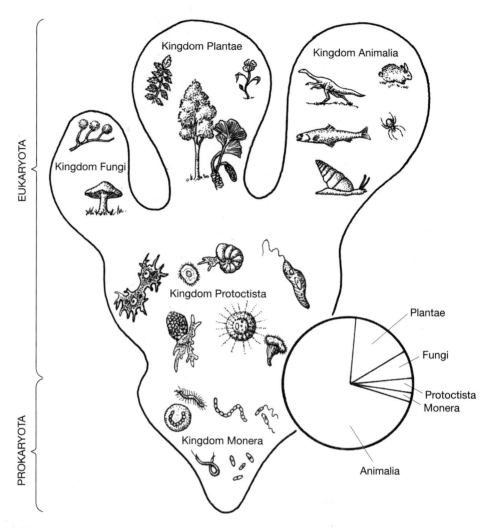

Figure 1.10 The five-kingdom system proposed by R.H. Whittaker in 1963, and the proportions of members of each kingdom living today. More recent classifications based on molecular structures classify organisms into three large categories termed *domains*, namely the Bacteria, Archaea, and Eukarya. Each domain includes two or more kingdoms. The Eukarya, for example, includes the Fungi, Plantae, and Animalia of Whittaker's classification.

EVOLUTION

"Every shape that's born bears in its womb the seeds of change," wrote the Roman poet Ovid. The phrase is prophetically apt when one considers the constant change of living things as revealed by the fossil record. Earliest life on Earth was microbial and unicellular. Change produced multicellular organisms, and these increased in complexity to form a vast array of naked and shelled invertebrates. For a time, invertebrates were the only members of the animal kingdom. Even today, they greatly surpass the vertebrates in number and diversity. Then fishes appeared, and with the further passage of time, amphibians, reptiles, mammals, and birds became established on Earth.

Within each of these great taxonomic classes, change continued, endlessly increasing the diversity of life. Yet throughout the millions of transmutations that led to diversity there has been a thread of unity, for all living creatures are descendants of forms that lived in the past. No matter how seemingly different, every organism exhibits basic similarities in its solutions to fundamental life processes. While studying the fossils in successively younger sequences of strata, the paleontologist can recognize the constant attributes of all life and also see changes unfold through time. Those changes constitute organic evolution.

PRE-DARWINIAN EVOLUTIONISTS

The concept that animals and plants have changed with the passage of time had been present long before Charles Darwin. Anaximander (611–547 B.C.) alluded to the idea 25 centuries ago. He taught that life arose from mud warmed in the Sun, that plants came first to the Earth, then animals, and finally humans. Reference to such changes was, however, rare during the Middle Ages (about A.D. 400 to 1400) because of the widely held belief in biblical Creation accomplished in 6 days. It was not until the eighteenth century that European intellectuals like Jean Baptiste de Lamarck, Georges Buffon, and Carl von Linné were able to provide convincing arguments that organisms have changed through time, even though the reason for the change was not understood. In fact, Lamarck presented a general theory of evolution that elicited considerable public attention and support from the scientific community. Lamarck suggested that all species, including humans, descended from other species. We would have no quarrel with that concept today. He further proposed, however, that new structures in an organism may arise because of the need or "inner want" of the organism, and that structures acquired in this way would somehow be inherited by later generations. Conversely, little used structures would gradually disappear in succeeding generations. Lamarck believed that snakes evolved from lizards that had a preference or need for crawling. Because of this inner need, certain lizards developed increasingly longer and more pliant bodies that were more efficient for crawling. Because legs became less and less useful in crawling, they gradually disappeared.

Although Lamarck's ideas were accepted by many during the nineteenth century, they were subsequently challenged. There was no argument about the concept that life changes through time, but Lamarck's belief that characteristics acquired during the life span of an individual could then be inherited was tested and shown to be invalid. We know that some organisms can adjust to environmental conditions in a narrowly limited way during their life span. For example, a fair-skinned person can acquire a protective tan. However, no amount of sunbathing will result in offspring born with a suntan. There is no way that our body cells (called somatic cells) can pass characteristics on to the next generation.

DARWIN'S THEORY OF NATURAL SELECTION

Charles Darwin (1809–1882) credited Lamarck for perceiving that there is change within a line of descent (a lineage) over many generations and across great lengths of time. He went further, however, when he added the important concept of *natural selection*. A colleague, Alfred R. Wallace, independently conceived of this same important idea about differences in survival and reproduction among members of a population that vary in certain traits.

The concept of natural selection was a logical outgrowth of two observations. First, there is always variation among individuals of the same species. Second, Darwin and Wallace noted that organisms have enormous reproductive capacity. They are able to produce more offspring than required for simple maintenance of their numbers. Yet even with this potential for providing huge numbers of progeny, populations do not increase infinitely. A population is limited by the survival of only a relatively small portion of its offspring. Competition for food and living space inevitably results in the elimination of weaker or less well adapted individuals and in the survival of individuals that vary in some way that makes them better able to cope with conditions in their environment. The survivors breed and provide the offspring from which the less well adapted members of subsequent generations are again culled. Thus, over many generations, useful variations will tend to prevail and to be maintained until such time as the environment itself changes. Inherent in Darwin's views about natural selection is the concept of adapation. An **adaptation** is a structural, physiological, or behavioral characteristic that promotes the probability of survival and reproductive success in a particular environment.

Processes of adaptation and natural selection are driven by more than changes in the physical environment. Conditions and changes in the biological environment also have a role. Organisms at all taxonomic levels constantly compete for food, shelter, living space, mates, and innumerable other necessities of life. Sometimes that competition is between individuals of a single population, and the individuals that compete successfully cause the perpetuation of useful adaptations by their success in breeding. Entire populations might also compete with other populations for a share of the requirements for life. Competition may be direct, as when two different predators compete for the same prey, or it may be indirect, as when organisms do not compete with one another, but rather there are too many individuals of different species dependent upon a finite supply of resources.

INHERITANCE

Although Darwin and Wallace both recognized the importance of variation in evolution, they were unable to explain the cause of the variability in a way that could be experimentally verified. The first step in finding the cause was made by a Moravian monk named Gregor Johann Mendel (1822–1884), who discovered the basic principles of inheritance. His findings, published in 1865 in an obscure journal, were unknown to Darwin, and indeed did not receive the attention of the scientific community until 1900. Mendel's report described a mechanism by which traits are transmitted from adults to offspring. Using pea plants as the basis for his experiments, Mendel demonstrated that heredity is determined by "character determiners" that divide in pollen and ovules and are combined in specific ways during fertilization. Mendel called these heredity determiners simply "factors." They have since come to be known as *genes*.

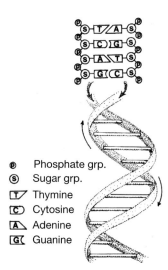

Figure 1.11   Part of a deoxyribonucleic acid molecule. The twisted side rails are composed of alternate sugar (deoxyribose) and phosphate molecules. Each rung is composed on one pair of nitrogenous bases. Of these, thymine normally links to adenine, and cytosine to guanine.

Ⓟ   Phosphate grp.

Ⓢ   Sugar grp.

T   Thymine

C   Cytosine

A   Adenine

G   Guanine

A **gene** is the basic unit of inheritance. Genes represent particular stretches of DNA (deoxyribonucleic acid) molecules. **DNA** is one of two kinds of nucleic acid found in cells (the other is ribonucleic acid, or **RNA**). Nucleic acids are long molecules composed of many similar molecular subunits. The repeating subunits in nucleic acids are called **nucleotides**, and each nucleotide consists of a five-carbon sugar, a phosphate group, and an organic nitrogen-containing base. In structure, the DNA moleule consists of two parallel strands coiled rather like the two handrails of a spiral staircase, hence the commonly used term "double helix" to describe DNA (Fig. 1.11). The handrails are constructed of the five-carbon sugar and phosphate subunits, and they are linked by the organic nitrogen-containing bases. There are four of these bases: adenine, guanine, thymine, and cytosine. In the DNA molecule, adenine (A) always attracts thymine (T), whereas guanine (G) attracts cytosine (C). In this respect DNA shows constancy, yet it also exhibits variability in that the sequence of bases in a segment will vary from one species to another. A species in the oyster family might have A-T followed by T-A, C-G, C-G, and T-A in a particular segment, while a corresponding sequence in a sea urchin might have a G-C, G-C, C-G, C-G, T-A sequence.

The vital property of DNA is its ability to replicate itself. The two nitrogenous bases forming the steps of the spiral ladder are joined only by weak hydrogen bonds. Enzymes within the cell are able to break the bonds readily, causing one side of the molecule to unwind from the other. When this occurs, nucleotides stored in the cell link to their partner nucleotides in each separated strand to reconstruct the double helix. As always, adenine will join thymine, and cytosine will link to guanine. In addition to replication, DNA is important because it controls the production of proteins by ordering protein synthesis through a series of rather similar but single-stranded ribonuncleic acid molecules. Many vital structures within organisms are formed of proteins, and the activities of organisms are precisely regulated by the specific catalytic proteins called *enzymes*. Thus, without DNA and the products it can produce there can be no life as we know it. DNA is the vital key and blueprint for life. Its ability to replicate itself precisely is the basis for heredity, and organic evolution ultimately results from structural changes in this remarkable molecule.

CELL DIVISION AND REPRODUCTION

When a cell prepares to undergo division, its DNA molecules appear extended like threads and have many protein molecules attached to them. These threads of DNA with attached proteins are called **chromosomes**. Genes are specific segments of chromosomes. In the formation of sex cells and in reproduction, chromosomes are recombined and redistributed to produce a variety of gene combinations in offspring. The result is the variability so essential to Darwinian natural selection. Except for prokaryotes, chromosomes are located in the nucleus of the cell and occur in identical (homologous) pairs. The kind and number of chromosomes are constant for any given species. In humans, for example, there are 46 chromosomes, or 23 pairs. Cells with homologous chromosomes are designated **diploid** cells. In animals and plants, new cells are produced constantly to replace worn-out or injured cells and to provide for growth. In asexual organisms, and in all somatic or body cells of sexually reproducing organisms, the process of cell division that produces new diploid cells with exact replicas of the chromosomal compliment of the parent cells is called **mitosis** (Fig. 1.12).

The amoeba is a familiar example of an asexual organism. When an amoeba has grown to a certain size, it reproduces by dividing in half to form two offspring. Before division, the organism makes a duplicate copy of its DNA, and one complete set of DNA is provided to each daughter cell. Because the DNA in the daughter cells is identical to that in the parent, each new amoeba resembles its parent, and there is little or no variability.

There is another highly important type of cell division that occurs in organisms capable of sexual reproduction. It is called **meiosis**, and it occurs when gametes (egg cells or sperm cells) are formed. Meiosis can occur in either unicellular or multicellular organisms. In the latter, it takes place only in the reproductive organs (testes or ovaries). There are two cell divisions in meiosis. In the initial division, the chromosomal pairs are divided in random fashion so that each of the potential new cells receives one member of each original pair. These daughter cells are termed **haploid** because they do not have paired chromosomes. The haploid cells then reproduce themselves exactly as in mitosis to form new haploid cells, which are gametes. When two gametes (egg and sperm) meet during sexual reproduction, the sperm enters and fertilizes the egg. Because two haploid gametes have combined into one diploid cell, there is now a full set of homologous chromosomes and genes representing a mix from both parents. The fertilized egg (zygote) then begins a process of growth by mitotic cell division that eventually leads to an adult organism.

The plant or animal that develops from the union of two haploid gametes will already have some variability relative to its parents because of the mixing of genes. There is, however, another aspect of meiosis that is important in producing variation. During the initial division while chromosomes still lie side by side, they may break at

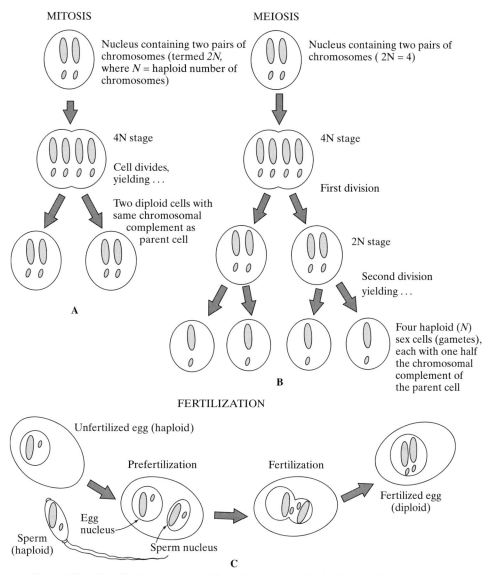

MITOSIS

MEIOSIS

Nucleus containing two pairs of chromosomes (termed *2N*, where *N* = haploid number of chromosomes)

Nucleus containing two pairs of chromosomes (2N = 4)

4N stage

Cell divides, yielding . . .

Two diploid cells with same chromosomal complement as parent cell

4N stage

First division

2N stage

Second division yielding . . .

A

Four haploid (*N*) sex cells (gametes), each with one half the chromosomal complement of the parent cell

B

FERTILIZATION

Unfertilized egg (haploid)

Prefertilization

Fertilization

Fertilized egg (diploid)

Egg nucleus

Sperm (haploid)

Sperm nucleus

C

Figure 1.12   Simplified comparison of the major features of (A) mitosis, (B) meiosis, and (C) fertilization. (From H. L. Levin, *The Earth Through Time*, 5th ed. Philadelphia: Saunders College Publishing, 1996.)

corresponding places and exchange their severed segments. The process is appropriately called *crossing over*, and it provides further mixing of genes.

We have seen that offspring commonly favor a particular trait of one parent over another. How does gene recombination explain this occurrence? We have used the term **phenotype** to refer to the outward appearance of an organism with respect to an inherited trait. Indeed, phenotype means "visible type." Mendel's plants, for example, produced both round and wrinkled pea phenotypes. The companion term **genotype** ("hereditary type") is used for the actual genetic makeup of an organism as distinct from its outward appearance. The genes brought together during fertilization may be identical or contrasting. If they are contrasting, one of the genes (the **dominant** gene) may mask the expression of its partner (the **recessive** gene). It is possible, however, for the recessive gene to be expressed if both of the combining genes are recessive. To illustrate with Mendel's experiment, we can call the gene for wrinkled peas *w* and the one for round peas *r*. The possible combinations are thus *rr*, *ww*, or *rw*. The *rr* genotype would produce a rounded pea phenotype, and *ww* would produce a wrinkled phenotype. But what of *rw*? If the offspring had rounded seeds, then the wrinkled pea gene expression was masked and would be termed recessive. The rounded trait would then be called dominant, and the combination recorded as *Rw*, with the dominant gene capitalized.

MUTATIONS

Although gene recombinations are an important source of variation in organisms, they do not account for the appearance of entirely new traits. Gene combinations involving an unchanging collection of genes are finite in number. For offspring to have a truly new trait, a process is required that will change the genes themselves. That process is **mutation**.

To understand the cause of mutations, one must look again at the structure of the DNA molecule. Recall that the base pairs adenine-thymine (A-T) form the steps of the helical ladder. Genes owe their specific characteristics to a particular order of these base pair steps. If this order is disrupted, a mutation will result. The disruption may occur during cell division when the twisted side strands of the ladder separate and attract new bases to rebuild the DNA molecule. As the rebuilding commences, there may be a mishap, such as an extra base being inserted, a base lost, or one base substituted for another in the sequence. One might, for example, have a guanine-cytosine (G-C) base pair positioned in what should have been the location of an adenine-thymine (A-T) base pair. When these events occur, the gene is altered and may result in a mutation.

Mutations may occur at more than one organizational level within genetic material. At the lowest level it may alter the building-block sequence of DNA in one or more genes as described above. At a higher level, the mutation might result from a structural alteration of entire chromosomal segments. At a still higher level, a mutation might result from an increase in the number of chromosomes. Mutations can occur in any cell, but their evolutionary impact is greatest when they occur in gametes, for then they will affect succeeding generations. They may be spontaneous and without recognizable cause, or may result from external effects such as ultraviolet (UV) radiation or radiation from the decay of radioactive isotopes. Many mutations are harmful, in that changing even a small segment of DNA can damage the steps required to produce some vital characteristic. Yet other mutations are beneficial or at least harmless.

In summary, mutations and gene combinations are the fundamental causes of variation among organisms. Without those variations there could be no evolution.

Natural selection acts upon the constantly appearing new traits, usually discarding those that have no benefit to the organism and preserving those that are useful.

EVOLUTION IN POPULATIONS

Evolution does not occur in individuals. It takes place in **populations**. A population is a group of individuals of the same species occupying a given area. Each of the individuals within the population has its own particular genotype, and the sum of all the genotypes within a breeding population constitutes the **gene pool** of the population. The gene pool is divided up in each generation and partitioned out to the new offspring. Inevitably, in each successive generation, new mutations and genetic combinations occur that manifest themselves phenotypically in the offspring. As natural selection comes into play, some of these traits will be passed on to the next generation in greater numbers, others in fewer numbers, so that ultimately the gene pool is altered. Thus, evolution results from the impact of natural selection on the gene pool.

The existence of a population depends upon the free flow of genes among its members. If part of the population becomes somehow isolated so as to interrupt the free flow of genes, new species may arise. Isolation can take different forms. In some cases it might result from a physical barrier, such as an arm of the sea that serves as a barrier between two groups of land snails and prevents their interbreeding; or a land barrier like the Panamanian land bridge that isolates Atlantic species of mollusks from sister species on the Pacific side. In other cases the isolation might be biological, as when one group changes the timing of its breeding cycle relative to the other, or for some reason becomes sexually unattractive to the other. Whatever the cause, various segments of the population become isolated for many generations. During such a period of isolation, a subpopulation may accumulate enough genetic differences so that interbreeding between the segments is no longer possible. Once this has occurred, the isolated segments have become different species. After a new species has appeared, its success will depend upon the advantages of its particular attributes relative to the environment. Some species encounter severe competition for food, shelter, living space, and other vital needs. Such conditions may shorten the life span of the species. If, however, it is able to cope well with its environment, it may increase and persist. With many subsequent speciation events, diverse organisms with varied living strategies emerge. The average duration for most species has been 5 million to 10 million years, a short time from a geological point of view.

PHYLOGENY

The term **phylogeny** refers to the historical development of groups of organisms so as to depict descent from other organisms. The depiction is usually a diagram called a **phylogenetic tree**. Branches on the tree are called **clades**. Most paleontologists today use one of two phylogenetic methods. The more traditional method is called **stratophenetic phylogeny**. In the 1960s, a newer method termed **cladistic phylogeny** was introduced.

In stratophenetic phylogeny, organisms are arranged in treelike fashion, with the most recently evolved species or groups on the upper branches and older, ancestral species on the lower branches and trunk. Thus, the tree includes the concept of change through time (Fig. 1.13A). The term *stratophenetic* is apt, for the tree relates the order of succession of life to the superpositional sequence of stratigraphic units. Stratophenetic phylogeny is dependent upon finding ancestral and descendant species in the fossil record. Unfortunately, this is not always possible. There are "missing links" where key forms are not found, so that parts of the tree may contain inferences about intermediary forms and ancestors. In some cases, cladistic phylogeny reduces some of the subjectivity inherent in stratophenetic phylogeny.

Cladistic phylogeny is a method by which organisms are analyzed objectively on the basis of shared characters to determine their ancestor-descendant relationships. The analysis is depicted on a diagram termed a **cladogram** (Fig. 1.13B). The establishment of such relationships requires recognition of two types of characters. On the one hand are **apomorphic** characters, which are advanced or derived, and on the other hand are primitive or **plesiomorphic** characters. For example, the cilia-encircled, pear-shaped trochophore larvae is a plesiomorphic character that appears in many groups of invertebrates, whereas eight-shell components is an apomorphic character found only in the molluscan group known as *chitons* (Polyplacophora). In the course of evolution, characters evolve from the plesiomorphic to the apomorphic state. In addition, apomorphic traits characterize narrow and more exclusive groups, whereas plesiomorphic traits characterize broad and more inclusive groups.

A unique apomorphic character can be termed an **autapomorphy**. It defines a group as distinctly different from all others. The rasplike radula of certain mollusks is an example of an autapomorphy. It is unique to Mollusca and does not provide a link to any other phylum. A companion term, **synapomorphy**, does not relate to a character that is unique to a particular group. The trochophore larva mentioned above is a synapomorph that occurs not only in the Mollusca, but in groups such as flatworms, nemerteans, bryozoa, phoronids, brachiopods, and even some annelids and arthropods. The trochophore is a synapomorphy that indicates some relationship among these organisms, but not necessarily a close relationship.

Yet another term used in cladistic phylogeny describes two groups that uniquely share apomorphic characters. Such groups are called **sister groups**. In Figure 1.13C, species *A* and *B* constitute a sister group. On the cladogram, three extant species (*A*, *B*, and *C*) are placed along the top. Both *A* and *B* share apomorphic character *X*, whereas *B* and *C* resemble one another in that both possess plesiomorphic character *Z* Thus, *A* and *B* are more closely related to one another than to *C*.

THE PACE OF EVOLUTION

Do changes in evolution occur gradually by means of an infinite number of subtle steps, or are there sporadic advances during which new species may appear over a short span of time? Gradual progressive change is referred to as **phyletic gradualism**. In this model of evolution, populations change from ancestors to descendants by slow degrees, and where apparent sudden changes do occur, they may be regarded not as true evolutionary breaks but merely what appear to be sudden changes, because transi-

Figure 1.13   Traditional stratophenetic phylogeny depicting extant orders of Bryozoa (A), a cladistic phylogeny of the classes of Mollusca (B), and a simple cladogram of three species in which *A* and *B* share an apomorphic characteristic and *B* and *C* share a plesiomorphic characteristic. The numbers on diagram (B) refer to the following characteristics:

1. Trochophore larvae; adult with gonopericardial complex.
2. Mantle with calcareous spicules; foot; retractor muscles; radula; 8 segments.
3. Foot with anterior guard; special radular muscles and support.

4. One shell with periostracum and nacre; trifold mantle edge, crystalline style; calcareous mantle spicules absent.
5. Eight shells.
6. Foot reduced to narrow keel.
7. Copulatory setae; internal fertilization.
8. Foot totally reduced; oral shield.

(Diagram (B) from C. Nielsen, *Animal Evolution*. Oxford: Oxford University Press, 1995.)

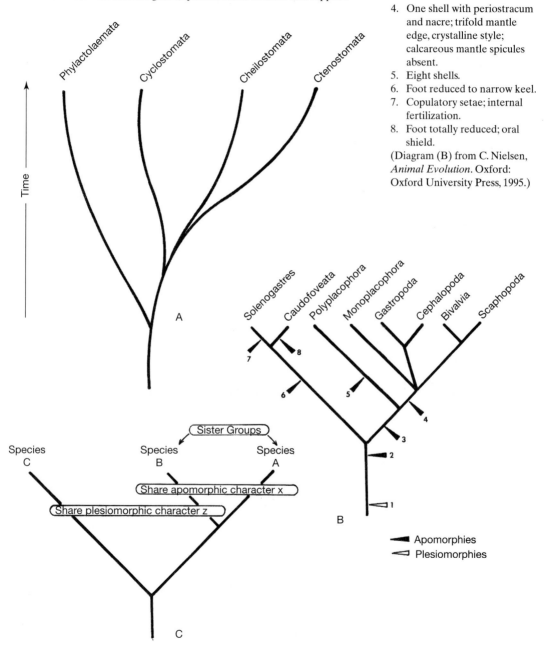

tional forms were not preserved, were destroyed by erosion after burial, or have not been discovered. To avoid these problems of possible gaps in the record and to show that phyletic gradualism is a valid concept, one would want to study a continuous sequence of sediments (no erosional gaps) containing abundant fossil organisms. Such a nearly ideal study situation exists when one examines the evolutionary changes seen in unicellular planktonic marine organisms known as foraminifera that occur in uninterrupted sections of well-dated, continuous deep sea cores. In such a study by B.A. Malgren and J.P. Kennett, based on deep-sea cores taken in the southwest Pacific, the gradualistic foraminiferal line of descent can be followed step by step, as *Globorotalia conoidea* progresses to *Globorotalia conomiozea*, *Globorotalia puncticulata*, and *Globorotalia inflata*. The complete sequence of gradual change occurred over a span of about 8 million years.

For evolution that includes sudden advances that "punctuate" long episodes of little evolutionary change (called **stasis**), the term **punctuated equilibrium** was introduced in the 1970s by paleontologists Niles Eldridge and Stephen J. Gould. It is interesting that Charles Darwin, who generally stressed gradualism in evolution, was also aware of stasis and the possibility that evolutionary advances did not always proceed in a uniform manner. In 1872 Darwin wrote, "Although each species must have passed through numerous transitional stages, it is probable that the periods during which they underwent modification, though many and long as measured in years, have been short in comparison with the periods during which each remained in unchanged condition."

The sudden morphological change that interrupts evolutionary equilibrium or stasis in the punctuated equilibrium model occurs primarily at the periphery of the area occupied by the species. The boundary members of the species are referred to as **peripheral isolates**. Gene flow is slow or nonexistent between peripheral isolates and the larger population, but is rapid among individuals of the peripheral isolates. Thus, changes in morphology or physiology leading to speciation may occur within a short span of time. The new species do not originate where their parental stock existed, but in peripheral boundary zones where they can be tested in new environmental settings. In many cases, the original parent population may suffer extinction, and the new species may move back into the old parental domain. They may also expand into a new territory, or they might move back into the parental domain and displace the parental population.

As one surveys the fossil record, it appears that examples of punctualistic equilibrium are more common, although it is likely that both styles of evolution have occurred in the past. The two modes of evolution, depicted as phylogenetic trees in Figure 1.14, need not be be mutually exclusive. The punctuated-equilibrium tree has short horizontal branches indicating traits that have changed rapidly around the time of speciation. Vertical continuation of the branch indicates stasis, or little subsequent change. In contrast, the gradualistic model shows gently inclined branches that suggest speciation occurred through gradual change in traits over geologic time.

LEVELS OF EVOLUTION

One may study evolution at two levels. The first, called **microevolution**, involves changes that occur within species that may lead to new species. Microevolution results largely from the interaction of variation, selection, and random changes in gene fre-

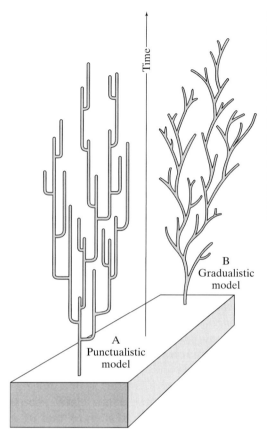

Figure 1.14   Phylogenetic trees expressing the differences between punctuated equilibrium (A) and phyletic gradualism (B). Changes in morphology or physiology are expressed by laterally directed parts of branches, and time is in the vertical direction. The abrupt horizontal branches of the punctualistic model depict sudden change, whereas inclined branches of the gradualistic model suggest slow and gradual change through time. (From H. L. Levin, *The Earth Through Time*, 5th ed. Philadelphia: Saunders College Publishing, 1996.)

quencies within a population. Thus, it is a generation to generation phenomenon that occurs over relatively short intervals of time. In contrast, **macroevolution** involves large-scale evolutionary changes marked by the appearance of higher categories of organisms, including families, orders, and classes. Also, macroevolution occurs over far greater intervals of time. "Geologic time" is involved, and the evidence of macroevolution is therefore found chiefly in the geologic record.

A common pattern in macroevolution is the rapid diversification of a parent group into many different new subgroups, each of which has become adapted to a different environmental situation or survival strategy. The process is called **adaptive radiation**. An example is the expansion and diversification of trilobites during the Ordovician Period. The radiation produced trilobite swimmers, burrowers, crawlers, and floaters, each adapted to exploit particular survival opportunities.

REVIEW QUESTIONS

1. What are fossils? How are they commonly preserved? What are some of the processes that might be expected to have occurred during the taphonomy of a fossil marine clam?

2. Compare the potential for preservation of marine invertebrates as compared to terrestrial invertebrates.

3. Distinguish between body fossils and trace fossils. What sort of information might be found in a stratum containing trace fossils of invertebrates, but lacking body fossils?

4. Why are fossils useful in the recognition of strata deposited during particular intervals of geologic time? When are fossils of particular value in biostratigraphic correlation?

5. A bed of limestone contains different marine invertebrate fossils in different geographic areas. If the limestone is the same age everywhere, what are some of the factors that would account for your finding different assemblages of fossils in different areas?

6. Define a species. Discuss the problems associated with the recognition of a fossil species. How does sexual dimorphism affect the recognition of fossil species? What rules must be followed in naming a fossil species not previously described?

7. Discuss the difference between the classification system employing five kingdoms and the system employing three domains.

8. What were the contributions of Darwin and Mendel to our understanding of organic evolution? What is the importance of mutations?

9. Distinguish between mitosis and meiosis, haploid and dipoid cells, and phenotypic and genotypic variation.

10. Distinguish between phyletic gradualism and punctuated equilibrium. In which process is stasis important? Explain the role of peripheral isolates in the concept of punctuated equilibrium.

SUPPLEMENTAL READINGS AND REFERENCES

Allen, K. & Briggs, D. (Eds.). 1989. *Evolution and the Fossil Record*. London: Belhaven Press.

Boardman, R. S., Cheetham, A. H. & Rowell, A. J. (Eds.). 1987. *Fossil Invertebrates*. Palo Alto, CA: Blackwell Scientific.

Briggs, D. E. G. & Crowther, P. R. (Eds.). 1990. *Paleobiology: A Synthesis*. Oxford: Blackwell Scientific.

Dawkins, R. 1986. *The Blind Watchmaker*. Harmondsworth, UK: Pelican.

Donovan, S. K. (Ed.). 1991. *The Processes of Fossilization*. New York: Columbia University Press.

Eldridge, N. 1991. *Fossils: The Evolution and Extinction of Species*. London: Aurum Press.

Eldridge, N. & Gould, S. J. 1972. Punctuated equilibrium: An alternative to phyletic gradualism. In T. M. Schopf (Ed.), *Models in Paleontology*. San Francisco: W. H. Freeman.

Grimaldi, D. A. 1996. Captured in amber. *Scientific American* 274(4):84–91.

Kauffman, E. G. & Kesling, R. V. 1960. An Upper Cretaceous ammonite bitten by a mosasaur. *Contributions from the Museum of Paleontology*. Ann Arbor: University of Michigan, 15:193–243.

Malgren, B. A. & Kennett, J. P. 1981. Phyletic gradualism in a Late Cenozoic planktonic foraminiferal lineage, D.S.P.D. site 284, southwest Pacific. *Paleobiology* 7(2):230–240.

Mapes, R. H., Sims, M. S. & Boardman D. R. 1995. Predation on the Pennsylvanian ammonoid *Gonioloboceras* and its implications for allochthonous vs. autochthonous accumulations of goniatites and other ammonoids. *Journal of Paleontology* 69(3):441–446.

Margulis, L. & Schwartz, K. V. 1988. *Five Kingdoms*. New York: W.H. Freeman.

Polnar, G. & Polnar, R. 1994. *The Quest for Life in Amber*. Reading, MA: Addison-Wesley.

Smith, A. 1994. *Systematics and the Fossil Record: Documenting Evolutionary Patterns*. Palo Alto, CA: Blackwell Scientific.

Runnegar, B. 1985. Molecular paleontology. *Paleontology* 29, pt.1:1–24.

Stearn, C. W. & Carroll, R. L. 1989. *Paleontology: The Record of Life*. New York: John Wiley.

Strickbergen, M. W. 1990. *Evolution*. Boston: Jones and Bartlett.

Stanley, S. M. 1979. *Macroevolution*. San Francisco: W. H. Freeman.

Wiley, E. O. 1981. *Phylogenetics: The Theory and Practice of Phylogenetic Systematics*. New York: Wiley Interscience.

Woess, C. R., Kandler, O. & Wheelis, M. L. 1990. Toward a natural system of organisms: Proposal for domains Archaea, Bacteria, and Eukarya. *Proceedings of the National Academy of Science* 87:4576–4579.

Fossils often provide information about the environments in which particular rock units were deposited. The slab of rock shown here (Coconino Sandstone from Grand Canyon National Park, Arizona) records the trails of millipedes. Millipedes inhabit continental areas, and therefore they provide evidence that the sandstone bearing these trails is a non-marine sandstone. (United States Geological Survey photograph)

The Value and Meaning of Fossils

In the high mountains I have seen shells. They are sometimes embedded in rocks. The rocks must have been earthy materials in days of old, and the shells must have lived in water. The low places are now raised high, and the soft material turned to stone.

Chu-Hsi, A.D. 1200

Long before geologists knew how to use radioactive elements to find the age of rocks, they used fossils. Fossils are uniquely suitable for this task, for they are different in rocks of each passing age. Older forms disappear, and once gone, they do not evolve again in later ages. It is for this reason that they are the geologist's signposts to particular intervals of geologic time. Their usefulness in determining the age of rock strata, however, is only one of the many uses put to fossils. Fossils provide the means for determining the equivalency or correlation of rocks at widely separated localities. They inform us about the way organic evolution has operated over vast spans of time, and they offer clues to ancient environments, climates, and geography.

FOSSILS AS GUIDES TO GEOLOGIC AGE

Life on Earth has changed with the passage of time. This is the basis for the use of fossils in recognizing strata deposited during particular intervals of the geologic past. Initially, it was recognized that in a vertical sequence of strata, the oldest beds are at the bottom and younger beds successively higher in the section. This relationship, termed the **law of superposition**, helped early geologists recognize groups of fossils from strata low in the geologic column as being older than those in overlying strata (Fig. 2.1). William Smith (1769–1839), an English surveyor and engineer, understood such relationships quite well. While surveying the route of the Somerset Coal Canal near Bath, England, he became adept at recognizing particular strata wherever they occurred by their color, composition, texture, and especially by the fossils they contained. He recorded how groups of fossils succeeded one another in the vertical sequence of strata. Supplied with this knowledge, Smith was able to predict the rock succession from place to place and to correlate strata across England and Wales. His work was a reflection of the **principle of faunal and floral succession**, which stipulates that animals and plants from each age of Earth's history were distinct, that the fossil remains of that life can therefore be used to recognize contemporary deposits around the world, and that fossils provide the means for assembling the scattered fragments of the sedimentary rock record into a composite chronologic sequence. The representation of that sequence is the standard geologic time scale (Fig. 2.2)

The geologic time scale is based on relative time. Although it has been an indispensable aid in organizing the physical and biologic events of the past, a method for determining the **actual** or **quantitative** age of rocks, and thereby of the events they record, is also needed. It is useful to be able to say that trilobites proliferated in the seas before oysters, but how much better to be able to state that they evolved about 300 million years before oysters. The development of a method for determining the absolute age of a rock followed the discovery of radioactivity in 1895. Called *radioisotopic dating*, the method is based on the observation that radioactive isotopes within rocks and minerals emit radiation and decay to more stable daughter isotopes. The rate of decay is constant and can be measured with precision. If one knows the rate of decay and has determined the proportions of the parent isotope to the daughter isotope, it is then possible to calculate how much time has elapsed since the rock formed. The rate of decay of a radioactive isotope is expressed as that isotope's half-life. **Half-life** is the time span in which one-half of an original amount of a radioactive isotope

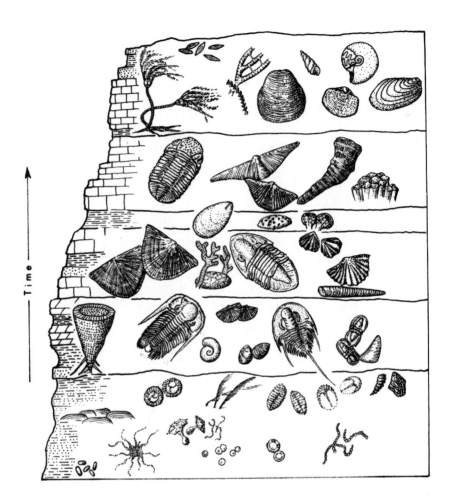

Figure 2.1   Conceptual illustration of how the superpositional sequence of strata and their contained fossils provided early geologists with a method for determining the relative age of both strata and fossil. Fossils could then be used in correlation and the formulation of a scale of relative geologic time.

decays to its daughter isotopes. Thus, at the end of the time constituting one half-life, 50 percent of the original quantity of parent isotope has undergone decay. After another half-life, half of what remains has undergone decay, leaving only 25 percent of the original quantity. After a third half-life, only 12.5 percent would remain, and so on. As an example, uranium-235 has a half-life of 704 million years. Thus, if a sample contains only 25 percent of the original amount of uranium-235 and 75 percent of the decay product (lead-207), two half-lives would have elapsed, and the sample would be about 1408 million years old (1.4 billion years).

Eons	Eras	Periods	Epochs
Phanerozoic	Cenozoic	Quaternary	Holocene Pleistocene
			—1.6—
		Tertiary	Pliocene
			— 5 —
			Miocene
			—24—
			Oligocene
			—37—
			Eocene
			—58—
			Paleocene
		— 66 —	
	Mesozoic	Cretaceous	
		—190—	
		Jurassic	
		—205—	
		Triassic	
		—250—	
	Paleozoic	Permian	
		—290—	
		Pennsylvanian (Late carboniferous)	
		—325—	
		Mississippian (Early carboniferous)	
		—355—	
		Devonian	
		—410—	
		Silurian	
		—438—	
		Ordovician	
		—510—	
		Cambrian	
		—544—	
Proterozoic		Late	
		—900	
		Middle	
		—1600	
		Early	
		—2500—	
Archean		Late	
		—3000	
		Middle	
		—3400	
		Early	
		—3800—	
Hadean			

Figure 2.2   The geologic time scale.

*Numbers are millions of years before present

THE GEOLOGIC TIME SCALE

Development of the geologic time scale did not proceed in an orderly manner. Early geologists had no way of knowing how many units would ultimately comprise the completed scale. Nor could they know for certain which fossils would be useful in correlation or which new sequences of strata might be discovered at a future time. Decades of research, fieldwork, and stratigraphic correlation were needed to piece together the time scale. It was, and continues to be, revised with each new discovery. The names for the units in the time scale are borrowed from geographic locations where one can examine rocks of the proposed unit, from the general character of life at the time, from the names of ancient inhabitants of an area, or even from the type of rock that was particularly characteristic of the time interval. As indicated in Figure 2.2, the largest divisions of the scale are the **Archean, Proterozoic,** and **Phanerozoic Eons**. Approximately 87 percent of all geologic time is encompassed in the Archean and Proterozoic Eons, which are simply referred to as the "Precambrian." To geologists, the phrase "in the beginning" usually alludes to the Archean, which began about 3.8 billion years ago. By Archean time, Earth had gathered most of its mass, possibly from a turbulent cloud of cosmic dust. Archean rocks reveal the characteristics of Earth during its early stages of development, and in a few places the rocks contain microscopic, unicellular remains of primordial life. Proterozoic rocks are generally less deformed and altered than are those of the Archean, and fossils of microbial life are more common. Near the end of the Proterozoic, multicellular animals are present in the fossil record.

All of the remainder of geologic time comprises the third great eon, the Phanerozoic. In contrast to preceding eons, the Phanerozoic has a rich fossil record. Based on this record, three divisions of the Phanerozoic are recognized. The oldest is the Paleozoic Era, which began 544 million years ago and lasted for 294 million years. The succeeding Mesozoic Era had a duration of 184 million years. The Cenozoic Era, in which we are now living, began 66 million years ago.

The geologic **eras** are divided into shorter intervals called **periods**, the periods into **epochs**, and the epochs into **ages**. Each one of these divisions represents intangible increments of time. They comprise **geochronologic units**. For the actual tangible body of rocks and fossils formed during a time interval, a parallel set of terms called **chronostratigraphic units** are employed. Thus, rocks deposited during the time span comprising a geologic period are known as a **system**. Similarly, the strata laid down during an epoch comprise a **series**, and those of an age are called a **stage** (Table 2.1). In using these parallel sets of terms it would be correct to speak or write of a snail that lived during the Cambrian Period (a geochronologic term), but its fossil remains would be found in rocks of the Cambrian System (a chronostratigraphic term).

Table 2.1 Terms used for the divisions of geologic time and their chronostratigraphic equivalents

Geochronologic Divisions	Chronostratigraphic Divisions
ERA	ERATHEM
PERIOD	SYSTEM
EPOCH	STAGE
AGE	ZONE

BOX 2.1   FOSSILS AS GEOCHRONOMETERS

In the early 1960s, John Wells described a method by which certain reef-forming corals could be used as geochronometers. The method is dependent on the presence of fine growth ridges on the outer surface of the coral skeleton. There are about 200 of these growth ridges per centimeter, each representing a single day's addition of calcium carbonate. In specimens suitable for geochronometry, the growth ridges are grouped into monthly bands separated by slightly constricted intervals. The constrictions may have been produced during monthly breeding cycles when levels of calcium carbonate secretion were reduced. Still broader bands contain a yearly increment of growth ridges. As a first necessary step in his study, Wells counted the growth rings on several species of extant corals and found the number "hovers around 360 in the space of a year's growth." Proceeding to the fossils of Paleozoic corals, he recorded far more growth ridges in an annual band. For example, there were about 400 growth ridges in corals known to be 370 million years old. Thus, there were about 400 days in a year in the late Devonian period. The finding substantiated calculations provided by astronomers indicating Earth's rotation has been slowing at a rate of about 2 seconds every 100,000 years. The cause is tidal friction.

As indicated on the graph, once the numbers of days in a year are known from growth ridge counts, it is possible to obtain the age of a specimen or the stratum in which it is found. Organisms other than corals, including bivalves, have employed the Wells's method. However, the method is not without problems. Only certain species lend themselves to growth ridge counts. Exceptionally well-preserved fossils are required, and large numbers must be scrutinized to obtain statistically valid results. Even specimens of suitable species may show spurious variations in growth-ridge production possibly resulting from adverse changes in the environment.

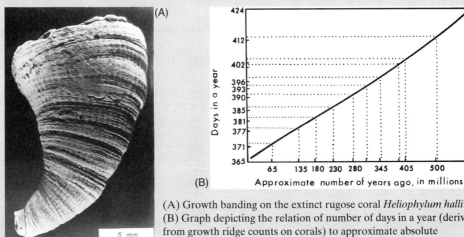

(A) Growth banding on the extinct rugose coral *Heliophylum halli*.
(B) Graph depicting the relation of number of days in a year (derived from growth ridge counts on corals) to approximate absolute geologic age. (Graph from J. W. Wells, *Nature* 197:948–950, 1963.)

LIFE THROUGH TIME: AN OVERVIEW

As yet, there is no fossil evidence of life on Earth during the planet's first billion years. The oldest direct indications of ancient life consist of bacteria and primitive algae discovered in rocks over 3.2 billion years old. These early fossils indicate that the long evolutionary march had begun. However, they also stand at the end of another long and remarkable period during which living things evolved from nonliving chemical compounds. We have no direct geologic evidence about how and when the transition from nonliving to living occurred. What we do have are reasonable hypotheses supported by experimentation. Some of these hypotheses will be examined in the next chapter, which deals with earliest life.

For most of the Precambrian, living things left only occasional and often obscure traces. Here and there, primarily in siliceous rocks, paleontologists have been rewarded with findings of bacteria or cyanobacterialike filaments (Fig. 2.3) and other forms of micro-algae. Some of these early microbes formed matlike structures called **stromatolites** (Fig. 2.4). The microbes responsible for the construction of stromatolites

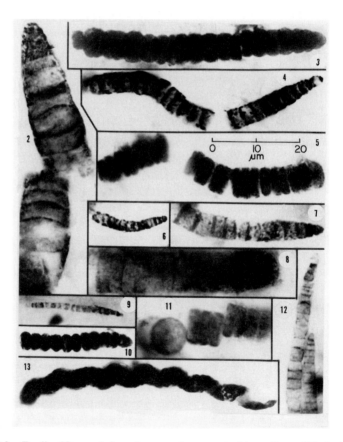

Figure 2.3   Fossils of Precambrian microorganisms seen in thin sections of black chert from the 900-million-year-old Bitter Springs Formation of Australia. The fossils are filamentous cyanobacteria similar in structure to certain species living today in algal mat communities of tropical, nearshore environments. (Courtesy of J. Wm. Schopf.)

Figure 2.4   Stromatolites from 2.2-billion-year-old beds of the Labrador Trough of northern Quebec.

are thought to have produced important quantities of oxygen by photosynthesis. An early indication of multicellularity is found in billion-year-old rocks of the Torrowangee Group of Australia. In somewhat younger rocks deposited about 0.7 billion years ago, fossil metazoans are known. The rapid expansion of shelled invertebrates, however, came at the approximate beginning of the Paleozoic.

Several generalizations can be made about the history of life during the Paleozoic. One notes that most of the principal phyla of invertebrates appear early in the era and that many of these are still represented by animals living today (Fig. 2.5).

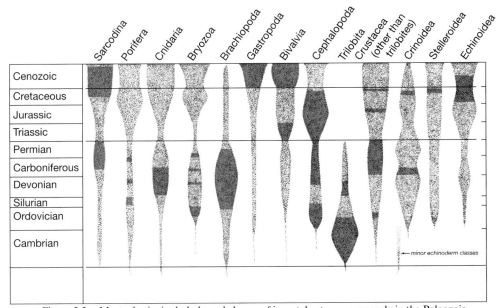

Figure 2.5   Most of principal phyla and classes of invertebrates appear early in the Paleozoic and have members that have survived to the present time. Shading indicates times when members of a taxon are particularly useful in biostratigraphy, and width of bands is a subjective estimate of the abundance of families. (Data from several sources.)

For the most part, we are not startled by sudden appearances of bizarre or exotic creatures that cannot be traced to earlier forms. We also find intervals when large numbers of families and lower taxa became extinct. Evidence of environmental adversity often coincides with these extinctions. Finally, there appears to have been a persistent increase in species diversity down through the ages. The evidence for increased diversity is obtained by compilation and scrutiny of decades of paleontologic reports and the subsequent statistical treatment of the data derived from these publications. Prominent in this area of research are J. J. Sepkoski Jr., R. K. Bambach, and D. M. Raup, who have reported a distinct increase in the diversity of invertebrate genera and families, as well as of trace fossils as one progresses through the Phanerozoic. Increases in diversity did not occur at a constant rate, however, for events such as extinctions caused fluctuations in the overall pattern.

Trilobites (marine arthropoda), brachiopods (bilaterally symmetrial bivalved invertebrates), nautiloid cephalopods (the chambered Nautilus is a living example), rugose corals, tabulate corals, bryozoans (attached and encrusting colonial marine invertebrates), and crinoids (stalked echinoderms) are the most frequently encountered fossil invertebrates in marine Paleozoic rocks. A brief referral to subsequent chapters will provide illustrations of these invertebrates. They lived in communities that changed in general composition and nature with the passage of time. With the exception of trilobites, one notices that many of the post-Cambrian invertebrates were sedentary suspension feeders. With time, these attached forms gave way to a greater number of mobile deposit and infaunal groups, which fed as they burrowed and churned the sediment of the seafloor. This is not to say that the sedentary forms were totally absent. They persisted here and there and did very well in areas where the sediment disturbers were rare or absent, such as along organic reefs or other areas inhospitable to burrowers.

Macroscopic invertebrate life of the Mesozoic would appear more familiar to us. The shells of bivalves, sea urchins, marine snails, and fragments of modern-looking (scleractinian) corals would have littered many Mesozoic beaches. Less familiar would be the shells of the ammonoid cephalopods (cephalopods having septa with crenulated margins), for these animals became extinct at the end of the Cretaceous. As expected, Cenozoic invertebrate faunas had a decidedly up-to-date appearance. Among the mollusks, marine snails, oysters, clams, scallops, squid, and octopods proliferated. Scleractinid corals continued as reef formers, and many families of crustaceans expanded. Also during the Cenozoic, echinoids, asteroids (starfish), and ophiuroids (serpent stars) were the most abundant members of the Phylum Echinodermata.

EVOLUTIONARY FAUNAS

Paleontologists have an advantage over biologists in being able to examine the entire history of life and thereby identify major changes in faunal dominance with the passage of time. John Sepkoski has identified three such large-scale changes, and has designated the faunas within each as **evolutionary faunas** (Fig. 2.6). Each evolutionary fauna dominated over a considerable span of geologic time; each had a similar history and achieved levels of diversity and ecologic complexity greater than the preceding fauna. Sepkoski defined each of the evolutionary faunas statistically by factor analysis

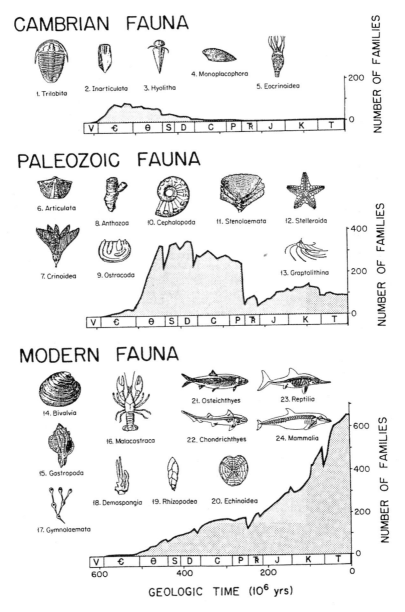

Figure 2.6   Family diversity in each of the three evolutionary faunas described by J. John Sepkoski Jr. (From J. J. Sepkoski, Jr. 1984. A kinetic model of Phanerozoic taxonomic diversity: III. Post-Paleozoic families and mass extinctions, *Paleobiology* 10(2):246–267.)

of diversity of families within classes. This kind of analysis permits one to reduce a complex mass of data to fairly simple form, and thereby to group together those classes that had reached their greatest diversity at about the same time. In each of the evolutionary faunas, there is an early period of rapid diversification, followed by a longer period of slow decline. Episodes of extinction separate the faunas. Sepkoski defined three Phanerozoic evolutionary faunas:

1. *Cambrian Evolutionary Fauna.* Trilobites, inarticulate brachiopods, monoplacophorans (small mollusks with cap-shaped shells), hyolithids (small probable mollusks with conical shells closed by opercula), archaeocyathids (conical to cylindrical double-walled animals), early echinoderms, and soft-bodied or chitin-shelled animals such as those in the Cambrian Burgess Shale characterize this fauna. From this list, it is apparent that the Cambrian evolutionary fauna was strikingly different from the subsequent Paleozoic fauna. Deposit and suspension feeding as well as grazing were the usual food-gathering methods characteristic of Cambrian animals.

2. *Paleozoic Evolutionary Fauna.* The predominant invertebrates in this fauna were articulate brachiopods, bryozoa, rugose (horn) and tabulate corals, cephalopods, crinoids, graptolites, ostracodes, and new classes of trilobites. By comparison to the Cambrian evolutionary fauna, the Paleozoic fauna existed in a more complex ecosystem.

3. *Mesozoic-Cenozoic Evolutionary Fauna.* This is a modern fauna in which predominant invertebrates are bivalves, marine gastropods, crustaceans, echinoids, modern bryozoans, and foraminifers. Brachiopods have lost their position of dominance to bivalves, tabulate and rugose corals to scleractinian corals, and stemmed echinoderms like crinoids to mobile echinoids. Once again, higher levels of diversity are achieved than had existed in the previous evolutionary fauna.

MASS EXTINCTIONS

Within the general continuum of life on Earth, species live for various lengths of time, become extinct, and are usually replaced by other species. There were, however, intervals when the planet was less hospitable or when catastrophic events occurred and caused the extermination of many taxa over a relatively short span of geologic time. Such biological destructive events are termed **mass extinctions**. Early geologists recognized the losses caused by mass extinctions, and they used that evidence to identify the boundaries between geologic systems. Although there have been fewer severe mass extinctions, five were particularly extensive. Of these, one occurred at the end of the Ordovician, one in the Late Devonian, one near the end of the Permian, one late in the Triassic, and one that marks the end of the Cretaceous (Fig. 2.7). Less severe extinctions occurred near the end of the Precambrian, the Late Cambrian, in the Eocene to Oligocene interval, and at the end of the Pleistocene.

Late Ordovician Extinctions Extinctions near the end of the Ordovician occurred in two phases encompassing a time span of about a million years. During the first phase, planktonic organisms such as graptolites as well as benthic trilobites and brachiopods fell victim. In the second phase, several trilobite families that had survived

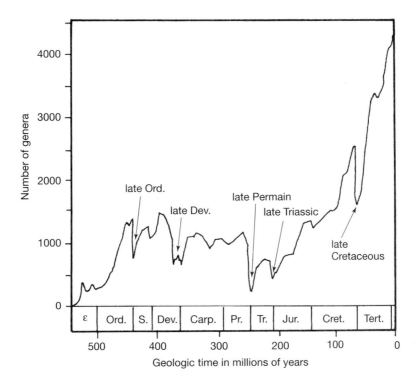

Figure 2.7   Diversity of marine animals as compiled by J. J. Sepkoski, Jr. from a database recording first and last occurrences of more than 34,000 genera. The graph depicts the five major episodes of mass extinction during the Phanerozoic. (Adapted from J. J. Sepkoski, *Geotimes* 39(3):15–17, 1994.)

the first wave of extinction perished. At the same time, corals, conodonts, and bryozoans were severely reduced in numbers and diversity. About 22 percent of invertebrate families disappeared. Both phases of extinction seem to be related to global cooling that was marked by the growth of ice caps in the present region of the Sahara Desert. In response to cooling at higher latitudes, the living zones of many families of invertebrates were shifted toward the equator. As ice accumulated on land, sea level declined, with consequent loss of shallow seas that flooded interior regions of the continent, as well as shallow marginal shelf environments. Many of these areas had been optimum habitats for invertebrates.

Late Devonian Extinctions   Following the late Ordovician crisis, the diversity of life again increased, building slowly in the Early Silurian, but attaining rich levels of diversity during the Devonian. By the end of the Frasnian stage of the Devonian, however, marine invertebrates were again confronted with adverse environmental change. A readily apparent result was the decimation of once extensive Devonian reef communities. Reef-building tabulate corals and stromatoporoids are rarely seen in rocks deposited during the remainder of the Paleozoic. Of the once ubiquitous rugose corals, relatively few groups survived. In addition, brachiopods, goniatites, trilobites, and conodonts were severely reduced in abundance and variety. Altogether, 21 percent of all families of marine creatures were killed off.

The Late Devonian extinctions occurred over an interval of several million years. Hence, a sudden catastrophic event such as the impact of a meteor or asteroid is unlikely to have been the cause. Nor is there evidence of fallout from heavy metals, which are the most important clues to such an event. As is the case for Late Ordovician extinctions, the basic cause was probably terrestrial. It appears to be related to cooling as indicated by continental glaciation in the Southern Hemisphere supercontinent Gondwana and the observation that tropical reef and shallow warm-water faunas were the most adversely affected. These faunas experienced a species loss of 96 percent. It has been suggested that during interglacial melting and rise in sea level, major reef ecosystems may have been too deeply inundated for survival of reef organisms. As suggested by the wide distribution of marine black shales, the seas that spilled onto low-lying continental tracts may have become oxygen deficient, contributing to the die off. In this regard, an association may have existed between the demise of marine creatures and land plants. Vascular land plants had spread widely by Devonian time, creating an abundance of nutrient and phosphorus-rich soils to be washed into rivers and the sea. Geochemist T. Algeo suggests that the nutrient-rich waters could have triggered an explosive growth of marine algae. On death, huge volumes of decaying algae consumed by bacteria may have robbed the sea of oxygen. The resulting anoxic waters might have been lethal to many marine invertebrates.

Late Permian Extinctions   The loss of families of invertebrates at the end of the Permian Period has been described by Smithsonian paleontologist Douglas Erwin as "the mother of mass extinctions." It exceeded all other extinctions, including the one involving the demise of the dinosaurs 180 million years later. Over an interval of only about 10 million years, an astonishing 57 percent of marine invertebrate families were exterminated. Rugose corals, many families of crinoids, spinose brachiopods called productids, lacy bryozoa, many families of ammonoids, and the once prolific fusulinids were among the victims of the mass extinction. Trilobites were not involved, for they had disappeared earlier in the Permian.

At the time of the Permian extinctions, the supercontinent Pangaea had completed its development. As is characteristic of large continents today, more rigorous climatic conditions existed across parts of the interior. Shallow inland seas (epeiric seas) were drained or restricted, reducing space favorable for shallow marine invertebrates. Pangaea blocked equatorial currents, disrupting the life zones of many organisms. Frigid polar regions lay to the north and south of Pangaea, forcing organisms accustomed to warmer waters toward lower latitudes. Cooling also inhibited the spread of organic reefs and lessened carbonate production. The Late Permian was also a time of extraordinary volcanism. One of the greatest episodes of flood basalt volcanism in Earth's history occurred in Siberia at the end of the Permian. Carbon dioxide released into the atmosphere during the volcanic activity may have triggered a greenhouse effect with consequent climatic stress on some organisms. Thus, it appears that many global changes were underway at the end of the Permian, and a combination of some of these changes might have disrupted ecosystems sufficiently to cause the extinction of life-forms evident in the fossil record.

Late Triassic Extinctions   Near the end of the Triassic Period about 20 percent of preexisting families of marine invertebrates had disappeared. Sponges, brachiopods, bivalves, crinoids, and ammonoid cephalopods suffered particularly heavy losses, whereas strophomenid brachiopods, conulariids, and conodonts disappeared completely. For this extinction, it is more difficult to find direct associations to changes in global conditions. We know that the growth of reefs in the Tethys Sea (which lay along the southern margin of Europe) essentially ended in the Late Triassic, and that there was a marked decline in carbonate production. Widespread karst conditions in parts of Europe at this time suggest increase in rainfall and runoff. This may have led to changes in ocean surface temperatures, ocean water pH and salinity, and consequent loss of carbonate habitats. The evidence, however, is tenuous, and the final explanation for Late Triassic extinctions requires further study.

Terminal Cretaceous Extinctions   The mass extinction at the end of the Cretaceous was not the most severe among Phanerozoic extinctions, but is the most popular because it involved the kill off of the dinosaurs. In the ocean, about 15 percent of invertebrate families were eliminated. Particularly striking were the extinctions of the ammonoid cephalopods, belemnites (cephalopods with straight internal shells), and rudistoid bivalves. Entire families of echinoids, bryozoans, and planktonic foraminifers were lost. The cause for this biologic crisis both on land and in the sea has engendered scores of extinction theories. Those that have the most credibility attempt to explain simultaneous extinctions of both marine and terrestrial animals, and seek a single or related sequence of events as the cause. The theories can be grouped into two categories. The first group relies on some sort of ruinous extraterrestrial event such as a bolide (meteorite, asteroid, or comet) striking Earth. The second group invokes extinction-causing events that occurred on the planet itself.

Bolide Impact   In 1977, Walter Alvarez sampled a thin layer of clay along the Cretaceous-Tertiary boundary near the town of Gubbio, Italy. Alvarez sent the samples to his father, Luis, a physicist, for analysis. The result was surprising, for the samples contained approximately 30 times more of the metallic element iridium than is normal for Earth's crustal rocks. Iridium is thought to be present in Earth's core and mantle, as well as in extraterrestrial bodies such as meteorites. It can also be brought to Earth's surface by volcanism. Alvarez, however, favored an extraterrestrial origin for the iridium in the clay layer, and he proposed that an iridium-bearing asteroid crashed into Earth at the end of the Cretaceous. The shattering blow from the huge body (presumed to be over 10 km in diameter) would have thrown dense clouds of iridium-bearing dust into the atmosphere. Transported by atmospheric circulation, the dust might have formed a lethal shroud around the planet, blocking the Sun's rays and thereby causing the demise of marine and land plants on which all other forms of life ultimately depend. As the dust settled, it would have formed the iridium-rich clay layers found at Gubbio and subsequently at 74 other localities around the globe.

In addition to the iridium-rich clay layer, other kinds of evidence support the terminal Cretaceous bolide-impact theory. One of these is the widespread occurrence of shocked quartz in sediment at the Cretaceous-Tertiary boundary. Grains of shocked quartz have distinctive sets of microscopic planes or shock lamellae produced when high-pressure shock waves, such as those emanating from the impact of a large meterorite, travel through quartz-bearing rocks. Often in the same stratum containing the shocked quartz one also finds tiny glassy spherules, called *tectites*, thought to represent droplets of molten rock thrown into the atmosphere during the impact event. The boundary clays also contain a rare silicate mineral known as *stishovite*, which forms only at extremely high pressures and temperatures, such as would be produced by the impact of an asteroid. Finally, sediments at the Cretaceous-Tertiary boundary often include a layer of soot that may be the residue of vegetation incinerated in firestorms caused by such a catastrophic impact.

In the course of Earth's 4.6-billion-year history, the planet has experienced many bolide impacts. The scars of most of these blows have been obliterated by erosion or destroyed by tectonic processes. The craters of a few, however, can still be discerned. One such structure that is thought to be associated with the terminal Cretaceous bolide impact lies buried in the northern part of the Yucatan Peninsula near the town of Chicxulub, Mexico.

Terrestrial Causes   Although the evidence for bolide impact as a cause for extinctions at the end of the Cretaceous is compelling, many questions are not yet fully resolved. Is the iridium truly of bolide origin, and if impact did occur, was it responsible for the mass extinction, or did it merely contribute to adverse terrestrial changes already in operation? Some argue that the iridium could have been derived from the mantle, from which it was carried to the surface in volcanic conduits and ejected into the atmosphere as iridium-rich volcanic ash. Shocked quartz can also result from explosive volcanism, and volcanism was especially prevalent during the late epochs of the Cretaceous. Especially significant were tremendous outpourings of lava in India. Known as the Deccan traps, these lava flows cover a large part of the Indian Peninsula. They were extruded about 65 million years ago at the same time that intensive volcanic activity was occurring in many other parts of the world, including the western United States, Greenland, Great Britain, Hawaii, and the western Pacific. Volcanoes produce dust, ash, and aerosols that block solar radiation and cause decline in temperatures. Sulfuric acid associated with eruptions may fall as acid rain, changing the alkalinity of the oceans and possibly killing plankton essential to animals higher in the food chain. It is noteworthy that the iridium at the Cretaceous-Tertiary boundary is often distributed across a sediment thickness of 30 to 40 centimeters, suggesting that it was deposited over a time span of several thousand years. Volcanic activity can persist for such time spans, but if the iridium was derived from a bolide impact one would expect it to be confined to a thin layer reflecting a short-lived event. Yet another aspect of the debate relates to the presence of abundant antimony and arsenic in the same beds containing the iridium. Although common in volcanic ash and lava, these elements are exceedingly rare in meteorites.

In addition to widespread volcanism, other conditions on Earth may have contributed to the biological catastrophe that closed the Mesozoic era. Before the time of the die-off, continents were extensively covered by shallow seas in which marine life proliferated. The interior seas contributed to the maintenance of moderate climatic conditions in many parts of the world. Near the end of the Cretaceous, these conditions began to dissipate. Stratigraphic sequences indicate lowering of sea level and withdrawal of epeiric seas. Such environmental changes spelled disaster to many creatures of the inland seas and broad, shallow marginal shelves. The demise of each group affected all dependent species within the ecologic system.

Those favoring terrestrial causes for the late Cretaceous extinctions emphasize that the die-off was not sudden as one might expect from a bolide impact. Families disappeared sporadically over an interval of up to 5 million years. This suggests that extinctions may have already been underway near the end of the Cretaceous. Perhaps the debilitated survivors of environmental stress were dealt a *coup de grâce* by bolide impact.

ECOLOGIC-EVOLUTIONARY UNITS

The record of Phanerozoic life reveals yet another interesting pattern in which diverse groups of organisms appear abrubtly at the beginning of a given time interval, then continue with relatively little change in overall aspect for a long span of time, and finally undergo extinctions. Arthur Boucot has proposed the name **ecologic-evolutionary units** for these relatively stable, co-appearing and co-existing groups. He has identified twelve such ecologic-evolutionary units (Table 2.2) among "level-bottom" marine invertebrates of the Phanerozoic. Each is characterized by minimum evolutionary innovation over intervals of tens of millions of years. When episodes of extinction and damage to the long-stable ecologic-evolutionary unit occur, the replacing groups undergo rapid diversification accompanied by many evolutionary innovations. Once these units are established, the new unit, like its predecessor, enters a lengthy interval of what might be called evolutionary monotony.

BIOSTRATIGRAPHIC CORRELA TION

When William Smith examined strata exposed by stream erosion in valleys, he was aware that the strata continued laterally beneath divides and soil cover and could be recognized again at distant locations. The process of identifying and determining the

Table 2.2

Ecologic-evolutionary Units	Time Intervals
I	Early Cambrian
II	Middle to Late Cambrian
III	Early Ordovician
IV	Middle to Late Ordovician
V	Two-thirds of Early Silurian
VI	One-third Early Silurian to one-half Late Devonian
VII	One-half of Late Devonian
VIII	Carboniferous and Permian
IX	Early Triassic
X	Middle to Late Triassic
XI	Jurassic and Cretaceous
XII	Cenozoic

equivalency of strata from one location to others is called **correlation**. Strata may be correlated by similarity in physical characteristics of the rock, by fossils, by position in the sequence of beds, or by their radioisotopic age. There are several different kinds of correlation. In some cases it is only necessary to trace the occurrence of a lithologically distinct unit, and the age of that unit is not critical to the solution of the problem. This type of correlation, which seeks to link lithostratigraphic units of similar lithology and stratigraphic position, is termed **lithocorrelation**. It is essential to the construction of geologic maps and cross-sections. Other problems can only be solved through the correlation of rocks that are of the same age. This kind of matching is termed **chronocorrelation**. Recognition of such chronologically equivalent units is essential for determining the relationships between contemporaneous, but often lithologically dissimilar, units that reflect different environments of deposition. Without chronocorrelation one would have difficulty both in recognizing the distribution of the environments inhabited by ancient organisms at particular times in the geologic past, and in deciphering how those environments changed or shifted with the passage of time.

Although the precise chronological equivalency of strata can sometimes be ascertained by using isotopic dating methods, more frequently chonostratigraphic correlations are based on fossils. As noted earlier, such correlations are possible because evolution has provided fossils that are different in rocks of different ages. Rocks of the same age from widely separated regions may contain similar fossils, thus making correlation possible. When they do not, the dissimilarity may reflect differing environments, or there may have been barriers that prevented dispersal or genetic exchange between contemporaneous faunas. A sandstone that originated on a floodplain would have quite different fossils from one formed at the same time in a nearshore marine environment. How might one correlate the floodplain deposit to the marine deposit? Often this can be done by physically tracing out the beds along a line of cliffs or valley sides. Occasionally, one is able to find fossils that transcend environments. Pollen grains, for example, can be carried into, and preserved within, sharply contrasting environments. In other cases, two dissimilar rock units may occur directly above a distinctive, firmly correlated stratum, such as a layer of volcanic ash. In this regard, ash beds are particularly good time markers because the ash was widely dispersed during a brief interval of time, and it is often possible to date the ash by radioisotopic methods.

Before the fossil remains of an organism can be used with precision in determining the age or correlation of a stratum, one needs to know when the species first appeared (its oldest occurrence) as well as the time of its final (most recent) occurrence. The interval between the first and last occurrence is the **stratigraphic** or **geologic range** of the species. Species that are found to have short stratigraphic ranges, and yet have achieved wide geographic and ecologic dispersal, serve as **index fossils** to rocks of a particular age. If in addition to their short range and wide distribution, biostratigraphically useful fossils should be abundant, easily recognized, and not confined to a restricted environmental niche.

Although individual fossil species may be biostratigraphically useful, many correlations are based on entire assemblages of fossils. Assemblages are less likely to cause error resulting from fortuitous absence of a single guide fossil. In particular, overlapping stratigraphic ranges of members of the assemblage can provide excellent correlations. The advantage of overlapping ranges is illustrated in Figure 2.8. It is apparent from the chart that strata containing both *Heterostegina* and *Assilina* must be

Late Eocene in age, whereas the occurrence of members of only one of these fossil groups would not provide as narrow a limit to the age of the rock.

In making correlations, paleontologists attempt to discover, define, and name biozones. A **biozone** is simply a body of rock characterized by one or more taxa of fossil organisms that permit one to distinguish that body from adjacent strata. Several different kinds of biozones exist. The simplest is the **taxon range zone**, which refers to the lowest (first) and highest (last) occurrence of a single species, genus, or higher taxon. Using Figure 2.8 as an example, the *Assilina* range zone is marked by the lowest occurrence of that genus at point *A* and its extinction at point *B*. **Concurrent range zones** are recognized by the overlapping ranges of two or more taxa. In Figure 2.8, the interval between *X* and *Y* would be designated the *Assilina-Heterostegina* concurrent range zone. **Assemblage zones** are bodies of strata characterized by three or more taxa in a natural association. Assemblage zones are named after an easily recognized and abundant member of the assemblage (although that member need not be present at every

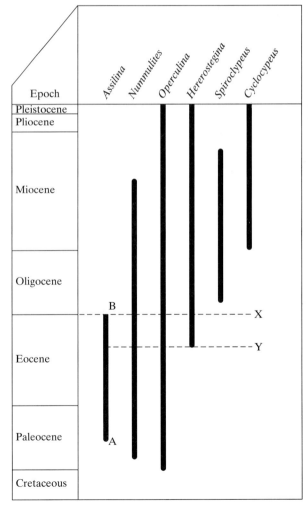

Figure 2.8   Geologic ranges of several genera of the family Nummulitidae (a family of foraminifers). *A* to *B* is the geologic range of *Assilina*. The *X* to *Y* interval represents the *Assilina-Heterostegina* concurrent range zone. (Data from R. S. Barnett, *Journal of Paleontology*, 48(6):1249–1263, 1974.)

location). In seeking the criteria for correlation, one may also designate **abundance**, **epibole**, or **acme zones** for an interval that contains the maximum number of a species, genus, or other taxon, but not necessarily its total range. Acme zones composed of larger fossils are readily recognized in the field, and hence often very useful in geologic mapping. However, all acme zones must be carefully evaluated, as their presence at any given locality may be influenced by intervals of slow deposition or exceptional ecologic conditions.

In practice, delimiting biozones is not an easy task. One must first accurately identify and record the precise sequence of fossils through the vertical section of rock. Tentative biozones can then be erected based on the first and last occurrences and ranges of the taxa recorded on a biostratigraphic range chart (Fig. 2.9). If these tentative divisions are then found in other localities, they can be elevated to the status of valid, confirmed biozones. Indeed, the validity and value of a biozone increases each time it is recognized at other and more distant locations.

As one examines the occurrence of fossils in a vertical section of strata, it is important to bear in mind that all changes need not reflect the appearance of a new species, nor the extinction of a species. They may merely indicate faunal migrations or shifts associated with environmental or geographic changes. Such a scenario is hypothetically illustrated in Figure 2.10A. It has been termed the **Lazarus Effect**, in that it gives the allusion of reappearance of dead (extinct) species. Paleontologists were aware of the Lazarus Effect long before it was named by David Jablonski. For example, in the sedimentary sequence of the southern San Joaquin Valley of California, foraminifera (shell-building unicellular protoctistans) diagnostic of the Saucesian stage of the Lower Miocene reappear in Middle Miocene strata. Micropaleontologists employed by oil companies in the area recognize the occurrence of the troublesome fossils in samples brought to the surface during drilling, and they informally designate the interval in which they appear as "pseudosaucesian."

In addition to being on the alert for faunal changes that are related to migrations (as opposed to changes resulting from evolution), one must also guard against misinterpretations caused by the occurrence of reworked fossils. Often in the geologic past, erosion has freed fossils from the rock that enclosed them. As with any particle freed in this way, if sufficiently durable, these fossils can be reworked into younger beds and the younger strata mistakenly assigned to an older geologic interval. There is also the possibility that small fossils from younger overlying beds can be leaked into older beds. J. D. Archibald has dubbed such occurrences the **Zombie Effect** (Fig. 2.10B). (Zombies leave their graves as walking dead, and the reworked fossils have left their burial site to reappear elsewhere.)

PALEOECOLOGY

Although inert and mute, fossils are vestiges of once lively organisms that nourished themselves, reproduced, grew, and interacted in countless ways with other organisms and their physical environment. The study of the interaction of ancient animals and plants with their environment is called **paleoecology**. Paleoecologists attempt to discover precisely where and how ancient creatures lived, the relation between function and their morphologic traits, and what the fossils reveal about the geography, climate, and other conditions of long ago. Various methods are used to accomplish these ends.

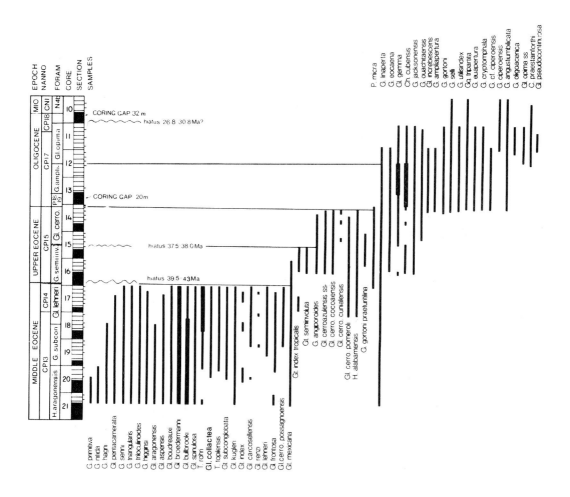

Figure 2.9   Example of a biostratigraphic range chart. The fossils were obtained from a deep sea core beneath the Gulf of Mexico. Geologic ranges of species of foraminifers are depicted by vertical lines, some of which are thicker to indicate exceptional abundance. The columns on the left show the foraminiferal biozones as well as biozones based on occurrences of microfossils known as nannoplankton (the column headed "NANNO"). (From G. Keller, *Journal of Paleontology*, 59(4):882–903, 1985.)

One can, for example, compare species known only from fossils with living species of generally similar form. The assumption is made that both the living and analogous species had approximately the same needs, defenses, feeding strategies, and tolerances. In addition, one can examine the anatomy of the fossil and attempt to identify structures that were likely to have developed in response to particular conditions. For example, broad flat shells of brachiopods may have evolved to prevent the animal from sinking into the soft ooze on the ocean floor, or possibly to increase the surface area of soft tissue for gas exchange.

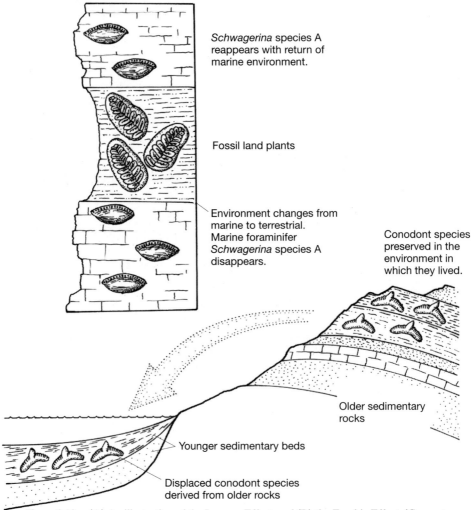

Schwagerina species A reappears with return of marine environment.

Fossil land plants

Environment changes from marine to terrestrial. Marine foraminifer *Schwagerina* species A disappears.

Conodont species preserved in the environment in which they lived.

Older sedimentary rocks

Younger sedimentary beds

Displaced conodont species derived from older rocks

Figure 2.10   (A) An illustration of the Lazarus Effect, and (B) the Zombie Effect. (Concept for Lazarus Effect from J. D. Archibald, 1996. *Dinosaur Extinction and the End of an Era.* New York: Columbia University Press.)

The terminology used in paleoecology is derived from **ecology**, the study of the present relationships between organisms and their environments. In ecologic studies, one tends to concentrate on the **ecosystem**, which is any selected part of the physical environment together with the animals and plants within it. An ecosystem can be as large as Earth or as small as a garden pond. Paleoecologists are particularly interested in the ocean ecosystem because of the richness of the fossil record of marine invertebrates. The physical aspects of the ocean ecosystem include the water itself, its dissolved gases (especially carbon dioxide and oxygen), salts (chlorides, phosphates, nitrates, and carbonates of sodium, potassium, and calcium), various organic com-

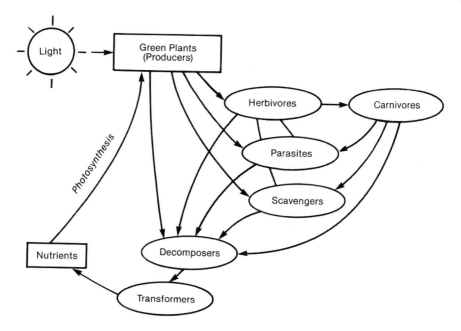

Figure 2.11   The movement of vital materials through an ecosystem. The organisms within the ellipses are consumers, and the highest level of consumers are the carnivores.

pounds, turbidity, light penetration, water movements (currents, wave action), and temperature.

The biological components of the ocean ecosystem are readily organized according to **trophic** or feeding levels (Figure 2.11). There are, for example, **producer organisms** such as green plants, which by means of photosynthesis manufacture organic compounds from simple inorganic substances. These producer organisms in the sea are mostly small (less than 0.1 mm in diameter) and include various forms of algae and bacteria. Producer organisms are eaten by **consumer organisms** such as mollusks, crustaceans, and fishes. The primary consumers feed directly on unicellular plants, and thus can be further designated as **herbivores**. The secondary consumers that eat the herbivores are termed **carnivores**. Tertiary and even quaternary consumers feed on carnivores from lower trophic levels. The ecosystem also contains **decomposers** and **transformers**, including bacteria and fungi, that are able to break down the organic compounds in dead organisms and waste matter and produce simpler materials that can be utilized by producers. Thus, in an ecosystem, the basic chemical components of life are continuously being recycled. The marine ecosystem also contains **parasites**, which feed on other organisms without necessarily killing them, and **scavengers**, which derive their nourishment from the dead.

Specific habitats for particular organisms occur within an ecosystem. In the ocean, the habitats range from the cold and dark realm of the abyss to the warm, illuminated waters of shallow seas. Within each habitat are **ecologic niches** in which particular organisms make their living. Each niche contains all the biological, chemical, and physical conditions that permit the organism occupying the niche to survive. A coral reef, for example, has many niches. Some are occupied by tentacled coral animals that build the framework of the reef and feed as carnivores on crustaceans and other animals. Certain marine snails occupy a niche along the surface of algal mats, where they graze on films of algae. Behind the reef in areas of soft mud, lugworms consume the soupy sediment for its content of organic nutrients. Each creature has its characterisic ecologic niche.

To facilitate study of ocean ecosystems, ecologists have developed a simple classification of marine environments. It begins with a twofold division of the entire ocean into pelagic and benthic realms. The **pelagic realm** consists of the water mass lying above the ocean floor. It can be divided into a **neritic zone**, which lies above the continental shelves, and an **oceanic zone**, which extends seaward from the shelves (Fig. 2.12). Within the pelagic zone are myriads of small animals and plants that float, drift, or feebly swim. These are **plankton**. **Phytoplankton** consist of plants and plantlike protoctista. Included here are the paleontologically important diatoms and coccolithophorids to be described in Chapter 4. **Planktonic** (often shortened to planktic) animals constitute **zooplankton**, and include radiolaria, foraminifera, certain tiny mollusks, small crustaceans, and the motile larvae of many different families of invertebrates that live as adults on the seafloor.

The pelagic realm is also the home of **nekton**, or true swimming animals. Nekton are able to travel where they choose under their own power. A swimming creature can search for its food and does not have to depend upon food particles carried in chance currents. It can use its mobility to escape predators, and it can move to more favorable areas when conditions become difficult. Nektonic animals include an immense diversity of invertebrates, as well as fishes, whales, and marine turtles.

The second great division of the ocean ecosystem is the bottom or **benthic** realm. It begins with a narrow zone above high tide called the **supralittoral zone**. Several

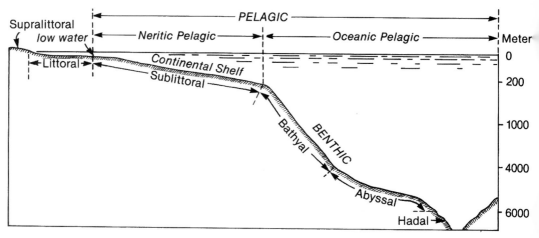

Figure 2.12   Terms applied to the various realms of the marine environment.

kinds of marine plants and animals have adapted themselves to this harsh environment where ocean spray provides vital moisture, but where dessication poses a constant danger. Seaward of the supralittoral zone is the area between high and low tide. This is the **littoral zone**. Animals living in the littoral zone must also be able to tolerate alternate wet and dry conditions. Some avoid dessication by burrowing into wet sand, whereas others have achieved adaptations that permit retention of body moisture when exposed to air. Seaward of the littoral zone is the **sublittoral zone**. It extends from low-tide levels down to the edge of the continental shelves, at about 200 meters. Depending on the clarity of the water, light may penetrate to the seafloor in the sublittoral zone, although the base of light penetration is usually somewhat less than 150 meters. In this well-illuminated environment, photosynthetic organisms thrive, as do abundant proto-zoans, sponges, corals, worms, mollusks, crustaceans, and echinoderms. These animals are not normally subjected to dessication. Their adaptations are mainly associated with food gathering and protection from predators. Some live on top of the sediment that carpets the seafloor and are called **epifaunal**. Others, termed **infaunal**, burrow into the soft sediment for food and protection. Those benthic animals capable of locomotion are further designated **vagile**, in contrast to immobile creatures termed **sessile**.

In the depths beyond the continental shelves, benthic organisms must be adapted to little or no light, generally cooler temperatures, and high pressures. Without light, photosynthetic microbes do not exist. This is the **bathyal** zone. It extends from the edge of the shelf to a depth of about 4000 meters. Still deeper levels constitute the **abyssal** environment. The term **hadal** is reserved for the extreme depths found in oceanic trenches. As might be expected, animals are less abundant in the abyssal and hadal environments. Most of these deep-water creatures are scavengers that depend on the slow fall of food particles from higher levels, or they are predators that feed on the scavengers. As will be described in Chapter 3, the abyssal environment also includes ecosystems associated with hydrothermal vents on the ocean floor. At such locations, heat-loving bacteria that derive their energy from chemosynthesis form the base of a food chain that supports tube worms, arthropods, and molluscs.

ADAPTATIONS

In paleontology, an **adaptation** is any aspect of form, function, or behavior that increases the probability of survival and reproductive success of an organism under a given set of environmental conditions. Paleontologists are keenly interested in adaptations because they provide clues, not only to the way ancient organisms lived but also to the nature of their physical and biological environment. The adaptations of a benthic animal may reveal if it crawled on the surface of the seafloor or burrowed beneath the surface, if that surface was well indurated (hardened) or consisted of soft sediment, if it was predator or prey, or if the environment was illuminated or dark.

A sampling of adaptive traits among animals known only as fossils would include the broad, spiny valve of the brachiopod *Marginifera ornata* (Fig. 2.13), which was an adaptation that served to support and anchor the animal on a seabed composed of soft mud. Reduction of the skeleton to mere spines in the Silurian trilobite *Deiphon* (Fig.

Figure 2.13   The spinose pedicle valve of the brachiopod *Marginifera ornata* from the Permian of Pakistan. The animal lived on a substrate composed of soft, very fine-grained sediment. (Courtesy of R.E. Grant.)

2.14) was probably an adaptation to provide bouyancy in an animal that fed near the surface of the sea. The zigzag form of the commissure (anterior gape) of many Paleozoic brachiopod shells is an apparent adaptation to prevent large sediment particles from entering the shell while simultaneously increasing the open area for currents carrying in food or eliminating waste (Fig. 2.15). The adaptation may also have protected against predators.

Coiling in living and fossil cephalopods appears to be an adaptation needed to bring the center of bouyancy above the center of gravity and thereby permit these creatures to keep an "even keel" while swimming. A list of additional adaptations

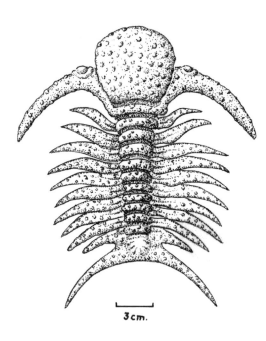

Figure 2.14   Although the majority of trilobites were benthic animals, the Silurian trilobite *Deiphon* illustrated here may have been pelagic. This is suggested by its spinosity, which may have promoted buoyancy. (Adapted from H. J. Harrington et al. In R. C. Moore (Ed.), *Treatise on Invertebrate Paleontology.* New York and Lawrence, KS: Geological Society of America and University of Kansas Press, 1959.)

3 cm.

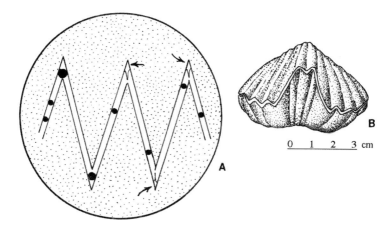

Figure 2.15   Costae (radial ridges) and plicae (radial ridges and grooves affecting both interior and exterior of a valve) in brachiopod shells result in a zigzag opening between the two valves (A). When valves were slightly opened, the zigzag space limited the number of unwanted larger particles that might otherwise enter the shell in inhalent currents. Some brachiopods developed delicate spines at the apex of each "V" in the zigzag pattern to further prevent entry of large particles (see arrows). (B) Zigzag opening as it would be seen in the slightly opened shell of the Ordovician brachiopod *Platystrophia*.

found throughout the fossil record would be immense, for every creature exhibits hundreds of adaptations. Not all of these are morphological. Biochemical and physiological adaptations also exist, although these are difficult to recognize in fossils. Moreover, not all structures are adaptations, for many features seem to have no apparent value or are even a detriment to the organism. Some may be relics of structures derived from ancestral forms, or they may simply be mutational novelties. Stephen J. Gould and E. S. Vrba have suggested that the term *adaptation* be restricted to those features of an organism that can be shown to have arisen as a result of natural selection. They recommend the use of the term **exaptation** for features that appear not to have been shaped by natural selection, but that nevertheless conveyed some advantage. Such exaptations may experience secondary adaptation to improve their effect further.

Adaptations have a purpose or function, and it is one of the more interesting tasks of paleontologists to discover that purpose. To do this one attempts to make functional morphological analyses of the kinds described below.

1. *Comparison with extant counterparts.* Where close living relatives of extinct species still exist, it is sometimes possible to relate analogous features in the living form to the fossil. For example, paleontologists have used primitive crustaceans called *cephalocarids* to reconstruct the musculature of extinct trilobites. The living nautiloid cephalopod *Nautilus* reveals the probable role of the cephalopod siphon in providing for fluid exchange in the chambers of the shell or conch.

2. Functional morphologic experiments. Often models of fossil animals can be constructed and used in experiments to test hypotheses about the function or probable benefit of an adaptive trait. Models of cephalopods of various shapes have been suspended in currents of water so as to measure the conch's resistance to the current. Such experiments provided information about the swimming efficiency of the various cephalods. M. J. S. Rudwick constructed models of aberrant Permian brachiopods called richthofenids, which have a large conical pedicle valve and a simple flat brachial valve (Fig. 2.16). By submersing the model in water containing visibly suspended oil droplets, Rudwick showed that the flat dorsal valve, by rapidly opening and closing, generated currents that carried food into the ventral valve and digestive system. In fact, he was able to film the process. Thus, the experiment supported the hypothesis that the flat dorsal valve was an adaptation for more efficient feeding, and it tended to discredit hypotheses stating its sole function was protection against predators.

   In another example of experimental functional analysis, J. G. Kingsolver and M. A. R. Koetel prepared models of Paleozoic insects. They placed the models in wind tunnels and tested theories about how the wings functioned aerodynamically, as well as in thermoregulation.

3. *Computer simulation.* David Raup has shown that it is possible mathematically to derive all the theoretically possible variations in coiling among coiled species of invertebrates such as cephalopods, gastropods, brachiopods, and foraminifers. Raup demonstrated that only four parameters are needed to accomplish this task, namely the shape of the cone in cross-section, the rate at which one complete turn or whorl expands, the distance and orientation of the curve of the conch relative to the axis of coiling, and the rate of translation around the axis. As a result of such analyses, paleontologists are able to understand better the constraints on design of shells and the range of variation that is possible.

4. *The paradigm method.* The paradigm method for deducing the relation of morphology to function was proposed by M. J. S. Rudwick in 1964. A **paradigm** is a model or form that can be invoked to explain the function of some morphological feature. In practice, the paradigm method includes many of the basic inferences used in experimental analyses. One first identifies and describes the structure under study. One then infers all the possible functions that can logically be related to the structure. Next, one formulates a paradigm to fulfill each of the

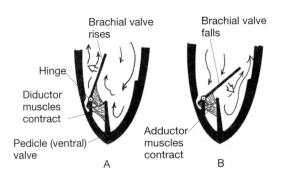

Figure 2.16   Conceptual diagram to show the action of the brachial valve of a Permian richthofenid brachiopod as it might have functioned in bringing food-bearing currents into the conical pedicle valve. (Concept from M. J. S. Rudwick, 1970. *Living and Fossil Brachiopods.* London: Hutchinson & Co.)

proposed functions. Finally, the paradigm with the closest fit to the real form is considered to represent the actual function of the feature.

FOSSILS AND PALEOGEOGRAPHY

The global distribution of present-day animals and plants is closely controlled by geographically related environmental conditions. Any given species has a definite range of conditions under which it can survive, and it is generally not found outside of that range. Ancient organisms, of course, conformed to their environments in a similar way. Therefore, it is possible to plot the locations of extinct organisms of a given age on a map and reconstruct the geography of the past. At the simplest level, locations of fossils of marine and terrestrial animals of the same age can provide information about the distribution of lands and seas for that age. Further study of the fossils of marine organisms may indicate which of the seas were deep, which were shallow, where reefs and shorelines were located, and where land barriers existed. The rise of the Panamanian land bridge during the late Cenozoic provides an example of the effect of land barriers on marine faunas. This land bridge between North and South America separated a formerly homogeneous marine fauna into two components having no possiblility for genetic interaction. The Pacific and Atlantic groups continued to evolve independently, producing a number of parallel but different species. The result was that the sum of species that existed after the barrier was established was greater than those that existed in the former homogeneous fauna. As the marine invertebrates were experiencing this divergence, the land bridge provided the opportunity for convergence among North and South American land animals, with fewer species present after the two continents were connected than had existed before.

In addition to providing clues to the locations of land bridges, seaways, and shorelines, fossils may also indicate locations of parallels of latitude, pole positions, and locations of drifting continents at various times in the geologic past. Today, diversity of species generally decreases at higher latitudes. In contrast to polar regions of low diversity (but abundant individuals of the fewer species), equatorial regions tend to have greater numbers of species with relatively fewer individuals within each species. The observation that species diversity increases from the poles to the equator can sometimes be used to infer the locations of global climatic zones during the geologic past.

The cause of the change in species diversity with latitude is somewhat problematic. It may be related to stress placed on evolving organisms living at increasingly higher latitudes by cold, seasonality, or more limited food sources. Whatever the cause, relatively fewer organisms appear able to adapt to the rigors of polar climates. Nearer the equator there is a stable input of solar energy, and a more assured supply of food and light. Of course, the generalization that biotas range from low diversity in high latitudes to high diversity in low latitudes can be upset in particular parts of the ocean by upwelling, currents, water depth, and hydrothermal activity on the ocean floor.

Ancient coral reefs provide yet another means to locate former equatorial regions (and therefore also the polar regions that are centered at 90° of latitude to either side). Nearly all living coral reefs lie within 30° of the equator (Fig. 2.17), and it is a reasonable assumption that ancient corals reefs existed at similar latitudes.

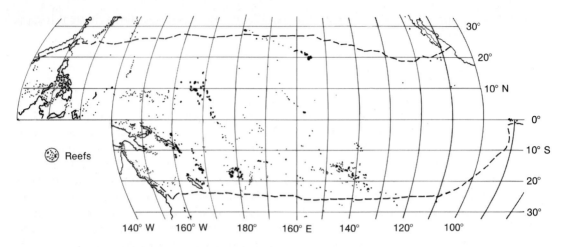

Figure 2.17 Distribution of living coral reefs in the Pacific Ocean. Latitudinal limits of reef growth are indicated by the broken lines. (After J. A. Wells, *Geological Society of America Memoir 67,* 1:609–631, 1957.)

FOSSILS AS PALEOCLIMATIC INDICATORS

Climate, and especially the temperature component of climate, is a major factor in determining the distribution of organisms. Fossils yield information about ancient climates in many ways. An analysis of fossil pollen may provide outstanding paleoclimatologic evidence. One can often compare fossils of extinct forms with living relatives having known climatic tolerances. As noted above, corals thrive in regions where water temperatures rarely fall below 18°C, and it is possible that their ancient counterparts were similarly constrained.

Climatic and other environmental conditions can sometimes be inferred from the size and architecture of the shells of marine invertebrates. For example, marine mollusks with well-developed spinosity and thick shells predominate in warmer regions of the ocean. *Globorotalia truncatulinoides* (Fig. 2.18), a planktonic foraminifer of the Pacific Ocean, develops a shell that coils to the left during episodes of relatively cold conditions (i.e., glacial stages), and to the right during warmer episodes (interglacial

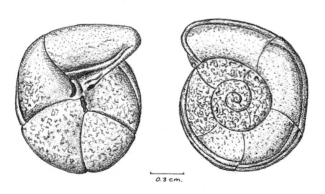

Figure 2.18 The two sides of a left coiling specimen of *Globorotalia truncatulinoides*. (Redrawn from F. L. Parker, *Micropaleontology* 8(2):219–254, 1962.)

0.3 cm.

stages). Another species, *Neogloboquadrina pachyderma*, exhibits similar changes in coiling directions, not only in the Pacific but also in the Atlantic Ocean. Because such reversals in coiling sometimes occur rapidly over broad regions of the ocean, they are noted carefully and used as biostratigraphic markers. Aside from these temperature-related morphological changes in foraminifers, entire assemblages of these protoctistans are used in the determination of paleoclimates. The ecology of many living foraminifera is well known, and this permits one to make accurate inferences about climate, as well as water depth and chemistry. Such comparisons, however, must be done cautiously, for certain species may have altered their ecologic ranges through time.

In cases in which the fossils themselves do not provide clues to temperature or climate, their skeletal composition may be instructive. Magnesium, for example, often substitutes for calcium in calcium carbonate invertebrate shells. For some marine invertebrates it has been found that high magnesium values correlate with a warmer water environment. In a somewhat similar way, the ratio of the two calcium polymorph minerals calcite and aragonite can be related to differences in water temperature. In colder regions, particular species will have higher calcite-to-aragonite ratios, whereas the ratio is lower in the same species living in warmer water. It would appear that secretion of aragonite is easier at higher temperatures. Interpretations based on calcite-to-aragonite or calcium-to-magnesium ratios, however, may present problems. Aragonite is relatively unstable and may revert to the more stable calcite. This probably explains the near absence of aragonite in older rocks. High magnesium calcite is also unstable, and in time is converted to its low magnesium counterpart. Error may also result from loss of magnesium due to weathering or dissolution after burial.

Oxygen isotope analyses offer yet another way to learn the temperatures of ancient seas. When water evaporates from the ocean, there is a fractionation of the oxygen-16 and oxygen-18 in the water. Oxygen-16, being lighter, is preferentially removed, while the heavier oxygen-18 tends to remain behind. When the evaporated moisture is returned to Earth's surface as rain or snow, water containing the heavier isotope precipitates first, often near coastlines, and flows quickly back into the ocean. Inland, the precipitation from the remaining water vapor is depleted in oxygen-18 relative to its initial quantity. If the interior of the continent is cold and supports growing ice sheets, the glacial ice will lock up the lighter isotope, preventing its return to the ocean and thereby increasing the proportion of the heavier isotope in seawater. As this occurs, the calcium carbonate shells of marine invertebrates such as foraminifera will also be enriched in oxygen-18, and thereby reflect episodes of cold climatic conditions associated with continental glaciation.

Even in the absence of ice ages, however, the oxygen isotope method may be useful as a paleotemperature indicator. As invertebrates extract oxygen from seawater to build their shells, there is a temperature-dependent fractionation of oxygen-18 between the water and the secreted calcium carbonate of the shells. Provided there has been no alteration of the shell, it is possible to find the temperature of the water at the time the shell was secreted. In an early study, the oxygen isotope ratio in the calcium carbonate skeleton of a Jurassic belemnite (a cephalopod with internal shell) indicated the average temperature in which the animal lived was 17.6°C ±6° of seasonal variation. Subsequent studies on both Jurassic and Cretaceous belemnites provided confirmation of the inferred positions of the poles during the Mesozoic, and indicated that tropical and semitropical conditions were globally more extensive during the late Mesozoic than they have been since that time.

FOSSILS AND PLATE TECTONICS

In the early 1960s, the revolutionary concept of plate tectonics was being formulated. Ideas about drifting and fragmenting continents and the opening and closing of ocean basins required the corroborative evidence that fossils provide. Paleomagnetic studies and various other lines of physical evidence indicated that global geography near the end of the Paleozoic was dominated by a large supercontinent roughly centered on the South Pole. Paleontologists were able to discern faunal similarities in the rocks of now widely separated continents that indicated those landmasses were once joined. More subtle indications of the supercontinent could be seen in the uniformity of nearshore benthic organisms that apparently resulted from a paucity of isolated, separate land-masses. A single supercontinent would also provide terrestrial organisms with relative ease of migration and interaction. The result would be cosmopolitan faunas with fewer families than would have been the case if there had been many smaller isolated conti-nents. Both geophysical evidence and paleontologic evidence indicate that the late Paleozoic supercontinent (called Pangaea) began to break apart in the Triassic Period. During the Jurassic and Cretaceous, the breakup and movement of continental frag-ments had produced a seaway that extended westward from the present site of Central America, across southern Europe, to the tropical eastern Pacific (Fig. 2.19). Called the Tethys Sea, it provided a circumglobal equatorial corridor for the spread of tropical marine invertebrates. Along this corridor from the Caribbean to southern Europe, the tropical marine invertebrate fauna of the Late Jurassic and Early Cretaceous was remarkably uniform. By the Late Cretaceous, however, plate movements had pro-duced barriers to dispersal in both the Mediterranean and the Caribbean segments of the Tethys. In addition, seafloor spreading had widened the Atlantic enough to inhibit dispersal of Eurasian and African species westward to America. As a result, Late Cretaceous tropical marine invertebrates on either side of the Atlantic no longer resembled one another.

In contrast to the Pangaean world of the Late Paleozoic, continents during the Cenozoic were exceptionally fragmented and widely dispersed across a considerable range of latitude. For the Cenozoic Earth, there existed, and continue to exist, a great variety of habitats, a large number of individual centers for adaptive radiations, and a correspondingly rich diversity of life.

REVIEW QUESTIONS

1. What are evolutionary faunas? Name three evolutionary fauna and indicate the invertebrate groups characteristic of each.
2. Distinguish between the following:
   a. geochronologic and chronostratigraphic units
   b. the Lazarus and the Zombie Effects
   c. taxon range zone and concurrent range zone

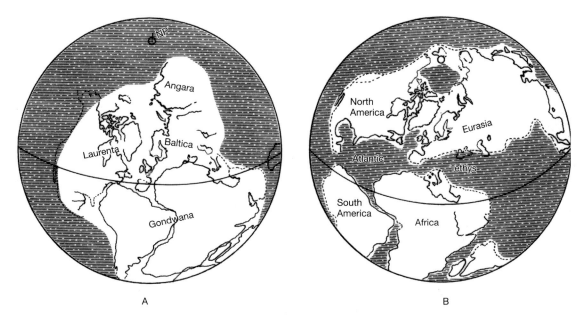

Figure 2.19   Global paleogeography during the Permian (A) and mid-Cretaceous (B). Note the Cretaceous circumglobal marine corridor just north of the equator. (After E. Irving. Fragmentation and assembly of the continents, mid-Carboniferous to present, *Geophysical Surveys* 5:299–333, 1983.)

   d. sessile and vagile invertebrates
   e. invertebrates of the littoral zone and those of the sublittoral zone
   f. plankton and nekton
   g. producer organisms and consumer organisms
3. What is a mass extinction? When did Earth's five greatest episodes of mass extinction occur? Compile a list of possible causes of mass extinctions.
4. Using the law of superposition and rule of cross-cutting relationships, sketch a hypothetical cross-section that illustrates how adjacent rocks yielding isotopic dates can define to some degree the actual age in years of fossiliferous sedimentary rocks.
5. What are ecologic-evolutionary units? What relation, if any, do such units have to the concept of punctuated equilibrium in Chapter 1?
6. What characteristics are usually present in a species of fossil invertebrate that causes it to be particularly useful in stratigraphy?
7. Discuss the relationship between changing levels of global biologic diversity and changes associated with the fragmentation and aggregation of landmasses.
8. What effect did the development of a land bridge between North and South America during the Late Cenozoic have on the diversity and distribution of marine invertebrates?

SUPPLEMENTAL READINGS AND REFERENCES

Ager, D. V. 1963. *Principles of Paleoecology.* New York: McGraw-Hill.

Boucot, A. J. 1981. *Principles of Benthic Marine Paleoecology.* New York: Academic Press.

Boucot, A. J. 1983. Does evolution take place in an ecological vacuum? II. *Journal of Paleontology* 57(1):1–30.

Boucot, A. J. 1990. *Evolutionary Paleobiology of Behavior and Coevolution.* New York: Elsevier.

Dodd, J. R. & Stanton, R. J. 1990. *Paleoecology: Concepts and Applications*, 2nd ed. New York: John Wiley.

Donovan, S. K. 1989. *Mass Extinctions.* New York: Columbia University Press.

Elliot, D. K. (Ed.). 1986. *Dynamics of Extinction.* New York: John Wiley.

Harland, W. B., Armstrong, R. L., Cox, A. V., Craig, L. E., Smith, A. G. & Smith, D. G. 1990. *A Geologic Time Scale.* New York: Cambridge University Press.

Lane, N. G. 1992. *Life of the Past*, 3rd ed. New York: Macmillan.

Sepkoski, J. J. & Raup, D. M. 1986. Periodicity in marine mass extinctions. In D. K. Eliot (Ed.), *Dynamics of Extinction.* New York: John Wiley.

Sepkoski, J. J. 1990. Evolutionary faunas. In D. E. G. Briggs & P. R. Crowther (Eds.), *Paleobiology: A Synthesis.* Oxford: Blackwell Scientific.

Stanley, S. M. 1979. *Macroevolution.* San Francisco: W.H. Freeman.

Stanley, S. M. 1987. *Extinction.* New York: Scientific American Library.

Valentine, J. W. 1973. *Evolutionary Paleoecology of the Marine Biosphere.* Englewood Cliffs, NJ: Prentice-Hall.

CHAPTER 3

Late Proterozoic frond and discoidal fossils from the Conception Group of Newfoundland. The frond fossil at the left is 20cm in length (Courtesy of S. B. Misra).

Early Life

I am, in point of fact, a particularly haughty and exclusive person of pre-Adamite ancestral descent. You will understand this when I tell you that I can trace my ancestry back to a protoplasmal, primordial, atomic globule.

W. S. Gilbert, *The Mikado*

The origin of life was the greatest of all events in biology and is of surpassing human interest. Small wonder that biologists and chemists find it such an intriguing topic for investigation. Presently, the answer to the question of how life arose on our planet has not been comprehensively answered, although recent discoveries in biochemistry, cell biology, and the behavior of viruses and genes appear to be converging on an answer. These studies indicate the strong probability that, on the primordial Earth, there were spontaneous organizations of carbon-containing molecules that were in time able to combine and form increasingly more complex molecules. Eventually viruslike molecules were produced that were capable of directing the synthesis of other units like themselves, and initiating a chain of events leading to the first unicellular organism. These remarkable transformations occurred nearly 3.8 billion years ago when Earth's atmosphere lacked molecular oxygen (O_2), when the planet was bathed in lethal radiation, and when volcanic activity was rampant.

THE PRIMORDIAL EARTH

DUST CLOUD TO PLANETARY HABITAT

Earth is the abode of life. How did it originate? What was it like "in the beginning"? The most adequate answers to these questions are provided by the so-called protoplanet hypothesis. According to this hypothesis, the formation of the solar system began when an immense cloud of dust particles and gases in space contracted and flattened as it took on a counterclockwise rotation. Turbulent eddies developed within the cloud, and some of these became collecting sites for matter brought from neighboring swirls. When the cloud had flattened to disklike form and had shrunk under the influence of gravity to a size somewhat larger than the present solar system, its denser state permitted particles to condense and grow. Small bodies merged to form boulder-size chunks, and swarms of these larger bodies swept up finer particles in their orbital paths. In this fashion they grew by accretion until some had reached the size of asteroids, moons, and planets. These events probably occurred over a span of about 10 million years.

If Earth developed in this accretionary manner, one must then account for the fact that it is not homogeneous from center to surface, but rather has internal zones of increasingly dense materials. Beneath the crust lies the mantle, which is composed of iron- and magnesium-rich silicate rocks. The mantle has an average density of about 4.5 g/cm^3. Beneath it lies the outer and inner cores, both believed to be composed of iron and nickel, and having an average density of 10.7 g/cm^3. These internal zones may have developed as a result of partial melting of the planet, which would allow liquid iron and other dense elements to settle toward the core, and lighter elements to migrate to the region occupied by the mantle. The heat needed for such an event could have been derived from the decay of radioactive elements, gravitational compression, and heat generated by the shower of incoming meteorites. Some planetary scientists, however, believe that a heating event was not needed to produce Earth's internal layering. They suggest the planet began its accretion when the dust cloud was hot, and that heavy, iron-rich compounds condensed in a mass that became the core. Subsequently, as the cloud was cooling, lower-density silicate minerals were swept up and added to the initial iron-rich body as an enclosing mantle.

ORIGIN OF THE ATMOSPHERE AND OCEAN

Although the question of how Earth's internal zonation originated continues to be debated, there is general agreement that our planet formed by aggregation of small and large bodies that were once part of a nebula of gas and dust. The particles and meteorites incorporated into the accreting planet contained significant amounts of gases. Water was also present in the hydrated minerals within meteorites. Both water and gases were released by heating and volcanically vented at Earth's surface. The process has been termed *outgassing*. Calculations indicated that the episode of outgassing that provided most of the water at Earth's surface was completed during the first billion years of Earth history. Since that time water has been continuously recirculated by evaporation and precipitation. Some is temporarily removed by being incorporated into sediments, but eventually these sediments are moved to subduction zones and melted into magmas that return water to the surface as volcanic exhalations. The early presence of an ocean is clearly indicated by marine metasedimentary rocks dating from as long ago as 3.9 billion years.

In addition to water vapor, the outgassed volatiles produced an atmosphere rich in carbon dioxide and nitrogen, with lesser amounts of carbon monoxide, hydrogen, and hydrogen chloride. The volatiles released by present-day volcanoes approximate those present in the early atmosphere, with the exception that primitive exhalations probably contained a higher amount of hydrogen and small amounts of methane and ammonia as well. Appreciable amounts of uncombined oxygen, however, were not discharged into the early atmosphere. Whenever small amounts of oxygen were released from the interior, it was quickly taken up by easily oxidized metals like iron, and thus it was never allowed to build up as an important component of the atmosphere.

After being vented at Earth's surface, the volatiles were subjected to a variety of changes. Water vapor was removed by condensation, and as Earth's solid surface cooled, an ocean began to form. The outgassed carbon dioxide and compounds of chlorine and sulfur made the ocean water considerably more acidic than it is today. Early acid rains promoted rapid chemical weathering, and thereby brought calcium, magnesium, and other elements into solution. When the seas were less acidic and oxygen had become an important component of the atmosphere, these elements were joined with carbon dioxide to form carbonate rocks, the calcareous structures of cyanobacterial mats, and the shells of myriads of marine creatures.

The earliest forms of life on Earth evolved in the anoxic environment described above. Only after life had begun could changes occur that would gradually produce our present oxygen-rich atmosphere. Free oxygen (O_2) entered Earth's atmosphere as a result of two processes. The one of lesser importance involved the dissociation of water molecules into hydrogen and oxygen. This process, termed *photochemical dissociation*, continues today in the upper atmosphere when water molecules are split by high-energy beams of utraviolet (UV) radiation from the Sun. Photochemical dissociation, however, does not produce oxygen at a rate sufficient to balance loss of the gas by dissipation into space. Another far more effective oxygen-generating process is required to attain and maintain oxygen levels similar to those that exist today in our atmosphere. That process is **photosynthesis**. It arrived with the advent of organisms that had evolved the capability of separating carbon from oxygen in atmospheric carbon dioxide.

Are there geologic clues to the change from an anoxygenic to oxygenic atmosphere? The most ancient rocks providing any sort of evidence are about 3.8 billion years old. Mineral grains within these old rocks show evidence of weathering, and there can be no weathering without an atmosphere. Further, the atmosphere under which these rocks were weathered must have lacked free oxygen, for minerals like hematite (Fe_2O_3), which form when iron is oxidized, are absent. Instead, one finds metallic sulfides that develop in oxygen-deficient environments. The lack of carbonate rocks such as limestone and dolomite are notably absent in older Precambrian sequences. An early atmosphere rich in carbon dioxide would produce water that is acidic, preventing the precipitation of carbonate minerals.

Evidence of the transition to an oxygenic atmosphere first appears in rocks about 2.5 billion to 1.8 billion years old. Called *banded iron formations* (BIF for short), these rocks have red iron-rich bands that alternate with thin gray bands. The red bands are colored by iron oxide. They reflect a periodic abundance of free oxygen in the atmosphere, whereas the gray layers suggest periods of low free-oxygen levels. Iron served as a collector of oxygen, precipitating the oxides as rapidly as atmospheric oxygen was generated. After much of the surface oxygen had been oxidized the oxygen supply became stable, so that rocks higher in the section are unbanded and uniformly oxidized. These unbanded rocks are about 1.8 billion years old, and they inform us of the prevalence of free oxygen in the atmosphere by that time. There is little doubt the oxygen was derived from cyanobacteria and marine algae, whose extensive development is documented by stromatolitic deposits in Precambrian rocks around the world. Also, with the attainment of a persistent oxygenic atmosphere, carbonate sediments began to accumulate in marine basins much as they do today.

THE BEGINNING OF LIFE

The favored hypothesis for the origin of life proposes that it developed on Earth from nonbiological (abiotic) chemical compounds about 3.8 billion years ago when the planet's surface temperature had fallen below 100°C. It occurred in the presence of an atmosphere deficient in free oxygen. As a result, there was no ozone screen to block UV radiation. With that radiation as a source of energy (and possibly electrical energy from lightning) there is a probability that water, carbon, hydrogen, and phosphorus at Earth's surface would chemically react and produce amino acids. Some amino acids may already have been present, as indicated by traces of this compound in meteorites known as *carbonaceous chondrites*. Amino acids are the building blocks of proteins, and therefore essential for life.

To test the hypothesis that amino acids could have formed on the early planet from abiotic precursors, Stanley Miller performed a now famous experiment in 1953. At the time Miller performed the experiment Earth's primordial atmosphere was thought to contain significant quantities of methane and ammonia, as well as water vapor and hydrogen. These gases were infused into an apparatus similar to the one shown in Figure 3.1. As the mixture was circulated, sparks of electricity (simulating lightning) were discharged into the chamber of the apparatus. At the end of 8 days, the condensed water in the apparatus had become turbid and deep red. Analysis of the crimson liquid showed that it contained a bonanza of amino acids as well as somewhat

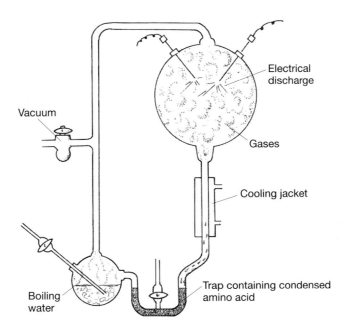

Figure 3.1   Diagram of the apparatus used by Stanley Miller to reproduce atmospheric conditions of early Earth. An electric spark was produced in the upper right flask to simulate lightning. The gases present in the flask reacted together, forming a number of simple organic compounds that accumulated in the trap at the bottom.

more complex organic compounds known to exist in living organisms. In subsequent experiments by other biochemists, similar compounds were synthesized in the presence of an atmosphere composed of carbon dioxide, nitrogen, and water vapor. The principal requirement for the success of the experiments was the absence of free oxygen. To the experimenters, it now seemed certrain that amino acids would have inevitably developed on Earth's primordial surface.

For amino acids to combine and form proteinlike molecules, they must lose water. This loss can be accomplished by heating the water containing amino acids to a temperature of at least 140°C. Thermal springs and volcanic emanations on the primitive crust could have provided such hot water. The reaction also occurs at temperatures as high as 70°C if phosphoric acid is present. In the laboratory, S. W. Fox was able to produce proteinlike chains from a mixture of 18 common amino acids. He termed these structures *proteinoids*, and reasoned that they may have been the transitional structures leading to true proteins. Subsequently, Fox found similar proteinoids in pools of hot water adjacent to active Hawaiian volcanoes. Apparently amino acids formed in the volcanic vapors and were combined to form proteinoids by the heat of lavas and escaping gases.

In brines and hot aqueous solutions, proteinoids tend to concentrate into tiny globules or microspheres that show certain characteristics of living cells. They exhibit a membranelike outer wall (reminiscent of a cell wall), are capable of osmotic swelling and shrinking, and have a kind of budding "reproduction" as they divide into daughter microspheres. Occasionally, they join into a chain in the manner of filamentous bacteria, and they exhibit a streaming motion of internal particles like that seen in true cells. These are significant similarities, yet there remain many important differences between microspheres and living cells. Microspheres do not truly replicate themselves, produce energy by feeding, or carry on photosynthesis.

THE NATURE OF EARLY LIFE

Although biochemists have not yet produced a living cell, they have provided evidence that abiotic synthesis of life in Earth's primordial environment would have been a likely occurrence. The sequence of events probably began with the synthesis of simple organic compounds in the manner suggested by Stanley Miller's experiment. The compounds thus created reacted to form larger structures called *polymers*, which are fabricated by the joining of repeating subunits. The polymers then reacted to form structures of increasingly greater complexity, leading eventually to living organisms. Even before those organisms appeared, however, the evolving organic structures probably competed with one another until those with the most stable and efficient characteristics remained. Perhaps some acted as templates upon which additional molecular components could be attached, giving rise ultimately to life at the level of a virus or bacterium. It is likely that natural selection was already an influence on these early transformations.

Many investigators believe that the earliest forms of life would not have evolved a means of manufacturing their own food, but rather assimilated small aggregates of organic molecules in their surrounding medium. Some life-forms, no doubt, would have even consumed some of their developing contemporaries. Organisms with this type of nutritional mechanism are termed **heterotrophs**. Food gathered by ancestral heterotrophs was externally digested by excreted enzymes before being converted into the energy needed for vital functions. In the absence of free oxygen, this can be accomplished by fermentation. There are several variations of the fermentation process. The most familiar involves fermentation of sugar by yeast organisms. It is a process by which organisms are able to disassemble organic molecules, rearrange their parts, and derive energy for life functions. A simple fermentation reaction can be written as follows:

$$C_2H_{12}O_6 \rightarrow 2CO_2 + 2C_2H_5OH + energy$$
Glucose   Carbon   Alcohol
dioxide

Animal cells are also able to ferment sugar in a reaction that yields lactic acid rather than alcohol.

By consuming organic compounds in their surroundings, the original fermentation organisms would eventually create an energy crisis, and the growing scarcity of food may have caused selective pressures for evolutionary change. At some point prior to the depletion of the food supply, organisms evolved the ability to synthesize what they required for sustenance from simple inorganic compounds. Organisms having this ability are called **autotrophs**. Some manufacture their food from carbon dioxide and hydrogen sulfide in a manner similar to that seen in today's sulfur bacteria. Others employ the mechanism of certain living bacteria that use ammonia as a source of energy and matter. However, of more importance than either of these kinds of autotrophs were the **photoautotrophs**. They had evolved the capability of dissociating carbon dioxide into carbon and free oxygen. In this familiar process of **photosynthesis**, the carbon combined with other elements to permit growth, and oxygen escaped to gradually change the composition of the atmosphere and thereby prepare the environment for the next important step in the evolution of Precambrian life. In its simplified form, the photosynthesis reaction is often summarized as follows:

$$6CO_2 + 6H_2O + \text{absorbed light energy} \rightarrow C_6H_{12}O_6 + 6O_2 + \text{chemical energy}$$

With the multiplication of oxygen-generating photoautotrophs, the Earth's primeval anoxic atmosphere was changed to one containing molecular oxygen. It was fortunate that the change was a gradual one, for if oxygen had accumulated too rapidly, it would have been lethal to developing microorganisms. Oxygen acceptors were needed to act as safety valves and prevent rapid buildup of the gas. As described earlier, iron in Earth's rocks and sediments provided the necessary oxygen acceptors, at least until organisms had evolved oxygen-mediating enzymes that would have permitted them to cope with the new atmosphere.

After much of the iron at Earth's surface had been oxidized and oxygen had begun to accumulate in the atmosphere, solar radiation acted on the gas so as to convert part of it to ozone. The ozone in turn formed an effective shield against harmful UV rays. Still-primitive and vulnerable life was thereby protected and could expand into environments that formerly had been subjected to lethal radiation. The stage was set for the advance of diverse aerobic organisms that use oxygen to convert their food to energy. The reaction, which can be considered a form of cold combustion, provides far more energy in relation to food consumed than does the fermentation reaction. This surplus of energy was an important factor in the evolution of more complex forms of life.

THE BIRTHPLACE OF LIFE

Charles Darwin suggested that the ocean was the birthplace of life. This traditional view has been with us for many decades. It has been nourished by our knowledge that ocean water serves as a solvent for a variety of vitally important organic and inorganic compounds, that ocean currents and waves ceaselessly mix and circulate these critical substances, and that the upper levels of the ocean are well illuminated for photosynthesis and serve as a partial shield against harmful ionizing radiation. Recent investigations, however, suggest that life may have originated in environments quite different and, to our view, far more hostile than that of the ocean. The ocean, rather than serving as the birthplace of life, may have been the site of an expansion of life that began elsewhere. Perhaps it arrived in a meteorite from another planet, or was born deep within solid Earth. We now recognize, for example, microbes of the Domain Archaea thriving in granite lying over 3 kilometers beneath the Earth's surface. These microbes not only

---

BOX 3.1   EXTRATERRESTRIAL LIFE IN THE SOLAR SYSTEM?

For centuries, people have wondered about the possibility that life exists or once existed on other planets. Is organic evolution a process unique to Earth? Might not the conditions that made the origin and evolution of life possible on Earth also exist elsewhere in the universe? And what are those essential conditions? To begin with, Earth is large enough that its gravitational attraction is sufficient to retain an atmosphere. The temperature across most of Earth's surface is low enough to provide an abundance of liquid water. Our planet also has all of the chemical elements required for life processes. Earth's temperature, which is a consequence of the size of the Sun and our distance from the Sun, is suitable for the chemical reactions required for life.

If extraterrestrial life exists within our solar system, it is likely to be microbial in nature. All evidence to date indicates that because of either size or distance from the Sun, conditions are currently too harsh on neighboring planets to permit the evolution of *higher* forms of life. Interest in organisms at the level of those in the Domains Archaea or Bacteria, however, was recently aroused by reports of organisms, or the chemical traces of organisms, within a Martian meteorite that landed here on Earth. The story of the Martian meteorite began about 16 million years ago when another meteorite struck Mars with sufficient force to blast fragments of its rocky surface into space. After traveling through space for millions of years, one of these fragments landed on an Antarctic ice field. It was found there in 1984 and labeled ALH 84001. If the meteorite was truly from Mars, it would contain certain chemical clues from that planet. In 1996, instruments aboard the Viking Lander probe had analyzed the atmosphere of Mars. ALH 84001 contained traces of gases identical to those identified in the Viking analysis. (Meteorites from the Moon or Venus lack these gases.)

Having confirmed the Martian origin of the meteorite, the search for evidence of life within it began. One interesting discovery were tiny carbonate globules within the meteorite. The carbonates were probably precipitated within the rock when it was part of the Martian

---

tolerate, but actually prefer, temperatures exceeding 60°C and pressures greater than 900 atmospheres. They are called *thermophiles* for their love of heat and *barophiles* for their fondness of high pressure.

Other examples of organisms of hostile environments include the acid-loving Archaea called *acidophiles*, which exist in sulfuric springs; the *halophiles*, which live in highly saline ponds and lakes; *psychrophiles*, which flourish on sea ice; and alkali-loving *alkaliphiles* of soda lakes. Life may have originated in one of these harsh environments. It may also have begun at great depths in the ocean, in total darkness, among jets of scalding water rising from submarine hydrothermal vents and volcanic fissures. Evidence for such a stygian beginning of life consists of bacteria called **hyperthermophiles**, which thrive in seawater at temperatures that exceed 100°C. These organisms are able to live deep within fissures below the actual vents, and they are often ejected in such great numbers that they cloud the surrounding waters. In the absence of light for photosynthesis, the bacteria derive energy from **chemosynthesis**, a

surface about 3.6 billion years ago. At that time the planet was wetter and warmer than at present. Water is essential for life, and the carbonate globules suggest the Martian temperature was once warm enough for the existence of water in its liquid form. In addition, investigators detected within the carbonate globules small amounts of organic molecules called *polycyclic aromatic hydrocarbons* (PAHs). Such compounds can be derived from the decomposition of bacteria. They may also, however, be produced inorganically. PAHs are common in factory smoke and engine exhaust and may have gotten into the meteorite as it entered Earth's atmosphere.

Did meteorite ALH 84001 contain direct evidence of organisms? With the aid of a high-resolution scanning electron microscope, investigators found ultramicroscopic, sausage-shaped bodies similar in shape, but not size, to certain bacteria known on Earth. But were these tiny structures really fossils? They were far smaller than Earth bacteria, and evidence of cell walls, cell division, or cell growth could not be discerned. Purely inorganic processes have been known to produce similar sausage-shaped objects.

From the above, it appears that confirmation of the former existence of life on Mars will require fossils clearly recognizable as those of true cells. A Mars lander is scheduled to examine rocks on the surface of the red planet in the year 2005 and deliver a sample to Earth about 2 years later. Without doubt, the rocks will be scrutinized in great detail for unambiguous evidence of life.

The satellite of Jupiter known as Europa is another planetary body in our solar system that may have the potential for harboring life. Data from the Voyager flybys of Europa suggest that the satellite has a liquid ocean covered by a thin, fractured crust of ice. That interpretation was confirmed in 1995 by close-up views of Europa sent back to Earth by the Galileo spacecraft. As noted above, water is required for life as we know it, and its presence beneath the icy surface of Europa suggests at least the potential for life on that satellite. By analogy, here on Earth we have discovered psychrophiles that live on sea ice, and species of algae thriving beneath impressive thicknesses of ice above the Arctic Circle.

process within the cell that causes hydrogen sulfide to react with oxygen and produce water and sulfur. As noted earlier, the chemosynthetic bacteria around submarine vents form the base of a food chain supporting an astonishing variety of invertebrates, including shrimplike arthropods, crabs, clams, and giant tube worms.

The environment in which present-day hyperthermophiles develop is a consequence of downward percolation of ocean water through the fracture system beneath mid-oceanic ridges. These waters eventually reach the zone of extremely hot rocks that supply the lavas extruded along the ridges. They are heated to about $1000°C$, yet high pressure prevents them from boiling. Such superheated water reacts readily with surrounding rocks, extracting elements needed to construct organic molecules. One can hypothesize that during the Archean similar molecules resulting from similar processes were conveyed by convection back toward the surface of the ocean floor, cooling as they ascended. Perhaps at that time, amino acids and other organic compounds would have been synthesized and combined to form the first protocells.

Growth, proliferation, and evolutionary processes operating on the protocells might then have led to chemosynthetic bacteria like those found today in hydrothermal vent environments.

PROKARYOTES, EUKARYOTES, AND SYMBIOSIS

There is no unequivocal evidence to date with precision the transition from chemical or prebiotic evolution to organic evolution. The most accurate statement that can be made is that it occurred at some time prior to 3.2 billion years ago. In rocks of that age, one finds the earliest direct fossil evidence of life. The fossils belong to a category of organisms called **prokaryotes** (Fig. 3.2), and are represented today by cyanobacteria and certain other bacteria. Primitive organisms such as these lack internal organelles and do not have membrane-bound nuclei containing genetic material arranged into discrete chromosomes. Modern prokaryotes do possess cell walls, and most are able to move about. Some, like the cyanobacteria, are capable of photosynthesis.

Prokaryotes are asexual and thus restricted in the level of variability they can attain. As noted in Chapter 1, sexual reproduction involves the union of gametes to form the nucleus of a single cell, the zygote. This involves recombination of parental chromosomes and leads to a great number of gene combinations among the gametes that contribute to the next generation. In asexually reproducing organisms, a cell divides from the parent and becomes an independent individual containing the same chromosome complement of the parent cell. For this reason, prokaryotes have shown little evolutionary change over nearly 2 billion years of Earth history. Mutations in prokaryotes can occur, but the pace of evolution caused by such changes is exceedingly slow.

Evolution proceeded from prokaryotes to organisms with a definite nuclear wall, well-defined chromosomes, and the capacity for sexual reproduction. These more advanced forms were eukaryotes. It is difficult to identify positively the first eukaryotes from fossil remains. This is not surprising, for internal structures in microbes are

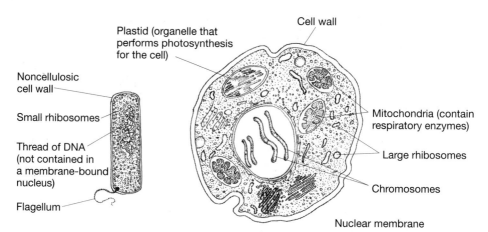

Figure 3.2   Comparison of a prokaryotic and eukaryotic cell. Note that the prokaryotic cell lacks a true nucleus, whereas the eukaryotic cell contains a true nucleus and an assortment of organelles. Also, prokaryotic cells are smaller. They range in size from 0.5 to 1.0 micrometers as compared to 10 to 100 micrometers for eukaryotic cells.

often not preserved or are vague in their altered fossil state. Nevertheless, the fossils do indicate that eukaryotes had begun their expansion about 1.4 billion years ago, and had become fairly common by about 750 million years ago.

Unlike prokaryotes, eukaryotes contain organelles, including plastids (which perform photosynthesis inside the cell) and mitochondria. Mitochondria function in cellular respiration and provide for the release of energy to support cell activities. Many biologists believe that organelles in eukaryotic cells were once independent microorganisms that entered the host cell and were then able to establish a symbiotic relationship. The environment within the host cell may have provided protection for the intruding microbe, or perhaps some were accidentally engulfed and subsequently proved beneficial. In any case, once they were within the host cell, they lost their ability to exist independently. In similar fashion, certain free-living photosynthetic bacteria may have been engulfed by host organisms, and then functioned in the host as a plastid organelle. A nonmotile host may have been provided with motility when it formed a symbiotic relationship with a whiplike spiral bacterium. Natural selection would favor such fortuitous, yet advantageous, affiliations.

ARCHEAN EVIDENCE OF LIFE

Among the oldest fossils on Earth are laminated structures produced by the activities of mat-building communities of microorganisms—principally filamentous photosynthetic prokaryotes such as cyanobacteria—as they trap or cause the precipitation of calcium carbonate. The structures are **stromatolites**, and they are known to have persisted from Archean time to the present (Fig. 3.3). Modern stromatolites are not as common as they were during the Precambrian when there were few molluscan (and other) invertebrates grazing on algal mats. Nevertheless, those that exist today reveal the way their ancient precursors developed. One observes an uppermost matlike layer, termed the *growth layer*, that is formed of a meshwork of filamentous cyanobacteria and associated aerobic microbes. Beneath this layer is a thin undermat containing so-called *facultative aerobes*, which can utilize oxygen if it is available, but are also able to

(A)

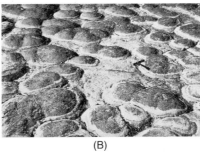
(B)

(C)

Figure 3.3   (A) Modern columnar stromatolites growing in the intertidal zone of Shark Bay, Australia. (B) Fossil stromatolites from the Transvaal Dolomite of Africa (about 2.3 billion years old). (C) Fossil stromatolites from the 2.0-billion-year-old Kona Dolomite of northern Michigan, showing the laminations that are the most distinctive feature of stromatolites. (Stromatolites at Shark Bay courtesy of G. Ross; others courtesy of J. Wm. Schopf.)

carry on by fermentation if free oxygen is not available. Photosynthetic bacteria that do not produce oxygen also live in this layer. The bottom layer, or *oxygen-depleted layer*, is the thickest. It contains a variety of predominantly fermenting anaerobic bacteria. J. William Schopf notes with interest that the upward progression from undermat to growth surface in modern stromatolites mimics the evolutionary sequence presumed to have occurred in the early development of life.

The oldest known stromatolites are found in rocks about 3.5 billion years old from Australia (Pilbara Supergroup), and from South Africa (Swaziland Supergroup). They are well-laminated and have simple undulatory, stratiform, domelike, or columnar shapes. Although filamentous and coccoid microfossils have been found in adjacent chert beds of the same age, as yet no microfossils have been discovered in the stromatolites themselves. Nevertheless, the structures are so similar to undisputed stromatolites in younger rocks that their organic origin appears certain.

During the Proterozoic, stromatolites attained their greatest abundance and diversity. Stratiform, domal, columnar, and branching forms expanded into all major habitable environments. Unlike their Archean precursors, these stromatolites contain sufficient fossil cyanobacteria to indicate that all modern cyanobacterial classes and orders were present by Late Proterozoic.

Remains of Archean fossil microbes that consist of bacterial-like filaments have been recovered from the black chert beds of the Onverwacht Group (Swaziland Supergroup) of Africa and the Warrawoona Group (Pilbara Supergroup) of Australia. These units are about 3.5 billion to 3.4 billion years old. The fossils exhibit microstructures identical to those in living filamentous bacteria and cyanobacteria. Individual cells in the filament have organic cell walls and are arranged in linear series. Reproduction by fission is apparent in partially divided cells along the filaments.

An additional indication of the early presence of life on Earth has been demonstrated by isotopic analyses of the carbonaceous matter preserved in some Archean rocks. During photosynthesis, an enzyme called Rubisco (short for ribulose biphosphate carboxylase/oxygenase) fixes carbon atoms from CO_2 into the organic compounds of an organism. The three most common isotopes of carbon (C) are ^{14}C, ^{13}C, and ^{12}C. The heaviest isotope, ^{14}C, is unstable and is lost after about 50,000 years, leaving ^{13}C and ^{12}C for analysis. An interesting property of Rubisco is that it preferentially reacts with $^{12}CO_2$ rather than $^{13}CO_2$. Rubisco prefers the lighter isotope of carbon. As a result, biologically derived carbonaceous matter will be enriched in ^{12}C relative to the amount of this isotope in atmospheric CO_2 and in rocks like limestone, which contain inorganically derived carbon. When the enrichment is measured with the use of the mass spectrometer, the sample can be identified as of biologic origin. This isotopic signature for biological carbon has been found in Archean rocks as old as 3.5 billion years.

PROTEROZOIC LIFE

Life of the early Proterozoic was not significantly different from that of the preceding late Archean. Blue-green scums of photosynthetic bacteria probably coated rocks and floated on seas and lakes. Anaerobic prokaryotes may have thrived in environments deficient in oxygen. Stromatolites, which were relatively sparse during the Archean, proliferated in the Proterozoic. The proliferation of stromatolites was a probable consequence of the expansion of shallow marine environments.

THE GUNFLINT FOSSILS

Extending eastward from Thunder Bay along the northern coast of Lake Superior are exposures of a rock unit known as the Gunflint Chert. Isotopic dating determinations indicate an age of about 1.9 billion years for the Gunflint Chert. The chert is the host rock for a variety of thread bacteria as well as cyanobacteria. Unbranched filamentous forms, some of which are septate, have been given the generic name *Gunflintia* (see Fig. 1.5). Finer septate filaments, such as *Animikiea* (Fig. 3.4), are remarkably similar to such living organisms as *Oscillatoria* and *Lyngbya*. Other Gunflint fossils, such as *Eoastrion*, resemble living iron- and magnesium-reducing bacteria. *Kakabekia* and *Eosphaera* are so different from any living microbes that their classification is uncertain. Classification difficulties aside, however, the important fact emerges that Gunflint and other stromatolite-associated organisms were proliferating during the Proterozoic and adding oxygen to the atmosphere.

EARLY EUKARYOTES

By about 1.8 billion to 2.0 billion years ago, evolution had progressed to the point where unicellular organisms resembling eukaryotes appear in the fossil record. At about this time, molecular oxygen was becoming a significant component of the atmosphere, and this would have enhanced the evolution of these more advanced organisms. Earliest eukaryotes may have lacked the capacity for sexual reproduction and reproduced by cell division. By about 1.1 billion years ago, however, sexual reproduction

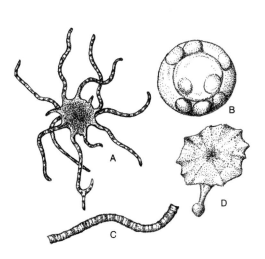

Figure 3.4   Microorganisms of the Gunflint Chert. (A) *Eoastrion*, (B) *Eosphaera*, (C) *Animikea*, and (D) *Kakabekia*. (*Eosphaera* is about 30 micrometers in diameter, and all four organisms are drawn to the same scale.)

was achieved, thereby providing for more rapid and diverse evolutionary change. The fossil evidence indicating sexual reproduction consists of cystlike structures that probably enclosed reproductive bodies. Uncertainty often attends the identification of fossil eukaryotes. This is not surprising when one considers the chances for preservation of such identifying characteristics as an enclosed nuclei and chromosomes. Size, however, provides a simple indicator that a fossil may be a eukaryote. Eukaryotes are much larger than prokaryotes. Living eukaryotes, for example, are nearly always larger than 60 micrometers in diameter, whereas prokaryotes rarely exceed 20 micrometers (see Fig. 3.2).

Fossils known as acritarchs provide another clue to the presence of eukaryotes in Proterozoic rocks. **Acritarchs** (Fig. 3.5) are unicellular, spherical microfossils with resistant, single-layered walls. The cell walls may be smooth or variously ornamented with spines, ridges, grooves, or papillae. Acritarchs are thought to be the resting stage of biflagellate marine algae known as *dinoflagellates*. Like dinoflagellates, acritarchs average about 60 micrometers in diameter and are therefore within the usual size range for eukaryotes. Earliest occurrences of acritarchs are in rocks as old as 1.6 billion years, but they do not become abundant until about 900 million years ago. During the remaining few hundred million years of the Proterozoic, they became common and diverse. As a result, they are widely used as guide fossils for Upper Proterozoic stratigraphic sequences. Their interval of expansion, however, was brief, for cooler climates signaled by glaciation at the end of the Proterozoic resulted in widespread extinctions. Only a few of the simple, smooth-walled forms survived into the Paleozoic.

THE ADVENT OF METAZOANS

Metazoans are multicellular animals that possess more than one kind of cell, and in which cells are organized into tissues and organs. They first appear in rocks that are about 630 million years old, and even at that early time they seem to have achieved global distribution. One of the best-known sites for the collection of early metazoans is in southern Australia at Ediacara Hills. One can consider the Ediacara Hills site a Lagerstätten, for fossils are abundant and provide a glimpse into an otherwise obscure ancient world.

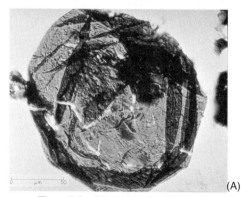
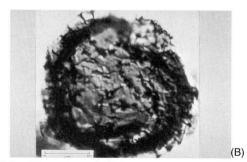

Figure 3.5   Examples of acritarchs from the 850-million-year-old Chuar Group, Grand Canyon of the Colorado River, Arizona. (A) *Kildinosphaera*, a finely striated form. (B) *Vandalosphaeridium*, a spinose acritarch. (Courtesy of G. Vidal.)

Figure 3.6   Fossil specimen of *Mawsonites* in the Rawnsley Quartzite, Pound Group, Ediacara Hills, southern Australia. Note the coarse nature of the host rock, which has greatly diminished the possibility that very small and delicate internal structures would be preserved. (Courtesy of B. N. Runnegar.)

The fossils at Ediacara Hills occur in consolidated quartz sands of shoreline deposits that comprise the Rawnsley Quartzite of the Pound Subgroup. The quartzite contains thousands of impressions of large soft-bodied animals. According to their shape, they tend to fall into three groups. There are discoidal (disklike) forms such as *Mawsonites* (Fig. 3.6), which have been considered to be jellyfish, and *Tribrachidium* (Fig. 3.7), which is difficult to relate to any existing phylum. It is considered an *incertae sedis* (of uncertain systematic position) by some investigators. M. A. Fedonkin would place *Tribrachidium* in a newly established class of the Phylum Cnidaria, which he has named the Trilobozoa.

The second group of Ediacaran specimens can be called *frond fossils* (Fig. 3.8). They resemble the living pennatulate cnidarians informally called *sea pens*. Living sea pens superficially resemble the fronds of ferns, except that tiny, tentacled polyps are aligned along the branches. The polyps capture and consume small organisms that drift

Figure 3.7   Plaster mold made from a fossil specimen of *Tribrachidium*. Diameter of the specimen is about 25 mm. (Mold courtesy of M.F. Glaessner.)

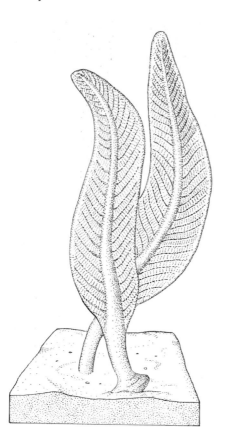

Figure 3.8   Reconstruction of the Ediacaran frondlike fossil *Charniodiscus*. This large organism attained heights of nearly a meter.

by. Frondlike fossils similar to those in the Rawnsley Quartzite are known from Africa, Russia, and England. In the English form, named *Charniodiscus*, the upright frond was attached to a concentrically ringed basal disk that apparently served as a base to stabilize the organism on the seafloor. The disks are often found separated from the fronds, suggesting to some investigators that many of the discoidal fossils common to Ediacaran faunas worldwide may be anchoring structures of frond fossils, and not jellyfish.

The third group of Ediacaran fossils are ovate to elongate forms that were originally regarded as impressions made by flatworms or annelids. Typical of these fossils is *Dickinsonia* (Fig. 3.9), a giant among Proterozoic animals, which attained lengths of up to a meter, and *Spriggina* (Fig. 3.10), which has a crescent-shaped structure at one end followed by a wormlike, slender body. Adolf Seilacher has suggested that these forms, even though superficially wormlike in shape, are related to the disk and frond fossils, and they belong within a single, heretofore unrecognized, taxonomic category.

The many puzzling features of Ediacaran fossils has fueled an ongoing controversy about their place in biologic classification. As suggested above, the traditional interpretation is that these creatures are Proterozoic members of such existing phyla as the Cnidaria and Annelida. Indeed, many paleontologists still retain this opinion.

Figure 3.9   *Dickinsonia costata* from the Ediacara Hills fossil site, southern Australia. Scale divisions are in centimeters. (Courtesy of B. N. Runnegar.)

Others, including A. Seilacher, maintain that most of the Ediacaran animals defy placement in the present taxonomic heirarchy. Seilacher believes the resemblance between living sea pens and frond fossils is superficial and misleading. In support of this view, he notes that the branchlets in frond fossils are fused together and do not have openings through which food-bearing water currents might pass. Living sea pens have such openings, which permit the polyps on the branchlets to capture food in the passing flow of water.

   With regard to the discoidal fossils traditionally deemed jellyfish, Seilacher notes that living jellyfish have radial structures at their centers and concentric structures

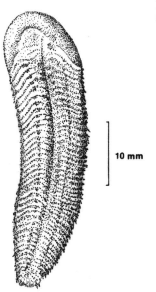

Figure 3.10   *Spriggina floundersi* from the Ediacara Hills fossil site.

10 mm

around the periphery. This arrangement is opposite to that found in the discoidal Ediacaran fossils. Finally, the resemblance of *Spriggina* and *Dickinsonia* to worms or arthropods is questionable in that the fossils reveal no trace of organs essential to such animals, such as mouths, oral cavities, guts, and anuses.

If indeed the Ediacaran organisms do not belong in extant (existing) phyla, then they may represent an early evolutionary radiation of metazoans that failed.

The occurrence of Ediacaran animals in quartzites that were originally sandy deposits also suggests that these creatures were not cnidarians or annelids, for such animals are rarely preserved in coarse granular sediments. For the Ediacaran animals to have left such distinct impressions in the enclosing sediment, they must have had tough body walls. The ribbed and grooved appearance of many Ediacaran fossil imprints suggests to Seilacher that these animals were constructed like air mattresses, and that this type of structure would have provided the firmness needed to make imprints in the sand.

Seilacher also proposes that the organisms may have contained photosynthetic and/or chemosynthetic symbiotic microorganisms, and that the symbiotic relationships would have provided the host animals with oxygen, nourishment, and a means of waste disposal. The large, broad, thin shapes of the Ediacaran animals would have facilitated such symbiotic relationships and provided ample surface area for the diffusion respiration critical to an animal that had not yet evolved the complex circulatory, digestive, and respiratory systems of thicker animals. The validity of Seilacher's hypothesis will probably be determined at a future date when fossils having better preservation are discovered.

Many investigators persist in their opinion that the Ediacaran organisms do not belong in a single taxon, but rather are a diverse group, some members of which can be placed in existing taxonomic categories. Perhaps, as G. J. Retallack proposed in 1994, some of the Ediacaran fossils are the remains of large lichens that were presumably widespread during the Late Proterozoic. Because of his conviction that the Ediacaran organisms cannot be placed in any existing taxonomic category, Seilacher once proposed that they be placed in a new kingdom, the **Vendobionta** (after Vendian, the final geologic period of the Proterozoic in Russia). Recently, however, Seilacher and L. W. Buss have suggested the alternative hypothesis that these enigmatic fossils were a phylum of cnidarianlike organisms (corals, jellyfish, etc.) that lacked cnidocytes (stinging cells).

The Ediacaran organisms made their appearance about 580 million years ago near the end of the Proterozoic. By that time, unicellular life had been thriving on our planet for over 3 billion years. One cannot but wonder why multicellular animals, like those of the Ediacaran fauna, appeared so late (and abrubtly) in geologic time. Perhaps their apparent sudden appearance was related to the accumulation of sufficient free oxygen in the atmosphere to support the evolution of large animals. Prior to the time during which Ediacaran animals flourished, atmospheric oxygen levels did not exceed 1 percent of present levels. Although low, this amount of free oxygen was apparently sufficient for the evolution of unicellular protoctistans and even multicellular marine plants. Large animals, however, require much higher levels of oxygen for their physiological functions, and thus would not have evolved until such levels had been attained. Recent estimates indicate animals like those in the Ediacaran fauna

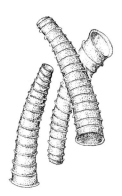

Figure 3.11   *Cloudina*, the earliest known organism bearing a calcium carbonate shell. *Cloudina* was first described from the Late Precambrian Nama Group of Namibia. The shells are about 2 mm long and are believed to be the remains of a tube-dwelling annelid worm.

would require oxygen concentrations equal to or greater than 6 to 10 percent of the level in today's atmosphere.

EARLY SHELLY FOSSILS

The Ediacaran animals were largely soft-bodied. There is, however, an interesting shell-bearing animal of Ediacaran age. First found in the Proterozoic Nama Group of Namibia, Africa, the fossil is named *Cloudina*. *Cloudina* (Fig. 3.11) has also been reported from several widely separated locations in the Southern Hemisphere. The fossil consists of a tubular shell constructed of short cylinders arranged one over the other like stacks of bottomless paper cups. *Cloudina* has been interpreted as a tube-dwelling annelid worm.

In rocks deposited during the final years of the Proterozoic and the initial stage of the Cambrian, one encounters the first truly diverse fauna of animals with skeletons. Most of the remains consist of tiny shell structures that rarely exceed a few millimeters in length. They are composed either of calcium carbonate or calcium phosphate. Some appear to be the shells of primitive mollusks (Fig. 3.12), whereas others are spiculelike and may be the supportive structures for sponges. Another group, the hyolithids (Fig. 3.13) are tubular or tusk-shaped. The peculiar anabaritids (Fig. 3.14) consist of tubes that expand along three grooves, each surmounted by a keel or flange. Perhaps the flanges provided support in soft mud or served to strengthen the shell. Another early shelly fossil is *Lapworthella* (Fig. 3.15), which has a conical shape. The shell, however, was probably not the single skeletal structure of an animal, but rather one of many similar elements that covered the body of a *Lapworthella*-bearing animal. This interpretation is indicated by discoveries of shells that are fused together in side-by-side arrangements, as well as by the observation that some fossils of *Lapworthella* lack a basal cavity that might have served as a living chamber.

Following the Tommotian stage (the second stage of the Early Cambrian), animals with skeletons became both abundant and diverse. More familiar fossils such as trilobites and brachiopods provide a fine fossil record. But there were also many soft-bodied creatures. Some of these are known from Lagerstätten such as the Chengjian fossil site in China and the Burgess Shale site in British Columbia.

Figure 3.12   Tiny shell-bearing invertebrates representative of those found in rocks of the Early Cambrian Tommotian stage. None of these fossils are more than a few millimeters in length. They are not drawn at the same scale. (A) *Pojetaia* is an early bivalve; (B) *Yochelcionella* is a distinctive mollusk because of the long snorkel projecting from the forward part of the shell; (C) *Heraultipegma* is considered to be a rostroconch mollusk; (D,E) are small gastropods; and (F) is a possible monoplacophoran mollusk. (Drawings of *Pojetaia*, *Yochelcionella*, and *Heraultipegma* are from M. A. S. McMenamin & D. L. S. McMenamin, *The Emergence of Animals.* New York: Columbia University Press, 1990.)

WINDOWS INTO THE PAST: THE BURGESS AND CHENGJIANG LAGERSTÄTTEN

On a ridge near Mount Wapta, British Columbia, there is an exposure of Middle Cambrian Burgess Shale that contains one of the most important faunas in the fossil record. The fossils consist of shiny, jet-black impressions on the bedding planes of the dark shale (Fig. 3.16). Most of the fossils are remains of animals that lacked shells. Over 60,000 specimens have been collected from the fossil site, including at least four major groups of arthropods, as well as sponges, onycophorans, crinoids, sea cucumbers, a probable chordate, and many forms that defy placement in established phyla.

Among the more unusual creatures in this richly diverse fauna are *Anomalocaris* (Fig. 3.17), *Opabinia, Hallucigenia,* and *Marella. Anomalocaris,* which exceeded 20 centimeters (cm) in length, was the largest of these genera. Its oval head had two lateral eyes mounted on short stalks, and it possessed a pair of curved feeding appendages used to capture prey and pass food to a ventrally located mouth. The circular mouth

Figure 3.13 *Hyolithes carinatus*, a hyolith from the Middle Cambrian Burgess Shale of British Columbia. The shell is subtriangular in cross section. A small subtriangular operculum serves to close the opening of the long, tapered shell. Between the operculum and the shell are two curved appendages called helens that may have helped to support the animal on the ocean floor. (Courtesy of the National Museum of Natural History, photograph by Chip Clark.)

was ringed with 32 teeth. Injuries seen in trilobite fossils appear to have been inflicted by the teeth of *Anomalocaris*, or by its close relative *Laggania*. *Laggania* also possessed 32 teeth, but its jaw was more rectangular in shape, its eyes were positioned differently, and it lacked the three pairs of vanes at the posterior end of the trunk.

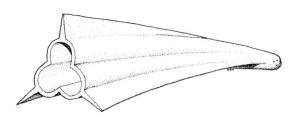

Figure 3.14 *Anabarites*. Rarely more than a few millimeters in length, this tiny fossil occurs as a phosphatic internal mold that can be dissolved from the enclosing rock with acetic acid. The vanes on the shell may have helped to stabilize the animal on the seafloor. (Courtesy of M.A.S. McMenamin & D. L. S. McMenamin, *The Emergence of Animals*. New York: Columbia University Press, 1990.)

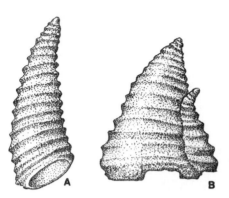

Figure 3.15   Sclerites of *Lapworthella schodackensis*. Sclerites are components of a larger external skeleton. Drawing (A) shows an arcuate sclerite having a shallow basal pit. Drawing (B) depicts two sclerites of different sizes that have been fused during development. Such fused specimens provide evidence that these are not the single shell of a solitary animal, but rather are parts of an external skeleton composed of many sclerites. The specimen on the left is about 0.5 mm in height. (Drawings based on photographs in E. Landing, *Journal of Paleontology* 58(6):1380–1398, 1984.)

If *Anomalocaris* and *Laggania* were among the largest animals of their time, *Opabinia* (Fig. 3.18) might qualify as the strangest. The animal had five eyes and a frontal, flexible nozzle that terminated in a spiny claw. The claw was apparently used in the capture of prey, which was then brought to the mouth by the proboscislike nozzle.

Hallucigenia (Fig. 3.19) is another intriguing Burgess Shale animal. It had a caterpillarlike annulated (ringed) body carried on stout paired legs that bifurcated at their distal ends. Seven pairs of plates were aligned along the body, and each bore a long tapering spine. Originally, the spines were thought to be legs, and the animal was reconstructed upside down. Only a single row of what we now know to be legs were visible in the fossil material, and so these were initially considered tentacles along the animal's back. Better specimens from China resolved the problem of orientation, and clearly showed that *Hallucigenia* is an early member of the phylum Onycophora, which includes the living velvet worm. *Aysheaia* (Fig. 3.20) is another onycophoran fossil found in the Burgess Shale.

Figure 3.16   The nature of preservation of Burgess Shale fossils can be seen in this photograph of the feeding appendages of *Anomalocaris*. Charles Walcott interpreted these fossils as the bodies of shrimplike animals (hence the name, which means "odd shrimp"). The discovery of additional fossils has shown that these appendages, which are about 7 cm long, were located in the front of the mouth of a huge swimming animal measuring nearly 60 cm in length. (Courtesy of the United States National Natural History Museum.)

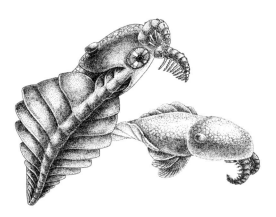

Figure 3.17 *Anomalocaris nathorsti* showing the ventral surface and circular mouth. At right is Anomalocaris canadensis in swimming position. (From S. J. Gould, *Wonderful Life*. New York: W. W. Norton & Co., Inc., 1989; artist Marianne Collins, used with permission.)

The small and elegant *Marella* (Fig.3.21) is a relatively abundant Burgess Shale arthropod. In *Marella*, two pairs of spines extend backward from a narrow head shield. The thorax is constructed of 24 to 26 body segments, each of which bears a pair of biramous (two-branched) appendages. Attached to the head shield are two additional pairs of appendages. The most anterior of these are long, thin, and many-jointed. Behind these appendages are a shorter pair having setae (chitinous bristles) on some of the segments.

The fossils of the Burgess Shale were discovered in 1909 by Charles B. Walcott, who also provided the original descriptions and interpretations of the fauna. Beginning in the 1960s, the fauna received additional study and reinterpretation over a period of about 15 years by Harry B. Whittington. The fauna clearly merited these long years of study, for along with the Chengjiang fauna to be described subsequently, it provides an unparalleled view of the diverse and complex biota present in the early epochs of the Paleozoic. The fossils demonstate that reconstructions of communities based on shelly fossils alone are likely to be deficient.

Figure 3.18   Model of *Opabinia* from the Burgess Shale fauna diorama at the Smithsonian Institution. (Courtesy of the United States National Natural History Museum.)

Figure 3.19   *Hallucigenia* as interpreted by Lars Ramskold. The creature was about 3.5 cm long. The animal is currently interpreted as a fossil onycophoran. (After L. Ramskold, The second leg row of *Hallucigenia* discovered. *Lethaia* 25:221–224, 1992.)

The Burgess Shale fauna is important for yet another reason. Sharing the seafloor with *Anomalocaris, Opabinia, Hallucigenia,* and *Marrella* were small, elongate animals that may be primitive early chordates. Chordates are animals that, at some stage in their development or throughout their lives, have a notochord (an internal, dorsal, longitudinal supportive rod), and a dorsally located nerve cord. Vertebrates are chordates in which the notochord has been supplanted by a series of vertebrae. The chordate animal of the Burgess Shale is named *Pikaia* (after Mount Pika near the Burgess fossil site). In addition to the apparent notochord in *Pikaia* one can discern the pattern of muscles arranged in a series of V-shaped segments as is characteristic of the musculature in certain primitive chordates and in fish (Fig. 3.22). The muscles, working with the flexible notochord, provided the sinuous body motion required for swimming.

Prior to 1984, the Burgess Shale offered the only extensive assemblage of animals having soft parts or coverings too delicate to allow easy preservation. In that year, researcher Hou Xianguang discovered a similar fauna in Chengjiang, China. The Burgess Shale animals are Middle Cambrian in age, whereas those at Chengjiang lived 30 million years earlier in the Early Cambrian. Over 80 species of invertebrates have been recovered

(A)

(B)

Figure 3.20   (A) Fossil remains of *Aysheaia* in the Burgess Shale. The animal had an elongate, annulated, cylindrical body about 3 cm long and had ten pairs of stocky limbs. Only the limbs on one side are visible in the fossil specimen. A model of *Aysheaia* (B) from the Burgess Shale fauna diorama at the Smithsonian Institution. Aysheaia was an onychophoran. (Photograph of model courtesy of the United States National Natural History Museum.)

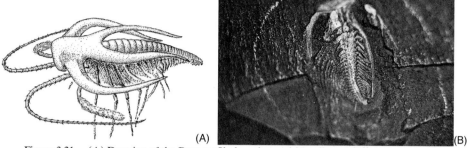

Figure 3.21   (A) Drawing of the Burgess Shale arthropod *Marella* and (B) photograph of a *Marella* fossil in a slab of Burgess Shale. (Photograph courtesy of the United States National Natural History Museum.)

from the Chengjiang site. The fossils exhibit extraordinary preservation. Jellyfish show the detailed structure of tentacles, radial canals, and muscles. Eyes, annulations, digestive organs, and ectodermal patterns are readily recognized in the fossil worms. Many Chengjiang organisms belong to such existing phyla as the Cnidaria, Porifera, Annelida, Brachiopoda, and Arthropoda. Arthropods dominate the fauna, with at least 30 species present. Some of these had been previously recognized in the Burgess Shale. Caterpillarlike animals that appear to be related to *Aysheaia* were found at Chengjiang. *Anomalocaris*, a prominent predator of the Burgess fauna, is also conspicuous in these fossil beds. At least as intriguing as *Anomalocaris*, however, is *Cathaymyrus*, which is currently considered the earliest known chordate. Like the Burgess Shale fossil *Pikaia*, the chordate affinities of *Cathaymyrus* are indicated by what appear to be an anteriorly directed notochord and a segmented musculature resembling that seen in the living cephalochordate *Branchiostoma* (commonly known as *amphioxus*).

The splendidly diverse animals exemplified by the Chengjiang and Burgess Shale faunas clearly had global distribution. They provide a vivid picture of the extraordinary range of creatures that inhabited the ocean at the beginning of the Paleozoic. The abruptness of their appearance and the rapidity of their evolution form the basis for the expression "the Cambrian explosion." That explosion, dubbed "evolution's big bang" in the popular press, lasted a mere 10 million years. During that geologically brief interval between 530 million and 520 million years ago, all the principal animal

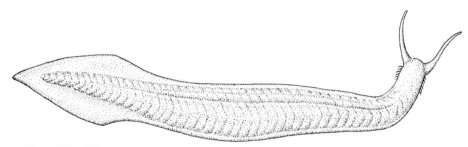

Figure 3.22   *Pikaia*. Note the rod along the back of the animal that appears to be a notochord. Fossils of *Pikaia* also reveal a zigzag pattern of muscles resembling the musculature in the primitive living cephalochordate *Branchiostoma*. (Length about 4.0 cm.)

phyla except the bryozoans appeared. Since that time, evolution has progressed primarily by modifications within phyla that had originated in the Cambrian. The reason for this will engage the interest of evolutionary biologists for years to come.

REVIEW QUESTIONS

1. What is the origin of Earth's hydrosphere (oceans, rivers, etc.) and original oxygen-deficient (anoxygenic) atmosphere?

2. How did Earth's atmosphere become oxygenic? What evidence suggests that free oxygen was beginning to accumulate in the atmosphere about 2.5 billion years ago? What was the probable role of stromatolites in the evolution of an oxygenic atmosphere?

3. What is the difference between chemosynthesis and photosynthesis?

4. How do heterotrophs differ from autotrophs? Name some unicellular members of each group.

5. How do eukaryotic organisms differ from prokaryotic organisms? In which of these groups do acritarchs belong?

6. When do metazoans appear on Earth? Describe these organisms. What is the basis for the suggestion that Ediacaran metazoans constitute a new taxonomic category of animals?

7. What phyla of organisms are represented in the Burgess Shale and Chengjiang fossil sites? How did the discovery of the Burgess Shale fauna change the way in which the history of early life on Earth was perceived?

8. Discuss the role of symbiosis in the origin of eukaryotes. What organelles may have originated by symbiosis?

9. What are hyperthermophiles, psychrophiles, acidophiles, alkaliliphiles, and barophiles?

10. Why have prokaryotes shown far less evolutionary change than eukaryotes over the last 2 billion years of biological history?

SUPPLEMENTAL READINGS AND REFERENCES

Bengston, S. (Ed.). 1994. *Early Life on Earth*. New York: Columbia University Press.

Briggs, D. E. G., Erwin, D. H. & Collier, F. G. 1994. *The Fossils of the Burgess Shale*. Washington, DC: Smithsonian Institution Press.

Buss, L. W. & Seilacher, A. 1994. The phylum Vendobionta: A sister group of the Eumetazoa? *Paleobiology* 20(1): 1–4.

Chen, J. Y., Edgecombe, J., Ramsköld, L. & Zhou, G. Q. 1995. A possible Early Cambrian chordate. *Nature* 377:720–722.

Collins, D. 1996. The "evolution" of Anomalocaris and its classification in the arthropod Class Dinocarida (nov.) and order Radiodonta (nov.) *Journal of Paleontology* 70(2):280–293.

Fox, S. W. & Dose, K. 1972. *Molecular Evolution and the Origin of Life*. San Francisco: W.H. Freeman.

Glaessner, M. F. 1984. *The Dawn of Animal Life.* Cambridge: Cambridge University Press.

Gould, S. J. 1989. *Wonderful Life.* New York: W.W. Norton & Co.

Jun-Yuan, C., Bergstrom, J., Lindstrom, M. & Xianguang, H. 1991. Fossilized soft-bodied fauna. *National Geographic Research and Exploration* 7(1):8–19.

Knoll, A. H. 1991. End of the Proterozoic eon. *Scientific American* 265(4):64–73.

Lipps, J. H. & Signor, P. W. (Eds.). 1992. *Origin and Early Development of the Metazoa.* New York: Plenum Press.

Margulis, L. 1981. *Symbiosis in Cell Evolution.* San Francisco: W.H. Freeman.

McMenamin, M. A. S. & McMenamin, D. L. S. 1990. *The Emergence of Animals.* New York: Columbia University Press.

Retallack, G. J. 1994. Were the Ediacaran fossils lichens? *Paleobiology* 20(4):523–544.

Runnegar, B. 1992. The evolution of earliest animals. In Schopf, J. W. (Ed.), *Major Events in the History of Life.* Boston: Jones and Bartlett.

Schopf, J. Wm. (Ed.). 1983. *Earth's Earliest Biosphere: Its Origin and Function.* Princeton, NJ: Princeton University Press.

Schopf, J. Wm. (Ed.). 1992. The oldest fossils and what they mean. In Schopf, J. W. (Ed.), *Major Events in the History of Life.* Boston: Jones and Bartlett.

Seilacher, A. 1989. Vendoza: Organismic construction in the Proterozoic biosphere. *Lethaia* 22:229–239.

Whittington, H. B. 1985. *The Burgess Shale.* New Haven, CT: Yale University Press.

CHAPTER 4

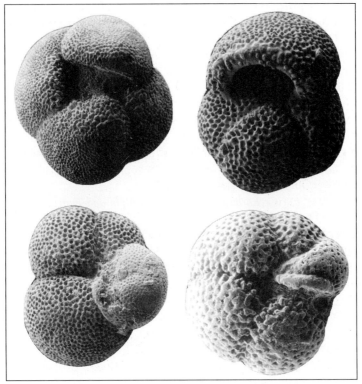

Electron micrographs of planktonic Oligocene foraminifera from the Gulf of Mexico, ×160.

Bodies of a Single Nucleated Cell: Protoctista

Dear God, what marvels there be in so small a creature.

Remark attributed to Anton van Leeuwenhoek's draftsman

The Kingdom Protoctista contains a bewildering variety of unicellular eukaryotic organisms. Some are plantlike autotrophs that have chloroplasts and are able to synthesize their food; others are heterotrophs that feed on other organisms; and some can switch their mode of nutrition, being autotrophic at certain times and heterotrophic at others. Although the plantlike protoctistans are not invertebrates, they are included here as a convenience to instructors and students wishing to examine the body of microfossils often included under the topic "micropaleontology." Protoctistans arose from prokaryotic ancestors during the last billion years or so of the Precambrian. Their early record is poor, probably because many lacked preservable coverings or were destroyed during episodes of erosion and metamorphism. As we progress through the Phanerozoic succession, however, fossil protoctistans become abundant. Many have been shown to be useful as biostratigraphic markers, and as paleoclimatic and paleogeographic indicators.

PLANTLIKE PROTOCTISTANS

PYRROPHYTES (DINOFLAGELLATES)

The flamelike reddish colors of pyrrophytes account for their name, which means "flame plants." The color results from a light-absorbing pigment that can impart a vivid red hue to extensive areas of the ocean. The Red Sea may have gotten its name from pyrrophyte blooms. Because many pyrrophytes produce nerve toxins, these harmful blooms or "red tides" kill large numbers of fish and other marine organisms. In 1996, a total of 146 manatees in Florida were killed by dinoflagellate toxins. Shellfish feeding on the toxic organisms become poisonous to humans as well.

The most abundant of the pyrrophytes are the dinoflagellates. The cells of dinoflagellates are between 20 and 150 micrometers in diameter and quite varied in form. All, however, have two distinctly placed flagella. One is longitudinal and occupies a furrow or **sulcus**. The other encircles the cell in a transverse or spiral groove, which is called the **cingulum** (Fig. 4.1). The name "dinoflagellate" is based on the Greek word *dinos*, meaning "whirling," and alludes to the spiral motion by which the flagella propel the organism through water. Dinoflagellates are found in marine as well

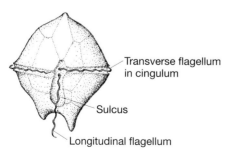

Transverse flagellum in cingulum

Sulcus

Longitudinal flagellum

Figure 4.1   The dinoflagellate *Peridinium* as it appears during the motile stage of its life cycle. The body wall is composed of plates of fibrous cellulose that fit together to form a theca. The resistant cysts of dinoflagellates form within the theca, and the fragile theca is subsequently discarded and is normally not preserved (× 300).

as freshwater habitats. In the ocean, they are second only to diatoms as primary producers in the food web.

Although plantlike in having chlorophyll pigments and cellulose in the cell walls, dinoflagellates are also capable of heterotrophic nutrition. The life cycle includes both motile and resting stages. During the resting stage a cyst (capsule) is formed within the formerly motile cell. The cyst walls are double and composed of a substance termed *sporopollenin*. Sporopollenin is highly resistant to chemical and bacterial attack. All dinoflagellate fossils are believed to represent the remains of cysts. In some species, the cyst develops in contact with the wall of the motile cell so that the shape and surface sculpture of the enclosing motile cell is, to some degree, duplicated. In others, the cyst has contact with the enclosing motile cell in just two or more places, or is separated by short supports. Such cysts do not exhibit traces of the motile cell's surface features.

Sexual reproduction has been reported in a few living species of dinoflagellates, but the majority reproduce asexually. Such reproduction consists of simple division of the cell into two halves (**binary fission**).

Fossil dinoflagellates have been reported in rocks as old as Precambrian. However, they are more frequently found in rocks of Permian to Holocene age. Their greatest expansion began during the Jurassic and continued into the Miocene. Species of *Hystrichosphaeridium* have been particulary useful as guide fossils. These so-called hystrichosphaerids are pyrrophyte cysts having spherical, subspherical, ovoid, or polygonal shapes from which radiate structures resembling tiny trumpets (Fig. 4.2).

CHRYSOPHYTES

Another phylum of protoctistans with photosynthetic pigments are the chrysophytes. Included here are golden-brown algae, silicoflagellates, diatoms, and coccolithophores. The remains of diatoms and coccolithophores occur in prodigious numbers in certain stratigraphic sections, and they are exceptionally useful in biostratigraphic studies.

Silicoflagellates   As indicated by their name, silicoflagellates are flagellated, unicellular organisms that construct a skeleton composed of opaline silica. Except for the fact that the rods comprising the skeleton are tubular rather than solid, the silicoflagellate skeleton resembles the skeleton in radiolaria (see discussion below). Silicoflagellates have a single flagellum (whiplike organelle) that is attached to the cell

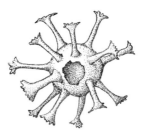

Figure 4.2   The dinoflagellate cyst *Hystrichosphaeridium*. These cysts are formed during the dinoflagellate resting stage of the life cycle. Trumpetlike processes seen on this drawing linked the cyst to the inner wall of the dinoflagellate theca. The large opening or *archaeopyle* serves as an escape hatch for the cell when it leaves the cyst and begins the motile stage of the life cycle.

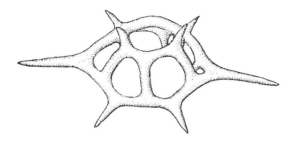

Figure 4.3   Skeleton of the silicoflagellate *Distephanus* (×1500). As in other silicoflagellates, the rods and spines are hollow and composed of opaline silica. The basic architecture of this form consists of a hexagonal basal ring having spines at each corner, and an overlying apical ring connected to the basal ring by inclined bars. (After an electron microscope image by W. W. Wornhardt.)

at its anterior end. Like dinoflagellates, the cells contain a photosynthetic pigment, but the organism can also consume prey, which they capture by means of threadlike pseudopodia. Most silicoflagellates range in diameter from 20 to 50 micrometers.

The silicoflagellate skeleton (Fig. 4.3) is secreted within the cell. It generally consists of a delicate circular, pentagonal, or elliptical ring that serves as the basal support for regularly spaced bars. Two to seven spines radiate outward from the central ring and bar structure. In addition to lending support for the protoplasm, the skeleton provides buoyancy by spreading the protoplasm for increased resistance to sinking. Buoyancy is also enhanced by the hollow (tubular) construction of skeletal elements.

All silicoflagellates are oceanic, and many have definite temperature preferences within the marine realm. Fossil forms have proven useful as paleotemperature indicators. They are found in sediments that range from the Early Cretaceous to the Holocene. Silicoflagellates are particularly abundant in siliceous sediment such as diatomite.

Diatoms   Diatoms are unicellular protoctistans with photosynthetic pigment and a silicified "shell" that is called a **frustule** (Fig. 4.4). The frustule is composed of two rimmed plates, one of which overlaps the other like the halves of a biologist's petri dish. Frustules range in size from 20 to 200 micrometers and are generally circular (**centric**) or oval (**pennate**) in shape. Perforations and opaline ridges of the frustule are often arranged in attractive geometric patterns (Fig. 4.5).

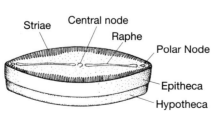

Striae
Central node
Raphe
Polar Node
Epitheca
Hypotheca

Figure 4.4   The extant pennate, freshwater diatom *Pinnularia*. The frustule consists of two overlapping valves. The outer part, which is comparable to the top of a box, is the epitheca. The inner part (lower part of a box) is the hypotheca. Extending down the so-called valve face is a unsilicified groove called the raphe. It is interrupted by a central node and two polar nodes. Rows of striae occur along the perimenter of the valve face (×500).

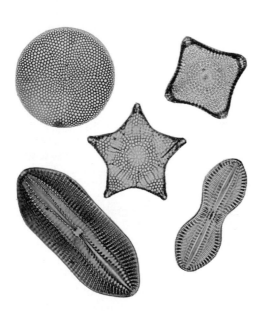

Figure 4.5 Diatom frustules collected off the coast of Crete. Note the variation in form and sculpture. (Courtesy of Naja Mikkelsen.)

Some diatoms do very well as solitary cells, whereas others attach to one another and form long chains. In either case, each cell contains photosynthetic granules, a large nucleus, and a central membrane-bound vesicle called a vacuole. Although flagellae are absent, the pennate forms can move slowly over a surface on a stream of protoplasm secreted along the underside of the frustule. In contrast, most of the round forms are planktonic and nonmotile. Fat droplets in the cell and delicate spines on the frustule help to maintain buoyancy.

Diatoms are found in a remarkable range of environments, including the ocean, lakes and streams, and even in wet leaves and soil. The pennate forms dominate in freshwater habitats, whereas centric types are most abundant in the marine environment. They reach their highest concentrations in subpolar and temperate seas. Because they require light for photosynthesis, diatoms are limited to the upper 200 meters (the photic zone) of the water column. In general, a great abundance of diatoms in a particular region of the ocean correlates with waters rich in nutrients such as nitrate and phosphorus. Such waters nourish a host of other forms of marine life as well, for diatoms are primary producers in the oceanic ecosystem.

Although there have been questionable reports of fossil diatoms in Jurassic rocks, their more substantiated geologic range is Cretaceous to Recent. At various times and places in the geologic past, the frustules of dead diatoms have settled onto the seafloor in such enormous quantities as to form diatom oozes. These siliceous sediments, on lithification, become the sedimentary rock known as diatomite. Diatomites in California are often hundreds of meters thick and contain over 6 million frustules in only a cubic centimeter of rock. Because of its low density and high porosity, diatomite is used as a filter and in the manufacture of lightweight construction materials. It is also used as an abrasive.

Coccolithophores   Coccolithophores are unicellular, planktonic, golden-brown algae. Although they occur in both freshwater and saltwater bodies, the majority are marine. Marine coccolithophores secrete an external covering composed of calcareous plates called **coccoliths**. Individual coccoliths measure only between 1 and 20 micrometers, and therefore the scanning electron microscope has become an essential tool for taxonomic work with this group of protoctista. Once familiarity with the various groups has been attained, a high magnification light microscope with polarizers and special illumination (phase contrast illumination) can be used for routine biostratigraphic work.

Reproduction in coccolithophores is predominantly asexual, although in some species there is an alternation between motile forms having flagellae and nonmotile forms that lack flagellae. The alternation can result in dimorphism.

Examination of live coccolithophores reveals that coccoliths are formed within the cell on organic matrices, and subsequently moved to the surface of the cell to form a sort of calcareous armor (Fig. 4.6). Coccoliths formed in this way may consist of one plate, or two superimposed plates. The plates are concave on one side so that they fit closely around the periphery of the spherical cell. Each is itself composed of smaller elements (crystalloids) arranged in astonishingly precise spiral, circular, imbricating (overlapping), or radial patterns (Fig. 4.7).

As photosynthetic organisms, coccolithophores must live within the photic zone. They are particularly abundant in open tropical seas having upwelling, nutrient-bearing currents. Distinctive assemblages occur in subpolar and temperate areas, and a few forms like the "geodesic" braarudosphaerids (Fig. 4.8) prefer shallow, nearshore environments. Like diatoms, coccolithophores are a major source of food for small, plant-ingesting planktonic organisms.

Upon death of the parent cell, coccoliths fall away and begin a slow descent (about 0.15 meters per day) to the ocean floor. Because of their small size and slow

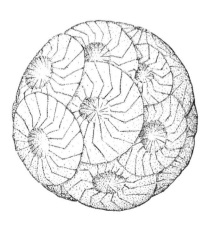

Figure 4.6   Coccosphere of *Cyclococcolithus*. Maximum diameter about 30 micrometers.

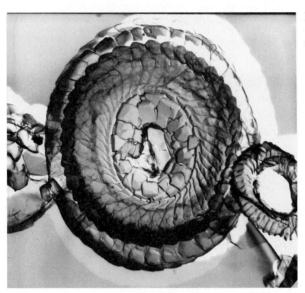

Figure 4.7   Electron micrograph of *Coccolithus*. Maximum diameter 15 micrometers.

descent, centuries may elapse before they come to rest. If coccoliths settle through a column of water no deeper than about 3,500 meters, they may accumulate to form a major part of the deep sea sediment known as *calcareous ooze*. Below that depth, however, the colder water holds more carbon dioxide. This increases acidity and causes the dissolution of the tiny skeletons. As a result, calcareous oozes are absent in the deepest tracts of the ocean. In such areas, siliceous oozes composed of the remains of radiolaria (see discussion in next section) and diatoms may predominate. The depth at which the rate of supply of calcite is balanced by the rate of its dissolution is called the **carbonate compensation depth**, or CCD.

Coccolithophores are members of a group of organisms more generally referred to as *calcareous nannoplankton*. The group includes a great variety of forms, some of which do not resemble the platelike coccoliths described above. Stellate, pentagonal,

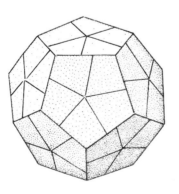

Figure 4.8   Coccosphere of *Braarudosphaera*. Diameter of coccosphere about 18 micrometers.

Figure 4.9   Electron micrograph of *Discoaster lodoensis*, an Eocene species. Diameter 14 micrometers.

wishbone, spiral, and stellate forms are common. The star-shaped **discoasters** (Fig. 4.9) are abundant in certain Tertiary strata. Discoasters, along with the many other varied forms of coccoliths, have been exceptionally useful in fixing the stratigraphic position of deep sea sediments recovered in Deep Sea Drilling Project cores of the ocean floor.

Although there have been a few questionable reports of coccoliths in Precambrian and Paleozoic rocks, they are primarily Mesozoic and Cenozoic organisms. They occur only rarely in Triassic rocks, but they become diverse and abundant in Jurassic strata. Mass extinctions of coccolithophores occurred at the end of the Cretaceous. The few genera that survived were the progenitors for later Tertiary forms, which steadily increased in diversity through the Paleocene and Eocene. Thereafter, diversity declined except for a small increase during the Miocene. Discoasters became extinct at the end of the Pliocene, but other forms of coccolithophores persist in modern seas.

ANIMAL-LIKE PROTOCTISTANS

RADIOLARIA

Radiolaria are planktonic, marine protoctistans that construct delicate, and often beautiful latticelike skeletons of opaline silica. The two groups that are most important paleontologically are the radially symmetrical **spumellarians** and the helmet-shaped **nassellarians** (Fig. 4.10).

The living radiolarian cell consists of an inner sphere called the **central capsule**, which is enclosed in an elastic membrane that is pierced by small pores. Within the

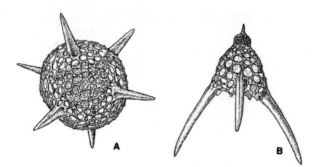

Figure 4.10   Among radiolarians, the spumellarians (A) have radial symmetry, whereas nasselarians (B) tend to be helmet or cap-shaped with apical spine. Diameter of the spherical wall of the spumellarian is 130 micrometers. The nasselarian is drawn to the same scale.

endoplasm of the central capsule are the nucleus and organelles required for reproduction, biochemical synthesis, and energy production. The central capsule is surrounded by a mantle of extracapsular ectoplasm from which radiate threadlike pseudopodia. Colorless, bubblelike vacuoles form a frothy, glutinous mass in the ectoplasm called the **calyma** (Fig. 4.11). Symbiotic zooxanthellid algae are also present in the ectoplasm. During daylight hours, carbon dioxide generated by radiolarian respiration is assimilated by the algae, thereby making the organism less buoyant, and causing it to slowly sink. At night, however, the algae are unable to photosynthesize, and carbon dioxide builds up in the calyma. As a result, the organism becomes more buoyant and rises in the water column. Fat globules, outstretched pseudopodia, and slender spines also promote buoyancy.

Most radiolarians (often nicknamed "rads") possess siliceous skeletons. Although the basic forms of the skeletons are either spheroidal with radiating spines

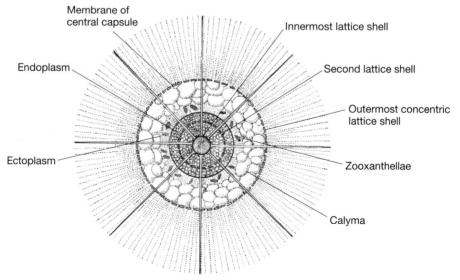

Membrane of central capsule
Innermost lattice shell
Endoplasm
Second lattice shell
Outermost concentric lattice shell
Ectoplasm
Zooxanthellae
Calyma

Figure 4.11   Schematic representation of a radiolarian depicting principal morphological features.

or helmet-shaped, a variety of other discoidal, pyramidal, and fusiform (spindle-shaped) forms also occur. Most skeletons are open and latticelike, but smooth forms are also found in which openings are filled with silica. In life, the radiolarian skeleton is enveloped in protoplasm, and pseudopodia extend outward as a covering around spines. It is likely that the protoplasmic cover protects the skeleton from dissolution.

Today's radiolarians are found from pole to pole throughout the ocean. Normally, they thrive in the upper 200 meters of ocean waters. As is the case for many kinds of marine invertebrates, they are most varied in tropical zones, and least diverse in polar seas. Where they are most prolific, their empty skeletons contribute to the formation of siliceous ooze on the seafloor. At depths in excess of about 8000 meters, however, the siliceous tests (shells) of radiolaria begin to dissolve.

Fossil radiolarians are known from strata ranging in age from Cambrian to the present. They are particulary abundant in marine deposits of the Mesozoic and Cenozoic. Dissolution and reprecipitation of radiolarian silica in rocks of these ages have produced a white, dense siliceous rock known as *radiolarian chert*. Under the effects of low grade metamorphism, these cherts have been altered to novaculite, a dense, even textured rock used in the manufacture of grinding stones.

FORAMINIFERA

Of all the protoctistans, none is more useful in biostratigraphy than the foraminifera (Fig. 4.12). Their remains are frequently found in marine sedimentary rocks, and they have experienced the kind of rapid evolutionary change that is highly useful for biostratigraphic studies. Foraminifera have a geologic range that extends from the

Figure 4.12   Foraminifera from the Miocene Monterey Formation of California. The two large coiled foraminifera near the center are *Valvulineria*, and these are about 0.2 mm in diameter. The spear-shaped forms are mostly species of *Bolivina* and *Siphogenerina*.

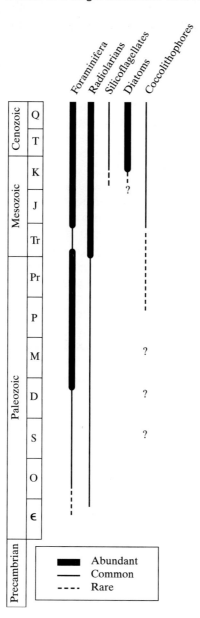

Figure 4.13   Stratigraphic ranges of foraminifers, radiolarians, silicoflagellates, diatoms, and coccolithophores.

Cambrian to the Holocene (Fig. 4.13). The **tests** (shells) of foraminifera are sturdier than the delicate skeletons of radiolaria, and these stronger structures are often obtained undamaged from small chips of rock brought to the surface in the course of drilling for oil.

Nearly all "forams" are marine and they include both benthic and planktonic groups. Benthic forams may live on or within the surface layer of sediment or cling to seaweed and the shells of other invertebrates. The bottom dwellers live in all environments from the intertidal zone to the deepest parts of the ocean, and from the tropics to the poles. Fewer species but large numbers of individuals constitute the planktonic group. Upon death or when abandoned during reproduction, the tests of these zooplankton rain down onto the seafloor and become components of calcareous oozes.

Foram tests consist of one or more chambers. Chambers in multichambered forms are added in a variety of patterns, including columns, coils, and spirals. The name *foraminifer* alludes to the openings or **foramina** located in the walls that separate chambers. A large opening, the **aperture**, is also present. The wall of the test may be organic (composed of a proteinaceous mucopolysaccharide), agglutinated (composed of foreign particles held together by an organic or mineral cement), or calcareous. However, an organic membrane forms the foundation for both the agglutinated and calcareous tests. In the classification of foraminifera, wall composition and microstructure are generally considered of greatest importance, followed by the mode in which chambers are added and characteristics of the aperture.

Allogromia, a form that lacks a mineralized test or shell, serves as a good specimen for the study of the living animal. In *Allogromia*, an organic membrane surrounds the endoplasm containing the cell nucleus and other organelles responsible for processing food and storing energy. A layer of frothy ectoplasm encloses the endoplasm. Branching pseudopodia radiate from the ectoplasmic covering (Fig. 4.14). These pseudopodia may extend outward to distances several times the diameter of the test. Within the pseudopodia one can observe conveyorlike movements of protoplasm referred to as *streaming*. A central stream moves away from the test, whereas outer

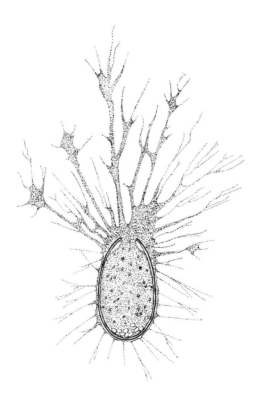

Figure 4.14   The extant foraminifer *Allogromia*. This primitive member of the family Allogromiidae has a thin, flexible, ovoid test composed of a chitinous substance. The Allogromiidae are not known as fossils. Diameter of test is about 0.4 mm.

streams move inward. The streams serve to bring in nutrients and expel waste. In general, food caught on the pseudopodia is chemically broken down, and the absorbed nutrients are then transferred by streaming to the endoplasm. Waste collects into small brown particles, is streamed out of the endoplasm, and is released. Food might consist of bits of organic matter, diatoms, bacteria, dinoflagellates, and even small copepods. Pseudopodia may also function in locomotion, for they are strong enough to pull the organism along from one part of a surface to another.

The mineralized test is of particular interest to paleontologists and is the usual basis for identification of species. As in the case of *Allogromia*, there are a few "naked" foraminifera that live within an organic membrane in lieu of a hard test. Agglutinated forms are more common. Often these agglutinated foraminifera are astonishingly selective in their choice of materials, using only sponge spicules, specific mineral grains, or even tiny black spheres known as *microtectites* (thought to be debris formed during meteorite impact). *Rhabdammina* and *Gaudryina* are two examples of foraminifera having agglutinated tests (Fig. 4.15).

By far the most abundant kinds of foraminifera tests are those made of calcium carbonate. These may have perforate walls as in *Globigerinoides* (Fig. 4.16), homogeneous microgranular walls as in *Endothyra* (Fig. 4.17), or smooth and either glassy (hyaline) or porcelaneous walls. With regard to variety in test shape, foraminifera display remarkable variety (Fig. 4.18). In some, the test consists only of a single chamber (**unilocular test**). More often the test is constructed of many chambers (**multilocular**). The chambers in multilocular forms can be arranged in a vertical series (**uniserial**), two ascending vertical columns (**biserial**), or even three columns (**triserial**). Rather like the coiling in some snail shells, the test may be coiled in a low helicoid spiral (**trochospiral**) or in a plane (**planispiral**). If the outer whorls of the test strongly overlap the inner ones, the test is termed planispiral **involute**. In contrast, where early whorls are not hidden by later whorls, the term planispiral **evolute** applies. The fusulinids have a test

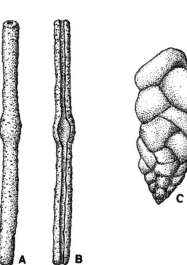

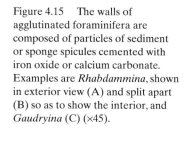

Figure 4.15 The walls of agglutinated foraminifera are composed of particles of sediment or sponge spicules cemented with iron oxide or calcium carbonate. Examples are *Rhabdammina*, shown in exterior view (A) and split apart (B) so as to show the interior, and *Gaudryina* (C) (×45).

Figure 4.16   Highly perforate calcareous test walls in two species of the planktonic foraminifer *Globigerinoides* (×90).

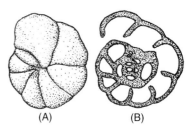

Figure 4.17   Side view (A) and section (B) of *Endothyra* (×80). The closely coiled test is involute, and the aperture is a narrow slit at the base of the apertural face.

shape resembling grains of wheat, and they are extended laterally along the axis of coiling. Other, often very large foraminifera added chambers in concentric rings in what is termed an **annular** pattern. Miliolid tests are composed of porcelaneous calcite and are imperforate. They typically have elongate chambers added in precise geomet-

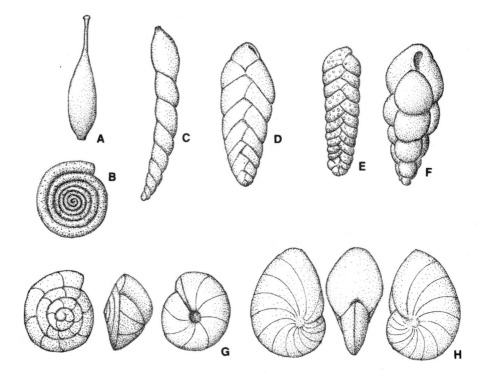

Figure 4.18   Common test forms in foraminifera. (A) Unilocular test of *Lagena*; (B) annular test of *Ammodiscus*; (C) uniserial test of *Dentalina*; (D) biserial test of *Bolivina*; (E) planispiral becoming biserial test of *Bolivinopsis*; (F) triserial test of *Bulimina*; (G) trochospiral test of *Gyroidina*; (H) planispiral involute test of *Nonion* (×200).

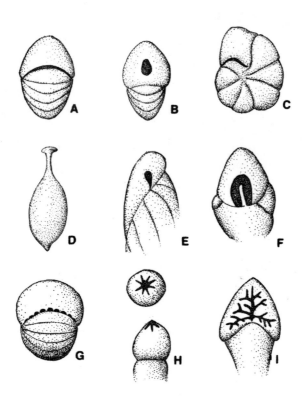

Figure 4.19 Several variations of apertures in foraminifera. (A) Aperture at base of apertural face; (B) middle of apertural face; (C) ventral along inner side of last formed chamber; (D) terminal with neck and lip; (E) comma-shaped; (F) with simple tooth; (G) multiple; (H) radiate; (I) dendritic. (After J. A. Cushman, *Foraminifera*. Cambridge, MA: Harvard University Press, 1948.)

ric angles, and whose ends are reversed with each chamber addition. One might think this is quite enough variety, but many other test shapes exist. Some foraminifera combine two or more chamber arrangements, with an early planispiral form becoming uniserial, or a triserial form reverting to biserial or uniserial.

The foraminifera aperture serves to connect the internal endoplasm with the ectoplasm outside the test. As with chamber arrangement, there appears to be a great variety in the shape and location of test apertures (Fig. 4.19). Apertures can be single or multiple, round or elliptical, radiate or comma-shaped, and with or without a "lip" or "tooth," or an elongate "neck." Correct identification of a species cannot be accomplished without determining apertural characteristics. Identification may also require a close look at ornamentation, which might consist of papillae, ridgelike costae, regularly arranged pits, raised bosses, or needlelike spines.

Foraminifera have one characteristic that can sometimes result in misidentifications. Many are known to exhibit test dimorphism. The dimorphism, which may involve differences in size or other morphologic characteristics, is related to the foraminiferal mode of reproduction, which alternates between asexual and sexual stages. The reproductive cycle involving an **alternation of generations** (Fig. 4.20) was originally observed in foraminifers by the famous British physician Joseph Lister in 1895. It begins with what can be termed the asexual stage represented by a foraminifer with a small (**microspheric**) initial chamber (**proloculus**) followed by many additional cham-

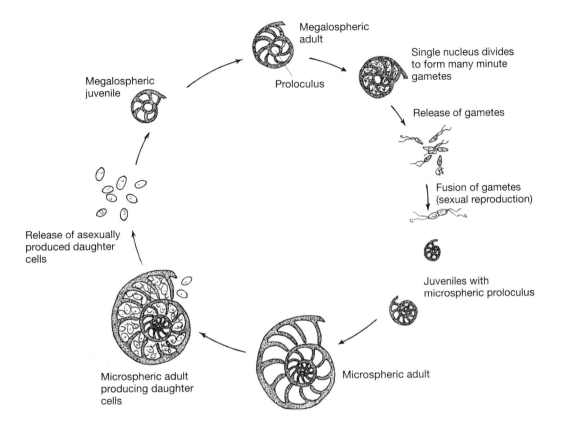

Megalospheric adult

Proloculus

Megalospheric juvenile

Single nucleus divides to form many minute gametes

Release of gametes

Fusion of gametes (sexual reproduction)

Juveniles with microspheric proloculus

Release of asexually produced daughter cells

Microspheric adult producing daughter cells

Microspheric adult

Figure 4.20   Diagram showing reproduction in foraminifera by alternation of generations. (Tests are shown as if thin-sectioned.)

bers. The microspheric form asexually produces tiny daughter cells, and these begin to secrete tests that have a large (**megaspheric**) proloculus, followed by few chambers. The megalospheric forms produce flagellated daughter cells that have a haploid number of chromosomes. They are actually sex cells that come together to form a fertilized cell or zygote. The zygote then begins to secrete the test of the microspheric form. Thus, the microspheric form is produced sexually but reproduces asexually, whereas the megalospheric form is produced asexually but reproduces sexually.

As with any large order of invertebrates, certain groups of foraminifera provide immediate recognition of strata of particular intervals of geologic time. For example, during the Mississippian Period the microgranular, coiled, multichambered endothyrids (such as *Endothyra*) were so abundant that they formed foraminiferal limestones. In the following Pennsylvanian and Permian periods, the fusulinid descendants of the endothyrids predominated (Fig. 4.21). Geologically rapid changes in their intricate internal structure and chamber arrangement have made them exceptionally useful in

Figure 4.21   Pennsylvanian limestone composed almost entirely of fusulinids. The limestone is part of the Modesto Formation, Peoria County, Illinois. (Courtesy of the Illinois Geological Survey.)

biostratigraphic studies. Like the endothyrids, fusulinids were sufficiently abundant to form foraminiferal (fusulinid) limestones. Yet by the end of the Permian, this once highly successful group had become extinct.

Foraminifera are sparse in Triassic rocks. With the success of lagenids and lituolids (Fig. 4.22) during the Jurassic, it is evident that the order had fully recovered

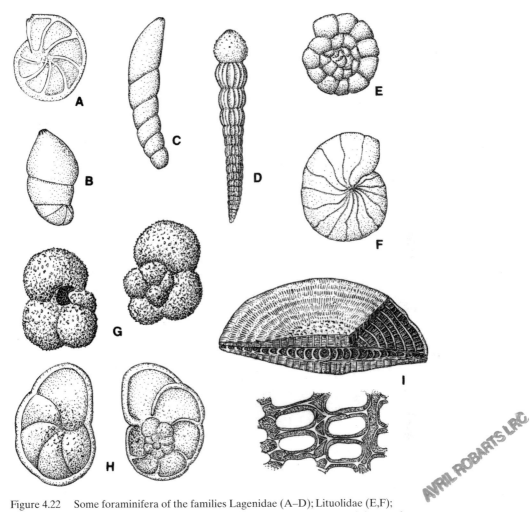

Figure 4.22   Some foraminifera of the families Lagenidae (A–D); Lituolidae (E,F); Globigerinidae (G,H); and Camerinidae (I). (A) *Robulus*; (B) *Marginulina*; (C) *Dentalina*; (D) *Nodosaria*; (E) *Trochamminoides*; (F) *Cyclamina*; (G) ventral (showing aperture) and dorsal views of *Globigerina*; (H) two views of *Globorotalia*; (I) *Cycloclypeus*, with enlarged section of median chambers showing septal canals. (*Cycloclypeus* is about 35 mm in diameter; average size of other foraminifera about 0.6 mm.)

from the terminal Permian episode of massive extinction. The first unquestioned planktonic foraminifera appear in the Cretaceous. They expanded dramatically during that period, as well as during the Eocene and Miocene when climates were somewhat warmer. During these warmer episodes, tests of *Globigerina* and *Globorotalia* accumulated by the billions in limy sediments that would become marl and chalk. These and other planktonic foraminifera are, in fact, the fossils of choice for mid-Cretaceous to Holocene stratigraphic studies. In addition, several lineages of large foraminifera evolved during the Eocene and proliferated in shallow tropical regions around the globe.

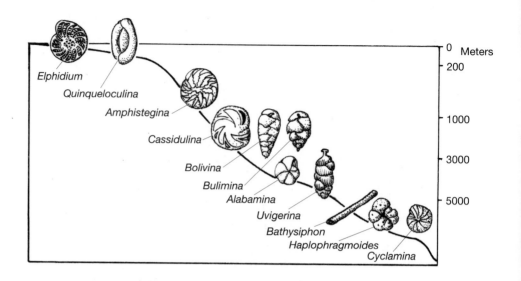

Figure 4.23   Depth occurrences of some benthic foraminifera at about 20° N latitude in the Pacific Ocean. Because deeper layers of ocean water are cooler than surface layers, the depth at which particular species live and reproduce is also strongly related to temperature. (Data from K. M. Saidova. *Progress in Oceanography* 4:143–151, 1967.)

Foraminifera are not only useful in stratigraphic correlation but they are also sensitive indicators of paleobathymetry (Fig. 4.23), paleosalinity, and paleotemperatures. Often, the first two of these environmental factors can be deduced by comparing fossil species with extant relatives and ascribing the environmental constraints of the living to their fossil counterparts. Such studies have demonstrated that certain benthic foraminifera show definite depth-related distributions. When mixtures of shallow and deep-water foraminifera are found in a given stratum, one may suspect downslope or current transport of sediment. Paleotemperatures can also sometimes be ascertained from foraminifera. An interesting example involves the direction of coiling in *Globorotalia truncatulinoides*. *Globorotalia truncatulinoides* builds a trochospiral test. Thus, all the whorls are visible on one side, and only the last formed whorl is visible on the other side. It has been observed that the majority of individuals of this species taken in warmer ocean water coil to the right, whereas left-coiling individuals prefer colder water. During the Pleistocene glacial stages when the temperature of the ocean declined, left-coiled forms predominated. These were replaced by right-coiled forms in subsequent warmer interglacial stages. The alternation of left- and right-coiled forms in deep-sea cores provides a record of the shifts in global temperatures that occurred during the Pleistocene Epoch.

REVIEW QUESTIONS

1. What are dinoflagellates? Describe their defining characteristics. Are they beneficial components of the marine environment? Under what circumstances are they harmful?

2. What are silicoflagellates? What function is served by the silicoflagellate skeleton?

3. Describe and prepare a labeled sketch of a typical diatom. Remains of what other protoctistans might you expect to find in a diatomite?

4. What are coccoliths? How do coccoliths differ from discoasters?

5. What is the carbonate compensation depth (CCD)? Why is it not at the same level throughout the ocean? What kinds of protoctistan skeletal remains are unlikely to be found below the CCD?

6. What are spumellarians and nassellarians? How do they differ?

7. What function is served by the calyma of a radiolarian?

8. Describe several different kinds of chamber arrangements in multilocular foraminifera. How does a planispiral test (shell) differ from a trochoid test?

9. What would be the geologic age of a rock containing abundant tests of fusulinids?

10. When did planktonic foraminifera appear?

11. Why are fossils of foraminifera generally more useful than fossils of radiolaria in the correlation of subsurface rock units encountered in the course of drilling for petroleum?

SUPPLEMENTAL READINGS AND REFERENCES

Bignot, G. 1985. *Elements of Micropaleontology*. Boston: International Resources Development Corporation.

Brasier, M. D. 1980. *Microfossils*. London: George Allen and Unwin.

Broadhead, T. W. (Ed.). 1987. *Fossil Prokaryotes and Protists: Notes for a Short Course* (Sponsored by The Paleontologic Society and The Cushman Foundation). Studies in Geology 18. Knoxville: University of Tennessee Department of Geological Sciences.

Haq, B. U. & Boersma, A. (Eds.). 1978. *Introduction to Marine Micropaleontology*. New York: Elsevier.

Lipps, J. H. 1992. *Fossil Prokaryotes and Protists*. Cambridge, MA: Blackwell Scientific.

Loeblich, A. R. & Tappan, H. 1964. *Sarcodina Chiefly "Thecamoebans" and Foraminiferida* In R. C. Moore (Ed.), *Treatise on Invertebrate Paleontology*. Lawrence: University of Kansas Press and Geological Society of America.

CHAPTER 5

Girtyocoelia dunbari, a Pennsylvania to Permian calcareous sponge from the Permian of Texas. This distinctive sponge forms bead-like chambers about 1cm in diameter that are "strung" on a tubular central canal. (Photograph courtesy of J.K. Rigby)

Porifera and Other Pore Bearers

There is found growing upon the rocks near the sea, a certain matter wrought together of foame or froth of the sea which we call sponges

Gerhard's Herball, 1636

Porifera is a phylum of aquatic animals known commonly as sponges. Sponges are simple, multicellular animals that live in aquatic, predominantly marine habitats. They range in size from about 1 to 200 cm. Multicellularity is a prerequisite for the attainment of large size in animals; it also permits diffusion of oxygen and metabolic products by individual cells that lie adjacent to intercellular fluids. In a single huge mass of protoplasm (the size of a sponge), such diffusion would be so slow as to preclude life. With their cellular, rather than protoplasmic, level of organization, poriferans are a higher form of life than are protoctistans. No single cell must carry on all the vital tasks, and there are distinctly different kinds of cells that are designed for specific functions. It is likely that sponges evolved from colonial protoctistans, as suggested by their larvae, which resemble a hollow, free-swimming colonial flagellated protoctistan. Sponges apparently did not give rise to any other animal group.

The cells within sponges differ among themselves in function and appearance, and to a degree they are dependent upon each other. Special food-gathering cells, for example, pass nutrients to other cells that are specialized for support, reproduction, or protection. Although there are such cellular interdependencies, sponges lack organ systems. In this regard they are among the more primitive of multicellular animals.

THE LIVING ANIMAL

The poriferan most familiar to us is the common bath sponge. When plucked from the seafloor, the bath sponge resembles a brown, slimy mass not unlike raw liver in appearance. This unappealing object is then dried, cleaned, and trimmed to provide the familiar flexible skeleton composed of tough elastic fibers. The fibers are composed of a proteinaceous substance called **spongin**. Because of their ability to hold water and protect against harmful impacts, processed sponges have been commercially important for centuries. They have been used for washing, as substitutes for drinking cups, and as protective linings in the metal helmets and leg guards of Roman legionnaires.

Although there are about 150 species of freshwater sponges, the majority of living and fossil sponges are marine. Species vary in appearance, and even a single species may have different shapes as individuals of that species adjust to the velocity and direction of currents, to turbulence, or to irregularities in the substrate. Regardless of differences in shape, however, the basic structure of a sponge is distinctive and is illustrated by the simple urn sponge named *Leucosolenia* (Fig. 5.1). The body of this small sponge has the shape of an elongated vase. The surface is perforated by many small openings or **incurrent pores**. These lead to a central cavity, a **cloaca** or **spongocoel**, that has a large opening at the top called the **osculum**. Some sponges have multiple **oscula**. By observing the movement of tiny colored particles added to the water near *Leucosolenia*, one can follow the path of water currents as they are drawn into the incurrent pores, pass into the cloaca, and stream out through the osculum. The cloaca may also provide a haven for a variety of small invertebrates.

The body wall of a sponge is rudely constructed of three layers, two of which are distinct cell layers. The outermost of these is the **ectoderm**, the cells of which lie side-by-side so as to form a surface rather resembling a pavement of floor tiles. Both at the outer surface and as a lining of canals, these flat ectodermal cells or **pinacocytes** pro-

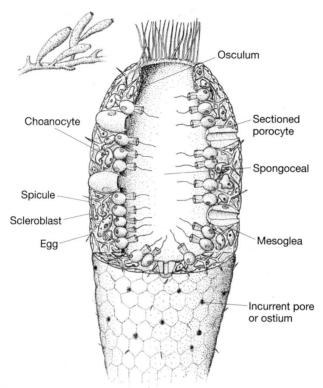

Figure 5.1   A simple (asconoid) sponge (×20) with a section cut out to show principal cells of the body wall. (After R. Buchsbaum, *Animals Without Backbones*. Chicago: University of Chicago Press, 1948.)

vide the protective equivalent of skin. In addition, those located near the base of the sponge may secrete cements that fix the organism to the substrate.

The inner cell layer or **endoderm** of a sponge may also contain some pinacocytes, but the most distinctive cells are the collared **choanocytes** (see Fig. 5.1). These face inward toward the center of the cloaca and along canals that lead into the cloaca. The flagellum on each of the choanocytes lash food particles such as bacteria, organic debris, and microplankton down into the collar where the food becomes trapped in protoplasm, is assimilated, and circulated to other cells. The vigorous beating of choanocyte flagellae provides currents within the sponge that assist in distributing food and oxygen and eliminating waste products. Sponges circulate water at an astonishing rate. A sponge the size of one's thumb can pump over 10 liters of water through its body in a single day.

The choanocytes of sponges have importance in evolutionary studies, for they demonstrate the relationship between the Porifera and the Protoctista. Protoctistans known as Choanoflagellida closely resemble the choanocytes of sponges. The similarity is so striking that some biologists propose that sponges arose from symbiotically associated aggregates of choanoflagellates and certain other protoctistans.

Between the ectoderm, with its prominent pavement cells, and the choanocyte-lined endoderm, lies a zone of gelatinous protoplasm called the **mesoglea**. Amoeboid cells within the mesoglea aid in digestion and transport waste products, whereas others called **scleroblasts** function in the secretion of spongin fibers or mineral spicules (see Fig. 5.4). In or near the mesoglea, one also finds pore cells called **porocytes**, which are tubular. Water can be drawn into the sponge through the porocytes as well as through incurrent canals. Porocytes are thought to have been derived from pavement cells that either developed an intracellular perforation, or enveloped space, to form a tube. **Contractile cells** which are able to constrict pores and small canals, are often associated with the pore cells, and they serve to close these openings against particles of sediment and other irritants.

Although sponges exhibit a variety of shapes and sizes, they can nevertheless be grouped into three general structural plans. At the simplest level is the **asconoid** plan exemplified by *Leucosolenia*, the urn-shaped sponge described earlier (see Fig. 5.1). Asconoid sponges tend to grow in clusters, rather than as solitary individuals. Next in structural complexity are **syconoid** sponges. Their plan is a modification of the asconoid pattern in that it includes folding of the body wall and reduction in the cloaca. These modifications provide more support for the syconoid sponge, more surface area for food gathering, and improved circulation and filtering. In syconoids, choanocytes line small radial chambers called **ampulae** that are developed between invaginations of the body wall (Fig. 5.2). The final and most complex structural plan is termed **leuconoid**. In the leuconoid plan, the body wall is thick with mesoglea, and choanocytes line numerous oval chambers that are connected by an intricate system of canals (Fig. 5.2C). A true cloaca is lacking, and there is ample surface area for food capture. The majority of living sponges, including the common bath sponge, have leuconoid construction.

REPRODUCTION

Sponges reproduce both sexually and asexually. In the sexual mode, sperm and eggs arise from choanocytes that are specifically modified for that function. Most species of sponges are hermaphroditic, but produce their eggs and sperm at different times. A milky cloud of sperm may be discharged suddenly through the osculum, may fertilize eggs in the external environment, or be drawn by currents into the chambers of nearby individuals, and then transferred to the mesoglea where fertilization occurs. The fertilized egg develops into a flagellated larva, which, if it settles on a suitable substrate, will begin its development.

The simplest method of asexual reproduction in sponges, called **budding**, occurs when a branch grows off the parent, constricts and separates at its base, and takes up an independent life. Sponges have remarkable regenerative powers, and this attribute is used to propagate sponges commercially. Cut pieces of sponges are attached to cement tiles and dumped into the sea where they regenerate and reproduce to provide a new crop of sponges in only a few years.

Budding is not the only means of asexual reproduction in sponges. All freshwater forms and a few marine forms reproduce asexually by forming structures called **gemmules**. A gemmule (Fig. 5.3) is a cystlike structure having a hard covering composed of spicules or spongin. A mass of amoeboid cells reside within this resistant covering. The gemmule is often able to survive drying and freezing that would be lethal to the parent sponge. Gemmules are set free from the parent by passing through the osculum or during disintegration of the parent. They are then dispersed by currents. On reaching a favorable environment, the amoeboid cells emerge through a small opening in the gemmule, aggregate, and grow into adult sponges.

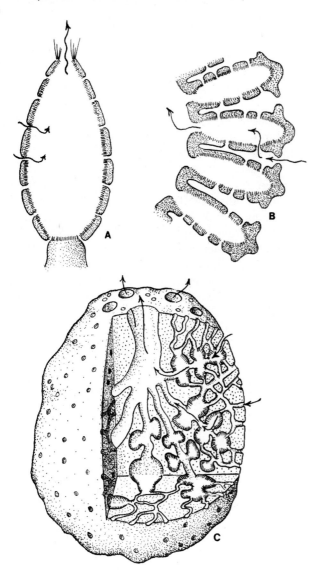

Figure 5.2 Sponge canal systems. In the ascon plan (A), the body may be tubular or vase-shaped. Water enters the cloaca or spongocoel through incurrent pores or ostia and passes out through the osculum. The sycon plan (B) differs in that the flagellated collar cells have withdrawn to small, radial chambers, each of which communicates to the cloaca. In the leuconoid plan (C), the small chambers lined with collar cells are deeply embedded in mesenchyme and connected by intricately branched incurrent and excurrent canals.

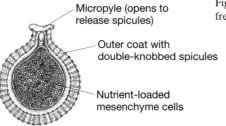

Micropyle (opens to release spicules)

Outer coat with double-knobbed spicules

Nutrient-loaded mesenchyme cells

Figure 5.3 Gemmule of a freshwater sponge.

THE SPONGE SKELETON

A framework of mineralized skeletal elements or a tough network of spongin fibers support the walls and passageways of sponges against the pull of gravity and currents. Spicules composed of calcium carbonate or silica improve the probablility of preservation and are also indispensable for identification of fossil species.

Depending on their size, sponge spicules are either **megascleres** or **microscleres**. The larger megascleres provide the principal framework of the sponge and are the spicules most frequently found as fossils. The smaller microscleres are distributed through the flesh of the sponge and are sometimes referred to as "flesh spicules." Spicules are designated according to the number of axes and rays they possess. The name of a spicule may include the root "actine" in reference to the rays, and "axon," which refers to the axis. Thus, a **triaxon** element possesses three axes that cross each other at a common point to form a six-rayed or **hexactin** spicule. Other basic forms (Fig. 5.4) include **monaxons**, which resemble straight or variously curved needles, and which may have pointed, knobbed, or hooked ends. **Tetraxons** have four axes radiating from a common intersection. The four axes may be of equal length, or as is often the case, there will be one major axon and three shorter ones. Such spicules are called **triaenes**. **Polyaxon** spicules have many short rods radiating from a central point.

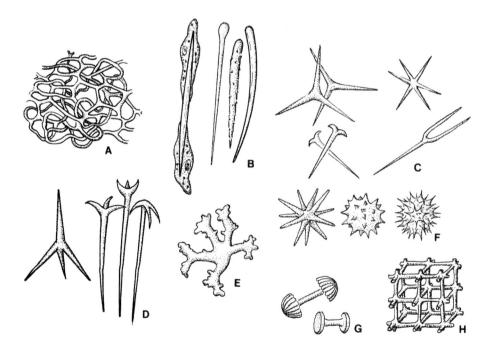

Figure 5.4   Sponge spicules. (A) spongin fibers; (B) monaxon spicules; (C) triaxon spicules. At the upper right of the triaxon group is a triaxon derivative called *hexactine*, in which three rays cross each other at a common point giving the appearance of six rays. The triaxon resembling a tuning fork is called a *pharetron*. (D) Tetraxons. (E) An aberrant form of spicule known as a *desma*. (F) Polyaxon, (G) birotules, and (H) the triaxonal lattice seen in many siliceous sponges.

Sponge spicules are produced by special amoeboid cells in the mesoglea. Most of the spicules remain in this inner layer, although they may project through the ectoderm. Often they are interlocked and fused together in distinct groupings that improve their supportive function. They are composed of either calcium carbonate or opaline silica. In general, sponges having calcium carbonate spicules prefer shallower and warmer parts of the ocean, whereas those with siliceous spicules prefer cold, deepwater environments. An exception is the common freshwater sponge *Spongilla*, which has siliceous spicules and thrives in Florida's Okefenokee Swamp.

MAJOR CATEGORIES OF SPONGES

As described above, the Porifera is a phylum of metazoans that possess choanocytes and a skeleton of spongin or mineralized spicules. There are four taxonomic classes within the phylum: the Demospongea, Hexactinellida (Hyalospongea), Calcarea (Calcispongea), and Sclerospongea. Members of the first three classes are found as fossils in rocks ranging in age from Late Proterozoic to Recent. The oldest sclerospongids are Ordovician in age.

DEMOSPONGEA

Demospongea (Fig. 5.5) have skeletons consisting of either siliceous spicules or spongin, or combinations of these materials. Nearly all fossil demospongids, however, have siliceous spicules. Characteristically, the rays of multirayed species meet at 120° or 60° angles. A relatively few species of demospongids possess monaxons, and many have lumpy, irregular spicules known as **desmas** (see Fig. 5.4E). Most demospongids are marine, although the only living freshwater sponges are also members of this class.

The boring or sulfur sponge *Cliona* is perhaps the most bothersome of living sponges. These creatures are the bane of oyster fisherman, for they bore into the shells of oysters causing them to crumble during harvesting. They also impart a repulsive sulfurous odor to the oyster's soft tissure. The bath sponges *Spongia* and *Euspongia* are examples of living marine sponges, whereas *Spongilla* and *Myenia* live in freshwater bodies. Demospongea are prominent members of the Cambrian Burgess Shale fauna. Common fossil demosponges include *Astylospongea*, *Caryomanon*, and *Microspongea* of Silurian age, the Mesozoic genera *Leiodorella* and *Siphonia*, and the Tertiary genus *Corallites*.

HEXACTINELLIDA

The Class Hexactinellida consists entirely of marine sponges having spicules composed of silica. The rays of hexactinellid sponges diverge from one another at right angles, in contrast to the rays of demosponges, which diverge at angles of 60° or 120°. Tetraxons with all four axes in the same plane are common in this class. In addition, five-, six-, and eight-rayed spicules occur. Nearly always, their rays join with those of adjacent spicules to form a firmly knit skeleton.

The hexactinellid families that have a significant fossil record are the Lyssakida, Dictyida, Lychniskida, and Heteractinida. The first of these families includes

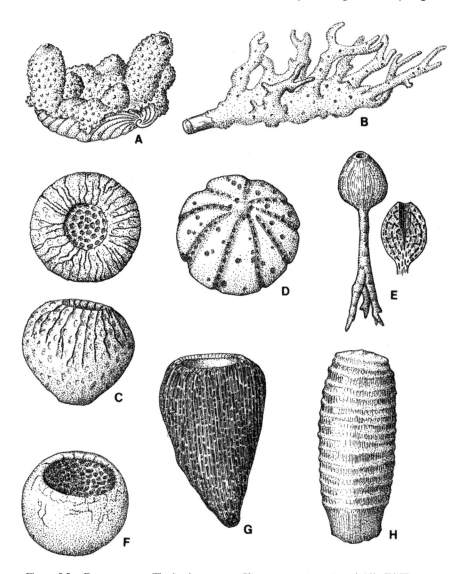

Figure 5.5   Demosponges. The boring sponge *Cliona* at a mature stage (×10). (B) The living freshwater sponge *Spongilla* (×0.35). (C) Top and side views of the Silurian sponge *Astylospongia*; (D) *Caryospongia* (Silurian); (E) *Siphonia* (Cretaceous to Tertiary); (F) *Palaeomanon* (Silurian); (G) *Calycocoelia* (Ordovician); and (H) *Nevadocoelia* (Ordovician). Sponges (C) through (H) are approximately natural size.

Euplectella (Fig. 5.6), a living sponge popularly called the "Venus flower basket." The skeleton of this tubular siliceous sponge is composed of delicate threadlike rays that form a beautiful and delicate polygonal network around the central cavity. Vertical opaline threads spiral around the perimeter of the skeleton. (Thus, the Greek derivation of the name, *eu* meaning "well," plus *plectos* meaning "twisted"). At the top of *Euplectella* there is a sievelike cap, and at the base there is a tuftlike holdfast that serves to anchor the animal on the seafloor.

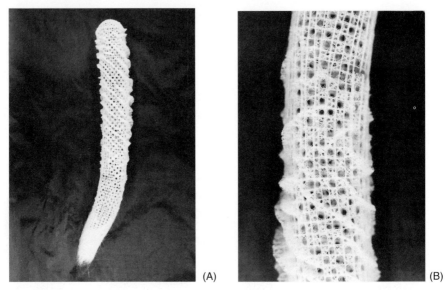

(A)   (B)

Figure 5.6   The Venus's flower basket sponge *Euplectella*. As depicted in the enlarged segment (B), the siliceous spicules in this sponge form a remarkably geometric lattice.

In Cambrian and Ordovician strata one may encounter walnut-size remains of the spherical lyssakid sponge *Protospongia* (Fig. 5.7). A single layer of spicules and an asconoid body plan characterize *Protospongia*. Like *Euplectella*, *Protospongia* attached itself to the seafloor by means of a root-tuft. It is likely that the protospongids gave rise to the more advanced lyssakids of the late Paleozoic. Among these are often well-preserved specimens of *Dictyospongia, Prismodictya* (see Fig. 5.7), and *Hydnoceras* (Fig. 5.8) found in Devonian rocks of eastern North America. *Dictyospongia* is an elongate form whose more rounded wall serves to distinguish it from *Prismodictya*, which takes the form of a slender, often expanding, prism. The spicules on each face of the prism outline quadrate openings. *Hydnoceras* has a similar wall structure, but in addition possesses blunt, hornlike protuberances. The fossils of these so-called glass sponges are internal molds formed in fine mud that settled into open areas of the skeleton. *Brachiostoma* was a differently shaped lyssakid. Like many living sponges it had thick walls, lacked a root-tuft, and had eight to twelve stocky radial protuberances.

The second major group of hexactinellids are the Dictyida. These sponges have a rigid skeleton composed of symmetrically arranged six-rayed spicules that are joined tip to tip so as to form a rectangular mesh. The spicular pattern can be seen in *Farrea*, a Holocene sponge (Fig. 5.9).

Lychniskid sponges (from *lychnos* meaning "lantern") are known for the distinctive geometry of their hexactine spicules. The spicules have short diagonal braces that connect adjacent rays in a manner suggesting an open-sided lantern. Lychniskids first appeared during the Jurassic, but did not become abundant until the Cretaceous. *Ventriculites* (Fig. 5.10) is a fairly common Cretaceous lychniskid. Rocks younger than Cretaceous rarely yield lychniskid sponges.

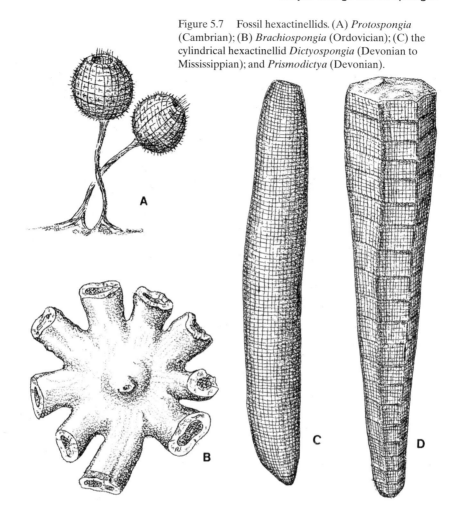

Figure 5.7   Fossil hexactinellids. (A) *Protospongia* (Cambrian); (B) *Brachiospongia* (Ordovician); (C) the cylindrical hexactinellid *Dictyospongia* (Devonian to Mississippian); and *Prismodictya* (Devonian).

Figure 5.8   The Devonian siliceous hexactinellid sponge *Hydnoceras* (Courtesy of J. Keith Rigby.)

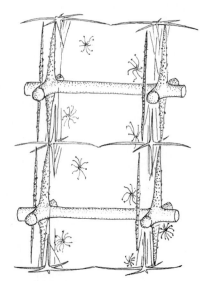

Figure 5.9   Unfused megascleres and associated microscleres of the hexactinellid sponge *Farrea*.

Figure 5.10   The funnel-shaped Cretaceous hexactinellid *Ventriculites*. This sponge was supported by six-rayed spicules and was attached to the substrate by rootlike holdfasts. (Average height 8 cm.)

The final order of the Class Hexactinellida are the Heteractinida. Members of this class are known only from Paleozoic strata. Their skeletons are constructed of polyactine siliceous spicules. *Astraeospongium* (Fig. 5.11) of the Silurian and Devonian is a biostratigraphically useful fossil sponge that is doubtfully classified as a heteractinid. The bowl-shaped wall of *Astraeospongium* is composed of a felted mass of eight-rayed spicules, six rays of which lie parallel to the surface of the sponge wall and the other two (shorter) rays normal to this plane.

CALCAREA

The Calcarea secrete only calcareous spicules. Commonly, these are either monaxons or three- or four-pronged spicules that resemble tuning forks (see Fig. 5.4). Such spicules are called **pharetrons**. Today, calcareans are found in tropical marine waters that do not exceed about 800 meters in depth. Among those living today, *Sycon* provides a good example of a sponge with a syconoid structural plan (Fig. 5.12).

Calcarea are rare in Cambrian rocks, but at least two species have been identified in the Burgess Shale. They become abundant by Ordovician time, and by the Permian they had become important in the construction of the basic framework of reefs. One of the most diverse late Paleozoic assemblages of calcareans occurs in the Permian reefs of South China. In North America, calcareans such as *Amblysiphonella* (Pennsylvanian and Permian) and *Girtyocoelia* (Pennsylvanian and Permian) are important in late Paleozoic biostratigraphy. *Cystothalamia* serves as a guide fossil for Permian strata in North Africa, and *Peronodella* is a well-known Cretaceous stratigraphic marker in Great Britain (Fig. 5.13).

Figure 5.11   The familiar North American fossil sponge *Astraeospongium*. (Diameter 5.5 cm.)

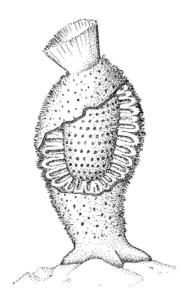

Figure 5.12   *Sycon*, with part of wall removed to show syconoid structure. (Height 2.5 cm.)

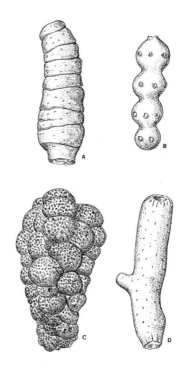

Figure 5.13   Calcareous fossil sponges. A) *Amblysiphonella* (Pennsylvanian to Permian); (B) *Girtyocoelia* (Pennsylvanian and Permian); (C) *Cystothalamia* (Permian); and *Peronodella* (Cretaceous). (Height of *Amblysiphonella*, *Cystothalamia*, and *Peronodella* about 12 cm. Height of *Girtyocoelia* about 4 cm.)

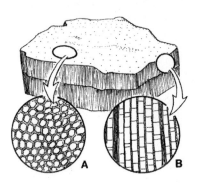

Figure 5.14   *Chaetetes* (Pennsylvanian). (A) Transverse section; (B) longitudinal section. Both sections ×10. Once thought to be a tabulate coral, *Chaetetes* is now placed in the Sclerospongea, sometimes called coralline sponges.

SCLEROSPONGEA

The Class Sclerospongea is the most recent addition to the Phylum Porifera. It is a taxon that contains only a small number of living species, most of which reside in dark crevices and caves within the walls of coral reefs. Their discovery did not occur until the invention of scuba-diving gear, which permitted marine biologists to examine closely the walls and tunnels of reefs. Sclerosponges have a distinctive structure in that the supporting skeletons of siliceous spicules and spongin fibers are underlain by an encasement of aragonite. In the living genus *Ceratoporella*, there are numerous raised oscula, each displaying starlike grooves made by excurrent canals that converge near the top of each oscular mound.

The most frequently encountered fossil sclerospongids are in the Order Chaetidae, represented by *Chaetetes* (Fig. 5.14). Species of *Chaetetes* are found as irregular bodies from the size of coins to large masses over a meter in diameter.

SPONGES THROUGH TIME

Today's sponges are found in a variety of marine and freshwater environments. Living bath sponges are most successful in tropical, shallow seas, whereas the glass sponges like *Euplectella* grow only in deep, cooler water. Fossils of ancient glass sponges, however, are often associated with sediments deposited in relatively shallow water, and their presence may have been associated with increased availability of silica from the dissolution of volcanic ash. As indicated earlier, Paleozoic demosponges and calcareans have been important contributors to the construction of reefs where they grew in dense but patchy aggregations.

The oldest strata known to contain sponge spicules are late Proterozoic in age. Although fossil sponges are more abundant following the Cambrian, their occurrence is sporadic, and they are generally used in combination with members of other phyla in biostratigraphic studies. Demosponges were common to abundant in both clastic and carbonate rocks throughout most of the Paleozoic and Mesozoic (Fig. 5.15). Paleozoic hexactinellids include the famous Devonian and Mississippian glass sponges of New York and Indiana. They probably reached their greatest diversity, however, during the Jurassic and Cretaceous. The calcareans experienced their greatest expansion during the Permian and Triassic when they were major components of extensive reefs.

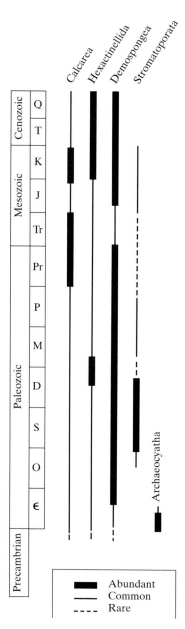

Figure 5.15   Stratigraphic ranges of Calcarea, Hexactinellida, and Demospongea, as well as Stromatoporata and Archaeocyatha.

ARCHAEOCYATHIDS

The cup or vase-shaped fossils known as archaeocyathids (Fig. 5.16) are interesting for several reasons. First, they are excellent fossils for biostratigraphic correlation of Lower Cambrian strata. Most species are Early Cambrian, but a few survived into the Middle Cambrian. Second, archaeocyathids are the first important reef-forming animals on Earth. An immense archaeocyathid reef in Australia extends parallel to a Cambrian shoreline for a distance of 200 kilometers. The spongelike porous walls of archaeocyathids, the presence of spicular elements within those walls, the similarity of

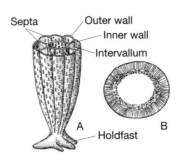

Septa   Outer wall
Inner wall
Intervallum

A   B
Holdfast

Figure 5.16   (A) Longitudinally fluted cup of an archaeocyathid, about 6 cm in height. (B) Transverse section of a nonfluted archaeocyathid having closely spaced parieties and a vesicular inner wall. (Diameter 4 cm.)

skeletal components of fossil to living calcareous sponges, and studies of functional morphology based on model and flume experiments all suggest a poriferan affinity for these biostratigraphically important fossils. However, they also possess a sufficient number of unique characteristics as to warrant their placement in a separate phylum, the Archaeocyatha.

Archaeocyathids constructed a double-walled conical to subcylindrical skeleton having an average height of about 10 cm. The outer wall is finely perforated and separated from the more coarsely perforated inner wall by vertical, radial partitions called **septa**. Septa are connected to one another by transverse rods called **trabeculae**. In addition, there are horizontal partitions called **tabulae**. The space between the inner and outer wall is referred to as the **intervallum**. A central cavity is present that is somewhat analogous to the spongocoel of poriferans, and there is a surmounting orifice corresponding to the poriferan osculum. Calcareous vesicles called **dissepiments** may be developed in the lower part of the central cavity. Whereas the upper part of an archaeocyathid is open, the lower tip is closed and is modified with rootlike **holdfasts** (anchoring structures). We know nothing about the soft parts of these long-extinct organisms, but we infer the presence of a fleshy wall surrounding the skeleton. Flume experiments with models of archaeocyathids indicated that particles of food were brought into the animal by currents moving through its perforate walls. A pumping mechanism for generating currents was apparently not needed, for sufficient flow would have been developed merely by differences in the velocity or direction of water movement across the top of the organism.

Before the end of the Middle Cambrian, all archaeocyathids, which at one time numbered more than 500 species, abruptly vanished. Possibly they were overwhelmed by the expansion of filter-feeding sponges.

STROMATOPOROIDS

Stromatoporoids are a group of marine lower invertebrates of uncertain affinity. Some investigators insist they may possess a sufficient number of distinctive traits to warrant their placement as a separate phylum. Others recommend the stromatoporoids be placed among the Porifera in the Class Sclerospongea.

Stromatoporoids first appeared in the Ordovician. Subsequently, they spread widely into the warm, shallow seas of the Paleozoic and Mesozoic. Particularly during the Paleozoic, they grew prolifically on the sea bottom and on parts of reefs where they formed domal, columnar, encrusting, or branching structures. The identification of stromatoporoids requires microscopic study of their skeletal structure in thin section. With such examination, one can recognize a framework of lamellae supported by connecting pillars, horizontal partitions, and curved panels (Fig. 5.17). Most of these features are well developed in such stromatoporoid fossils as *Stromatopora* (Ordovician to Permian), *Actinostroma* (Silurian to Jurassic), and *Labechia* (Silurian to Devonian) (Fig. 5.18).

One of the most distinctive features of stromatoporoids are conspicuous small protuberances called **mamelons** and tiny rosette patterns of grooves termed **astrorhizae** (see Fig. 5.18A) that occur regularly across the upper surface of the colony. Mamelons and astrorhizae are thought to be skeletal traces of parts of a water-conducting system.

Figure 5.17   Vertical section of the Devonian stromatoporoid *Stylostroma sinense* showing conspicuous mamelon columns with intervening concave-upward laminae. Relief of mamelons at surface up to 6 mm. Width of field 5.5 cm. (Courtesy of G. W. Stearn.)

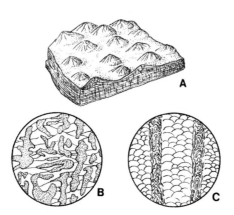

Figure 5.18   (A) Small block of *Actinostroma* (Silurian to Jurassic) with well-developed mamelons and astrorhizae. (B) Vertical section of Stromatopora (Ordovician to Permian). (C) Vertical section of *Labechia* (Silurian to Devonian) (sections ×12).

The astrorhizae relate to an interesting recent discovery of encrusting sponges in the Pacific Ocean as well as the Caribbean. The sponges secrete layers of calcium carbonate in which soft tissue and siliceous spicules are embedded. Tiny canals radiate from pores at the outer surface, and these are strikingly similar to astrorhizae, prompting the belief that most stromatoporoids are Porifera and can be classed as Demospongea.

REVIEW QUESTIONS

1. In what way is a multicellular animal such as a sponge a more progressive form of life than a colony of protoctistan cells?
2. What is the primary function of pinacocytes, choanocytes, scleroblasts, and contractile cells in a sponge?
3. What are the three general plans on which sponges are constructed? What was the advantage to the organism in the increasingly more complex structural plans?
4. What are the most apparent differences among the Demospongea, Hexactinellida, and Calcarea? Sketch a fossil representative of each.
5. Although freshwater sponges exist, why have sponges been more successful in marine environments?
6. Most modern members of the Hexactinellida live in very deep, cold parts of the ocean. Yet the Devonian–Mississippian hexactinellids were dwellers in shallow water. How do you account for this apparent contradiction of the concept of uniformitarianism?
7. What characteristics of archaeocyathids suggested to early investigators that they were sponges? Suggest some possible causes for the rather abrupt demise of the archaeocyathids.
8. What are stromatoporoids? When were they most abundant? What evidence suggests they are members of the Phylum Porifera?

SUPPLEMENTAL READINGS AND REFERENCES

Bergquist, P. R. 1978. *Sponges.* London: Hutchinson & Co.

Debrenne, F., and Wood, R. 1990. A new Cambrian sphinctozoan sponge from North America, its relationship to archaeocyathans and the nature of early sphinctozoans. *Geological Magazine* 127(5):435–443.

Hill, D. 1972. Archaeocyatha, Part E (revised). In Teichert, C. (Ed.), *Treatise on Invertebrate Paleontology.* Boulder, CO and Lawrence, KS: Geological Society of America and University of Kansas Press.

Rigby, J. K. 1987. Phylum Porifera. In Boardman, R. S., Cheetham, A. H. & Rowell, A. J. (Eds.), *Fossil Invertebrates.* Palo Alto, CA: Blackwell Scientific.

Rigby, J. K. 1971. Sponges and reef and related facies through time. In Rigby, J.K. (Ed.), *Reefs Through Time*, Pt. J. North American Paleontological Convention, Chicago, 1969. Lawrence, KS: Allen Press.

Savarese, M. 1992. Functional analysis of archaeocyathan skeletal morphology and its paleobiological implications. *Paleobiology* 18(4):464–480.

Stearn, C. W. 1975. The stromatoporoid animal. *Lethaia* 8:89–100.

CHAPTER 6

The rugose coral *Lithostrotion*, ×1.

The Cnidaria

Along the shores and beneath the seas of warm and tropical climates, the coral animalcules are laying the foundations of new islands and continents, and producing reefs of rock hundreds of miles in extent, which when elevated above the waters of future ages, will rival in magnitude and extent the mountain chains of Europe

G.A. Mantell, *Medals of Creation*, 1844

The Phylum Cnidaria (formerly Coelenterata) includes such animals as hydras, sea anemones, jellyfish, and corals. Because of their radial symmetry and the often bright coloration, many cnidarians are noted for their beauty. Their name is derived from specialized cells called **cnidocytes** (Fig. 6.1) which assist the animal in capturing prey. Cnidarians possess an internal space for digestion called an **enteron**, which opens to the outside at one end to form a mouth. The existence of a mouth and enteron permits cnidarians to utilize a much greater range of food sizes than is possible in porifera. A circle of tentacles that represent extensions of the body wall surrounds the

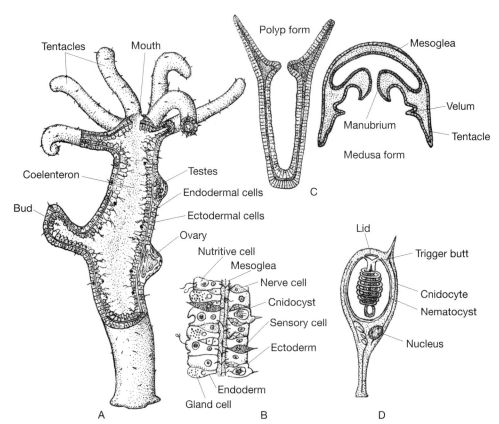

Figure 6.1   (A) Drawing of *Hydra* with section cut away. This tiny, tentacled cnidarian (3 to 10 mm in height) lives in freshwater bodies. As with many other cnidarians it is radially symmetrical. (B) Enlargement of the body wall of *Hydra* to show types of cells. (C) The polyp (left) and medusa (right) body plan of cnidarians. (D) A cnidocyte containing an undischarged nematocyst.

mouth and assists in the capture of food. Cnidarians differ from poriferans in having well-developed nerve nets, better muscular abilities, and coordinated associations of cells to form true tissues able to perform specific functions. In fact, cnidarians are regarded as the first animals to attain a tissue level of organization. They are, however, relatively simple in their embryology and in their lack of true organ systems. This has little to do with their success, for they have flourished on Earth for about 700 million years.

THE LIVING ANIMAL

POLYP AND MEDUSA

Cnidarians have two different body types (see Fig. 6.1). The polyp form is typically cylindrical, with the upper end containing the mouth with its one or more rows of encircling tentacles. These may be extended during feeding, or retracted and folded over the oral surface when the animal is threatened. Food brought to the mouth by tentacles is passed directly into the enteron, although in some groups it may first move through a short constriction of the body wall that is referred to as the esophagus. Most polyps are sessile; that is, they are attached at their aboral ends to objects on the seafloor. A polyp that is perhaps familiar to many of us is the freshwater hydra. This creature is actually able to change locations, either by slowly sliding along on its base, or by bending over and somersaulting to a slightly more advanced position (Fig. 6.2).

The medusa (Fig. 6.3), exemplified by the jellyfish, is the second body type found in cnidarians. Although they seem very unlike polyps, medusae are really inverted polyps with umbrella shaped bodies. The mouth is located on the underside, and the tentacles hang down around the ventral margin. Some cnidarian groups are known only as medusae, others only as polyps, and still others develop both body forms at different stages in their life cycles.

THE BODY WALL

Regardless of whether it is a polyp or a medusa, the body wall of a cnidarian is composed of three layers: the **ectoderm**, **endoderm**, and **mesoglea**. The external ectodermal layer is paleontologically important because it contains the cells that secrete

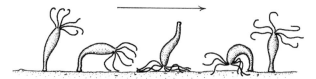

Figure 6.2   The most common type of movement in *Hydra* is by somersaulting. The body bends, attaches its tentacles to the substrate, releases its base, and somersaults to a new position. A second somersault brings its base back to the substrate. *Hydra* may also glide on its basal disk or float near the surface by suspending itself upside-down from a gas bubble secreted by basal disk cells.

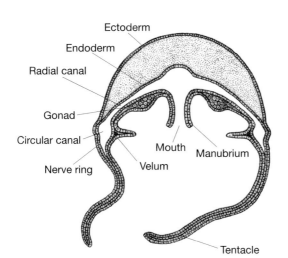

Ectoderm

Endoderm

Radial canal

Gonad

Circular canal

Nerve ring

Mouth

Manubrium

Velum

Tentacle

Figure 6.3   Diagrammatic section illustrating the anatomy of a medusa.

skeletal parts in those cnidarians that secrete a mineralized skeleton. Also within the ectoderm are special muscle cells that allow for movement, interstitial cells that give rise to egg and sperm, sensory cells, nerve cells, and cnidocytes. The unique **cnidocytes** (see Fig. 6.1) are ovoid cells containing stinging structures called nematocysts. **Nematocysts** are flask-shaped capsules containing coiled, hollow threads attached to the capsule at one end. They are most numerous in the tentacles of both polyps and jellyfish. When a small crustacean or other prey organism comes near the cnidoblast-laden tentacles, the nematocyst capsules compress and evert. The threads, often barbed like harpoons, are ejected with great force, so that the unfortunate victim finds itself riddled with poisonous barbs and threads. Paralysis quickly follows as the tentacles move the captured prey to the mouth, which opens widely to receive the nourishment.

The endoderm or inner layer of the cnidarian body wall has as its principal function the assimilation of food that is brought into the enteron. Gland cells in the endoderm then begin the process of digestion and soon reduce the prey to a soupy mass. Special nutritive cells engulf parts of this material and complete its digestion. Nutrients are then transferred from cell to cell by diffusion. Movements of the body wall and currents produced by flagellated cells carry bits of undigested debris and waste to the mouth where they are eliminated.

The third layer of the cnidarian body wall is the mesoglea. It functions chiefly in circulating fluids between the ectoderm and endoderm. Suspended within the mesoglea of many cnidarians are amoeboid cells, and at times the mesoglea may also contain various cells of the ectoderm and endoderm.

REPRODUCTION

Cnidarians reproduce by both sexual and asexual means. Many that have only the polyp form throughout their lives reproduce asexually by dividing to produce new individuals or developing lateral offshoots or branches from the parent. The first process is a form of fission, and the second is called *budding*.

Most cnidarians that have only the medusoid form reproduce sexually. The gametes (eggs and sperm) are released into the enteron or surrounding water where fertilization occurs. The fertilized egg or zygote then develops into a mature medusoid individual.

In those cnidarians that have both polyp and medusoid forms during their life cycles, reproduction is asexual during the polyp stage and sexual during the medusoid stage. An example of this **alternation of generations** is provided by *Obelia* (Fig. 6.4). The polypoid stage in *Obelia* consists of branching colonies of polyps specialized either for feeding or reproduction. The reproductive polyps asexually produce tiny, saucer-shaped medusae, which escape through an opening at the top of the reproductive polyp. The primary function of the medusae formed in this asexual manner is to produce gametes for the sexual stage. The male medusae produce sperm and the female provide eggs. Both kinds of gametes are shed into the surrounding water where fertilization takes place. Once fertilized, the egg develops into a larva, which swims about for a short time before settling onto a surface and developing into a polyp. The polyp, in turn, may bud asexually.

CNIDARIAN CLASSES AND THEIR CHARACTERISTICS

The three major taxonomic classes of the Phylum Cnidaria are the **Hydrozoa, Scyphozoa**, and **Anthozoa**. If the "frond fossils" and other still questionable fossils of the Ediacaran fauna are truly cnidarians, then at least the Scyphozoa and Anthozoa had representatives in the Precambrian, and all have survived to the present day. They have been largely marine creatures throughout their history, but some scyphozoans are able to tolerate the brackish waters of estuaries. A few hydrozoans have become adapted to freshwater bodies.

THE HYDROZOA

Individual hydrozoans may be medusoid, polypoid, or manifest both of these forms in the course of alternation of generations. The *Hydra* (see Fig. 6.1) is a well known polypoid form, whereas the Portugese man-of-war (*Physalia*) has the medusoid form. Like all hydrozoan jellyfish, *Physalia* possesses an inwardly projecting shelf or **velum** (see Fig. 6.4).

Hydrozoans are considered the least advanced of the three classes of Cnidaria. Unlike the scyphozoans and anthozoans, they lack cells in the mesoglea, and their cnidocytes are confined to the epidermal layer. (In the Scyphozoa and Anthozoa, amoeboid cells are present in the mesoglea and at least some cnidocytes are internal.)

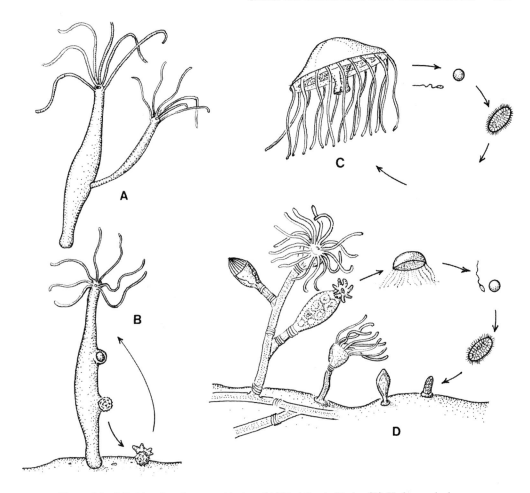

Figure 6.4   Life histories of some cnidarians. (A) Budding in *Hydra*. (B) *Hydra* producing eggs and sperm, with fertilized egg developing into a new polyp. (C) Eggs and sperm are released from medusa, unite to form a zygote that develops into a ciliated, motile, planula larva. The larva develops through a pre-medusa stage before becoming a new medusa.
(D) Reproductive cycle in *Obelia* during which a reproductive polyp asexually produces medusae that mature and produce eggs and sperm. Fertilization produces a zygote that becomes a planula. The planula positions itself on the substrate and develops into the polypoid form.

The taxonomic orders within the Class Hydrozoa that have some paleontologic importance are the Hydroida (Precambrian to Holocene), Trachylina (Jurassic to Holocene), Siphonophora (Ordovician to Holocene), Milleporina (Cretaceous to Holocene), and Spongiomorphida (Triassic to Holocene).

The Hydroida includes a few naked solitary forms such as *Hydra*, but more common members of the order are colonial (live in colonies). *Obelia* (see Fig. 6.4) is representative of the colonial forms. Within the bushy colony of *Obelia*, one can recognize tentacled feeding polyps (**hydrozoids**), as well as distinctive reproductive polyps

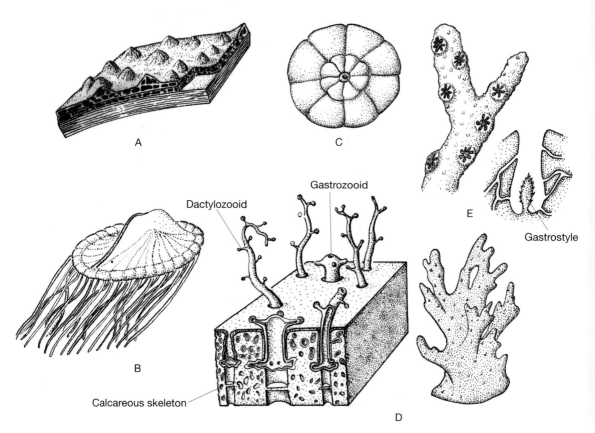

Figure 6.5   Hydrozoans. (A) Skeletal mat of *Hydractina* (Order Hydroida) encrusting a fragment of a bivalve shell, ×20. (B) Reconstruction of *Plectodiscus* (Order Siphonophora). (C) *Kirklandia*, a Trachylina medus, ×0.50. (D) Cut block of *Millepora* (Order Milleporina) exhibiting dactylozooids (defensive polyps) and gastrozooids (feeding polyps). The drawing to the right of the block shows a *Millepora* colony. (E) External view of *Allopora* (Order Stylasterina) with an adjacent enlarged section of a gastropore with its distinctive gastrostyle.

(**gonozoids**). Each one of these specialized polyps resides in a cuplike skeletal **theca** (pl. thecae), the hydrozoid in its **hydrotheca**, and the gonozoid in its **gonotheca**. Although the lengthy Precambrian to Holocene geologic range of hydroids is impressive, they are rarely found as fossils. An exception is *Hydractina* (Fig. 6.5A), the roots of which accumulate as a thin mat of chitinous or slightly calcified crust on shells and other objects. *Hydractina* has survived since the Jurassic.

Hydrozoans in the Order Trachylina are entirely medusoid, yet they differ from true jellyfish or Scyphozoa in having dissimilar larval stages, and tentacles that arise from the upper part of the medusoid umbrella rather than from the rim. Trachylinids are rarely preserved. The best known ancient genus is *Kirklandia* (see Fig. 6.5C), from Cretaceous and possibly Middle Jurassic strata of Europe and North America.

Next among the hydrozoan orders are the Siphonophora. These cnidarians occur as large floating colonies. Because they lack mineralized skeletal parts, fossils are rare.

Sporadic discoveries indicate that the siphonophorids lived from the Middle Ordovician to the Holocene. A representative is *Plectodiscus* (see Fig. 6.5B) from rocks of the Late Devonian. *Plectodiscus* possessed a large elliptical disk that was surmounted by a sail. Tentacles and polyps were suspended from the ventral side of the disk.

The Milleporina (Cretaceous to Holocene) have always been colonial hydrozoans having sturdy skeletal structures like those of corals. The skeleton is composed of aragonite, a calcium carbonate mineral having the same composition as calcite, but differing in crystal structure. Individual dimorphic polyps live in pores that penetrate the outer surface. As seen in *Millepora* (see Fig. 6.5D), feeding polyps or **gastrozoids** occupy the larger pores (**gastropores**), whereas the smaller pores hold special defensive polyps called **dactylozooids** that are well supplied with cnidoblasts. The occurrence of abundant cnidoblasts accounts for the informal designation of Millepora as the "stinging coral." The Milleporina also have pores called **ampulae** in which small medusae are produced and set free. All of these different kinds of structures are connected to one another by tubules located beneath the calcareous surface of the colony. As milleporina polyps grow upward, they seal off early portions of their pores with horizontal partitions (**tabulae**), and the underlying deeper tissue is allowed to disintegrate.

Like the Milleporina, the Stylasterina are capable of secreting sturdy calcareous skeletons. Polyp forms are similar in the two orders, but the pores in which the polyps reside are often stellate (star shape) in plan view. Also, dactylopores are arranged more uniformly around the larger gastropores. The living genus *Allopora* (see Fig. 6.5E) is distinctive in that gastropores have an axial column (**gastrostyle**) that extends upward from the base for a short distance.

The final order of the Class Hydrozoa is the Spongiomorphida. These hydrozoans form thick colonies containing vertical, radial pillars. They lack gastropores, and bear radial structures resembling the astrorhizae of stromatoporoids. A representative genus is *Stromatomorpha*, which was common during the Triassic Period.

THE SCYPHOZOA

The medusa is the dominant scyphozoan form. These are true jellyfish that consist largely of mesoglea, and that differ from the medusae of Hydrozoa in their generally larger size, lack of a velum, and the possession of four pouchlike depressions on their ventral, concave surfaces. The adult medusae reproduce sexually. The fertilized egg develops into an oval, ciliated, actively swimming larva called a **planula**. Planula larvae settle to a substrate and form small sessile polyps. From this sessile form, medusae bud off one at a time (Fig. 6.6) and grow into adult jellyfish. Some scyphozoans develop directly from a zygote without passing through the polypoid stage. Most scyphozoans have bell-shaped forms that range from 2 to 40 cm in size. The grand prize for size, however, would go to *Cyanea arctica*, which grows to a diameter of over 2 meters. At the present time, about 200 species of living scyphozoans have been described. They live at all latitudes, from polar seas to the equator.

As expected in animals that lack mineralized skeletal parts, scyphozoans are rare as fossils. Usually, preservation takes the form of impressions in shale, although

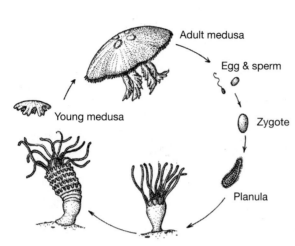

Figure 6.6 The life cycle of the
scyphozoan *Aurelia*. The polyp stage
produces young medusae by
budding.

impressions in coarser sediment, such as the Rawnsley Quartzite at Ediacara Hills in
Australia, are also found. Fifteen species of presumed scyphozoans are among the
Ediacaran fossils. Cambrian strata in Alabama have yielded many fossils of a medusoid organism named *Brooksella*; however, it is sufficiently unlike living forms to warrant placement in a separate class, the Protomedusae.

THE ANTHOZOA

Anthozoans are either solitary or colonial polypoid cnidarians. Medusae do not occur
in this cnidarian class. About two-thirds of all cnidarians are members of the Class
Anthozoa, including most corals, sea anemones, sea fans, and sea pens. Because so
many members of this class build skeletons of calcium carbonate, they are important in
the construction of reefs and provide a rich fossil record.

The anthozoan polyp differs from the polypoid form seen in hydrozoans in that
the mouth leads into an ectodermally lined cylindrical pharynx (Fig. 6.7). The pharynx
characteristically bears one or more flagellated grooves called **siphonoglyphs** that
direct water into the body. The siphonoglyphs are useful in determining the orientation
of the polyp. Also important in orientation, as well as in systematics (taxonomy), are
the mesenteries that extend radially from the body wall toward the center of the polyp.
Mesenteries are thin sheets of tissue that function primarily in increasing the surface
area across which digestion can occur. Hydrozoans, which lack mesenteries, can only
increase the digestive area by increasing the diameter of the enteron. They would
quickly reach a limiting size where the volume of water within the enteron would be so
great as to reduce the effectiveness of digestive juices. Also, with increasing size, a
point would soon be reached where the weight of the animal would exceed the ability

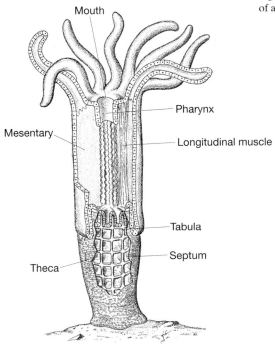

Figure 6.7   Diagram of the relation of a coral polyp to its skeleton.

Mouth

Pharynx

Mesentary

Longitudinal muscle

Tabula

Septum

Theca

of the body to provide support. Mesenteries are a solution to these problems. They provide absorptive surfaces on both sides, do not require an increase in the size of the polyp, and displace water that might otherwise dilute digestive secretions.

The two subclasses of anthozoans that have provided a significant fossil record are the Octocorallia and Zoantharia. A third subclass, the Ceriantipatharia, is virtually unknown as fossils and will not be considered here.

OCTOCORALLIA (ALCYONARIA)

The Octocorallia are small, dimorphic, colonial anthozoans with eight tentacles and eight primary mesenteries. The latter differ from the mesenteries of all other Anthozoa in having all muscle bands on the mesenteries facing the side of the pharynx where a single siphonoglyph is located. The tentacles are **pinnate**, meaning they have small side branches. By convention, the siphonoglyph defines the ventral side.

Included in the Subclass Octocorallia are gorgonian corals such as sea fans, sea feathers, the red organ pipe coral (*Tubipora*), the blue coral (*Heliopora*), the precious red coral (*Corallium*), and sea pens (*Pennatula*) (Fig. 6.8). Some octocorals lack mineralized skeletons. In those with skeletons, septa (mineralized partitions) are lacking, and the thecae are composed of either a horneous proteinaceous material called **gorgonin**, calcareous spicules embedded in gorgonin, calcareous fibers, or cemented spicules.

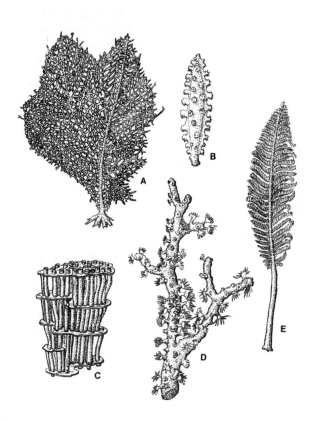

Figure 6.8   Octocoralla.
(A) *Gorgonia*, the "fan coral," about 30 cm in height. (B) Spicule of *Gorgonia*, 1 mm long. (C) The "organ pipe coral," *Tubipora*, diameter of theca 3 mm. (D) Branches of the "red coral," *Corallium rubrum*, average width of branch 4 mm. (E) *Pennatula*, the sea pen, about 20 mm in height.

When the colony dies, spicules are freed from the soft tissue they once supported and accumulate as components of seafloor sediment.

Compared to the Zoantharia, the Octocorallia have only a modest fossil record. If the frond fossils of the Ediacaran fauna are confirmed as sea pens, they would extend the geologic range of the octocorals into the Late Proterozoic. Aside from the Ediacaran occurrences, unquestioned fossils of octocorals have been found in Cambrian, Ordovician, Cretaceous, and Tertiary strata. They are prominent members of living reefs.

ZOANTHARIA: THE TRUE CORALS

Within the Subclass Zoantharia are solitary and colonial cnidarians whose polyps (unlike those of octocorals) have *more* than eight tentacles. Some, like the anemones, lack skeletons. Ancient sea anemonies are known primarily from impressions and carbonized films. The majority of zoantharians secrete calcium carbonate around the ectoderm to form a cup that is termed a **theca** or **corallite** as well as several kinds of skeletal structures within the theca. Among these so-called stony corals are those of the orders Rugosa, Tabulata, Scleractinia, Heliolitida, Heterocoralla, and Cothoniida.

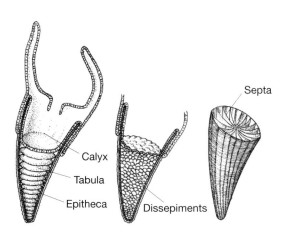

Figure 6.9   Relation of tabulae, dissepiments, and septa in the calyx of a rugose coral. Some taxa of Rugosa have only dissepiments, others only tabulae and septa, and others possess all three skeletal structures.

ORDER RUGOSA (MIDDLE ORDOVICIAN TO PERMIAN)

The Order Rugosa takes its name from wrinkled ridges or **rugae** on the thecae. As noted above, the theca is the "cup" in which the polyp resides. In rugose corals, the theca is surrounded by another calcareous layer, the **epitheca** (Fig. 6.9) secreted by the ectoderm at the base of the polyp. A prominent feature of the Rugosa are radiating vertical plates or **septa** that extend inward from the outer wall. They are constructed of vertical rods or **trabeculae** that are aligned like the logs of stockade fences. Septa are also secreted by the ectoderm as it is folded inward along the base and sides of the polyp. Because septa are inserted at only four locations during mature stages of growth, rugose corals are sometimes also called *tetracorals*.

Two kinds of septa, major and minor, can usually be recognized. **Major septa** extend inward the farthest, whereas **minor septa** are usually shorter and thinner. Septa may be thickened by secretion of additional calcium carbonate. When thickened only at their inner margins, they appear club-shaped in sections cut at right angles to the vertical axis (saggital sections) and are termed **rhopaloid** (Fig. 6.10). **Carinate septa** have shallow flanges along their sides, and **contratingent septa** are minor septa that lean against neighboring major septa. Such septal characteristics are important for the identification and classification of rugosan corals.

Other skeletal components of the rugose corals are the calyx, fossula, columella, tabulae, and dissepiments. The **calyx** is merely the basin-shaped depression at the top of the theca, which was the final residence of the polyp. A **fossula** is a groove in the calyx occupying a position where a septum is absent or reduced. It is useful in orientation. The **columella** is an axial structure that rises upward from the base of the theca as a column. It is secreted by the ectoderm as it domes itself upward from the bottom of the polyp. The principal function of the columella is to reduce the volume of the enteron. A similar function is served by the **streptocolumella**, which is an axial column composed of the twisted inner margins of the longer septa. The **tabulae** are transverse partitions that extend across the inner area of the theca. They can be flat or curved, and can either extend from wall to wall or be confined to the central axial region. Tabulae

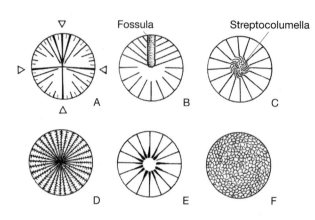

Figure 6.10 Transverse sections of rugose corals showing some skeletal structures within the theca. (A) Section showing primary or first-formed septa meeting at the center, and the order of subsequently appearing septa indicated by progressively decreasing length. Triangular symbols identify the cardinal septum (at top), counter septum (bottom), and the two lateral alar septa. Short, thin septa are those of the second cycle. (B) Transverse section with fossula. (C) Transverse section with streptocolumella. (D) Carinate septa. (E) Rhopaloid septa. (F) Cystiphylloid structure in which dissepiments fill the theca.

are secreted in succession by the basal ectoderm as the polyp draws itself upward during growth.

Finally, the **dissepiments** are small curved plates usually formed between septa near the periphery of the corallite. They resemble bubbles, with their convex upper surfaces facing inward and upward. Like the columella, dissepiments were important in reducing the volume of the enteron. When sectioned longitudinally, the frothy mass of dissepiments are seen in a broad marginal zone called a **dissepimentarium**. The central area occupied by tabulae is called the **tabularium**.

To provide uniformity in formal descriptions of rugose corals, they are described with respect to quadrants delineated by certain septa and fossulae. In the conventional scheme, the first four septa to be formed are called first order or **directive septa**. One of these is further designated the cardinal septum. Usually the **cardinal septum** lies in the fossula and may be longer than the other three directive septa. On the side of the calyx opposite the cardinal septum is the **counter septum** (Fig. 6.11). The two remaining directive septa, called **alar septa**, arise on opposite sides of the cardinal septum. Usually one can recognize an extra wide space on the counter side of each alar septa. These are termed the **alar pseudofossulae**. They provide the final feature needed to identify the four quadrants of the coral. The right and left cardinal quadrants extend from the cardinal septum past the alar septa. The regions from the alar pseudofossula to the counter septa constitute the right and left counter quadrants. A further orientation aid may be provided by the shape of the theca. In solitary horn-shaped corals, the cardinal side is concave, as a consequence of its slower rate of growth.

Rugose corals occur both as solitary forms resembling horns or cylinders, or as compact colonial masses composed of an aggregate of individual corallites. In life, the solitary forms were variously attached to the seafloor by rootlets or rudderlike talons. Most were attached at the tip or apex of the theca, but those shaped like Persian slippers probably lay horizontally on the seafloor. The twisted shapes of some horn corals clearly indicate that they had an initially vertical position, but had fallen over and changed their direction of growth in order to "right" themselves.

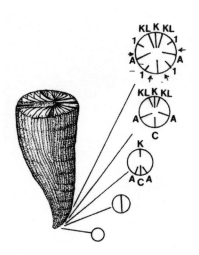

Figure 6.11   Successive transverse sections through a solitary rugose coral showing the order of insertion of septa during coral growth. Arrows indicate locations at which new septa are inserted. *C* indicates cardinal septum; *K* counter septum; *KL* counter lateral septum; *A* alar septum.

The sheer number of families of rugose corals makes classification difficult. There are nearly a hundred families, and their recognition is commonly dependent upon examination of transverse and longitudinal sections. In a simplified classification, which provides a format of description for some of the more familiar groups, the Order Rugosa is divided into the suborders Streptelasmatina, Columnariina, and Cystiphyllina. The Streptelasmatina can then be further divided into the superfamilies Cyathaxoniicea and Zaphrenticae.

Suborder Streptelasmatina   The solitary or colonial corals of the Suborder Streptelasmatina (Ordovician to Lower Permian) are characterized by domed tabulae and either a marginal dissepimentarium composed of tiny globular dissepiments, or by the presence of septa that are thickened at their inner edges to form a well-defined marginal zone called a **stereozone**. Within the suborder, the Superfamily Cyathaxonia (Ordovician to Triassic) consists of small, conical, solitary rugosids that usually do not have dissepiments, but possess a narrow septal stereozone. The major septa in cyathaxonids usually extend inward all the way to the columella. *Cyathaxonia* (Fig. 6.12A,B), a representative genus with a prominent columella, carinate septa, and distinctly ribbed surface, is used worldwide for the correlation of certain Mississippian stratigraphic units.

The Superfamily Zaphrenticae (Ordovician to Permian) contains both solitary and colonial forms. A dissepimentarium or stereozone is usually present, as well as domed or conical tabulae. The superfamily contains many well-known guide fossils (see Fig. 6.12) including *Zaphrenthis, Heliophylum, Streptelasma,* and *Lithostrotion. Zaphrenthis,* a Devonian fossil, has carinate septa, a narrow dissepimentarium, and numerous irregular tabulae. Heliophyllum, another Devonian form, has particularly prominent carinate septa. *Heliophyllum halli* is the species used by J.W. Wells in a famous study that confirmed calculations that tidal friction has been slowing Earth's rotation at a rate of about 2 seconds every 100,000 years (see Box 2.1, Fossils as Geochronometers).

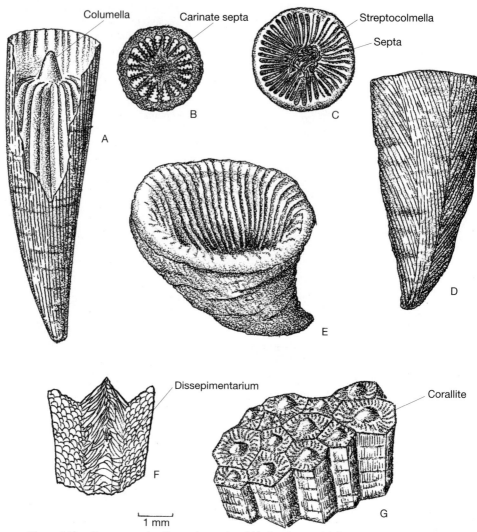

Figure 6.12 Representative Streptelasmatina. (A) *Cyathaxonia*. (B) Transverse section of *Cyathaxonia*. (C) View into cup of *Streptelasma*. (D) Lateral view of *Streptelasma*. (E) *Zaphrenthis*. (F) Longitudinal section of a solitary *Lithostrotion*. (G) *Lithostrotion* colony.

Streptelasma, an Ordovician rugosan, takes its name from its prominent strepto-columella. Species of this genus lack dissepiments and have short minor septa. They are the earliest abundant solitary rugose corals to evolve. In *Lithostrotion* (Mississippian), the tabularium is well developed, and a sharp upward doming of tabulae produces a distinct columella.

Suborder Columnariina (Ordovician to Permian) Corals of the Columnariina are largely colonial, although solitary forms do occur rarely. The individual corallites

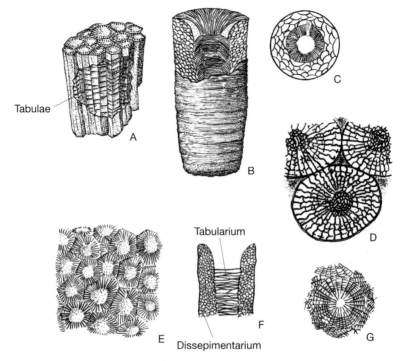

Figure 6.13   Representative Columnariina. (A) *Columnaria*, ×1. (B) *Blothrophyllum* with part of the thecal wall broken away to reveal dissepiments and tabulae, ×0.7. (C) Transverse section of *Blothrophyllum*, ×0.7. (D) Transverse section of *Waagenophyllum*, ×2.5. (E, F, and G) Surface of colony, longitudinal section, and transverse section of *Hexagonaria*.

within the colony may be loosely grouped in parallel units (**phaceloid**) or tightly packed to resemble hexagonal prisms (**ceroid**). Most genera have thin septa and flat or axially depressed tabulae. Examples (Fig. 6.13) include the Ordovician genus *Columnaria*, which has the honeycomb appearance of many ceroid rugosans; *Blothrophyllum*, a solitary Devonian form with a wide dissepimentarium and septa withdrawn from its outer margin; *Waagenophylum*, a Permian genus with slender fasiculate corallites and a dissepimentarium of small interseptal plates; and *Hexagonaria* of the Devonian, which is noted for its attractive hexagonal pattern. Amateur collectors know *Hexagonaria* as "Petosky stone."

Suborder Cystiphyllina (Ordovician to Permian)

This third suborder of the Rugosa is composed of colonial and solitary corals in which the trabeculae are large and complex. Often in these corals dissepiments are extraordinarily prominent, as in the Silurian genus *Cystiphyllum* (Fig. 6.14). *Calceola* is an interesting Devonian slipper-shaped cystiphyllid with a triangular base and semicircular corallites that are capped by opercula. Its distinctive form indicates it lay on the seafloor with its convex side down and its lidlike operculum poised at an angle of 90° so as to close quickly on the approach of a predators.

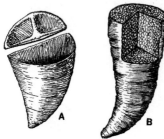

Figure 6.14 Two members of the Cystiphyllina. (A) *Calceola*, with operculum. (B) *Cystiphyllum*, a horn-shaped coral with interior almost completely occupied by dissepiments. (Both drawings ×1.)

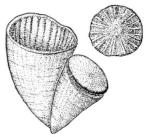

Figure 6.15 The small, operculate, Middle Cambrian rugose coral *Cothonion*, ×5. Note the septal ridges in the open calyx (Cothoniida had weak septa). An underside view of the operculum is shown at the upper right.

ORDER COTHONIIDA

The cothoniids are a rare but interesting group of colonial and solitary corals that are known only from the Middle Cambrian. They are characterized by small, originally calcitic thecae having opercula and fragile septa. The septa are added in a manner similar to that observed in rugosids, but cothoniids appear too early and are too advanced morphologically to be encompassed by that group. They are viewed as an evolutionary side branch of coralline evolution that ultimately failed. The most representative genus is *Cothonion* (Fig. 6.15) from the Middle Cambrian of New South Wales, Australia.

ORDER HETEROCORALLIA

Although only five genera of heterocorals have been recognized, they are sufficiently distinctive to warrant their placement in a separate taxonomic order. Heterocorals have a peculiar order of septal insertion. Four primary or prosepta are inserted first, and these are joined at the axis. Subsequent additional septa are added in each of four quadrants, and these branch symmetrically. A typical heterocorallid rarely exceeds a few centimeters in length or more than 6 mm in diameter. An example is *Heterophyllia* (Fig. 6.16) from the Lower Carboniferous of Europe.

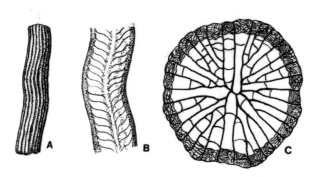

Figure 6.16 *Heterophyllia*. (A) Exterior of fragment, ×1. Usually only such short fragments are recovered. (B) Longitudinal section, ×2. (C) Transverse section of mature stage, ×8.

ORDER TABULATA

All tabulate corals are colonial. Fossils are found in rocks ranging in age from Ordovician to Late Permian. The group takes its name from the tabulae that are its most prominent feature. As with rugose corals, the polyp rested on the last formed tabula, and successive tabulae were added to provide for upward growth. Corallites in the colony are either loosely held, or tightly bundled in a prismatic arrangement. In the tightly bundled forms, the walls of the corallites may be solid or perforated by small openings termed **mural pores**. The mural pores might have been openings for the passage of tissue or nutrients between adjacent individuals. Where the corallites do not lie directly next to one another, the intervening space is spanned by horizontal plates, occupied by tubular processes, or filled with spongy or vesicular skeletal material called **coenenchyme**.

The Tabulata can be divided into the following suborders: Lichenariina, Sarcinulina, Favositina, Auloporina, Syringoporina, and Halysitina.

Suborder Lichenariina (Early Ordovician to Middle Silurian)   This suborder consists of tabulates having prismatic corallites bundled into massive colonies. The corallites of most lichenariids contain 16 or more septa, although some such as *Lichenaria* (Fig. 6.17A) lack septa. Mural pores are commonly present and arranged in horizontal rows.

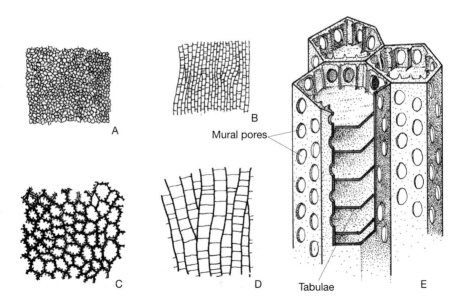

Figure 6.17   Transverse (A) and longitudinal (B) sections of *Lichenaria* (Early Ordovician). *Lichenaria* is among the earliest of corals and lacks septa. Transverse (C) and longitudinal (D) sections of *Favosites*. (E) Diagram depicting the basic internal structure of *Favosites*. (A–D, ×1.)

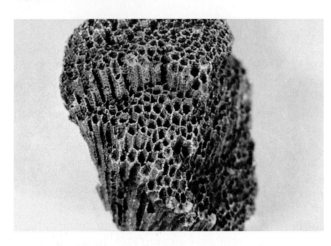

Figure 6.18 Colony of the tabulate coral *Favosites* (Ordovician to Permian). The prismatic corallites are about 5 mm in diameter and are pierced by mural pores.

Suborder Sarcinulina (Early Ordovician to Middle Ordovician) Sarcinulinids form massive colonies composed of corallites that may have up to 24 septa. Extensions of septa and tabulae form spongy or vesicular tissue called **coenenchyme**, readily seen in thin sections. An example is *Sarcinula* (Middle Ordovician to Early Silurian).

Suborder Favositina (Middle Ordovician to Permian) The sturdy, massive colonies of this large suborder of tabulates are composed of prismatic corallites that may contain septa formed by longitudinal rows of short spines. Tabulae and mural pores are prominent, and they can be readily examined with the aid of a hand magnifier. The most familiar genus is *Favosites* (Figs. 6.17 and 6.18), species of which are informally called "honeycomb corals." Although the total range of *Favosites* is Late Ordovician to Permian, species are most useful biostratigraphically in Silurian and Devonian rocks. Other representatives are *Palaeofavosites* (Ordovician-Silurian), and *Alveolites* (Silurian-Devonian).

Suborder Auloporina (Ordovician to Permian) These tabulates generally form low encrusting networks of tubes, rather than massive colonies. Septal spines are present, and tabulae are either widely spaced or absent. An example is *Aulopora* (Fig. 6.19A), a simple form with trumpet shaped corallites. *Aulopora* was an epifaunal coral, and characteristically grew on the shells of larger Paleozoic invertebrates.

Suborder Syringoporina (Late Ordovician to Permian) The corallites of syringoporinids are cylindrical and very loosely bundled. They are joined at intervals by lateral tubes. Most species are nonseptate, but some have rows of small septal spines. The syringoporinids are sometimes referred to as fossil organ pipe corals, although they are quite distinct from the familiar modern organ pipe coral (*Tubipora*). *Syringopora* (see Fig. 6.19B) is a representative genus that ranged from Silurian through Pennsylvanian.

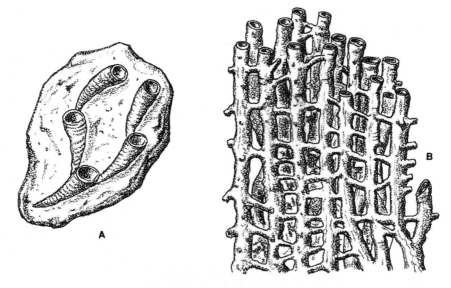

Figure 6.19   The tabulate corals *Aulopora* (A), about three times natural size, and *Syringopora* (B), about two times natural size. Both genera range from Silurian to Permian.

Suborder Halysitina (Middle Ordovician to Devonian)   The halysitinids are commonly called *chain corals* because their corallites appear as chainlike strands when the colony is viewed from above. Actually, the colonies more closely resemble short tubes arranged like logs in a stockade fence. Corallites in halysitinids are mostly elliptical in cross section, but circular and quadrate forms also occur. They tend to be uniform in length and are distinctly tabulate. Corallites may have 6 to 12 short septa. In some species, smaller tubes called *microcorallites* alternate between larger ones. Mural pores are not developed. Species of the suborder were most abundant during the Ordovician and Silurian. *Halysites* (Fig. 6.20) is an example that is seen by most beginning paleontology students in their study of common invertebrate guide fossils.

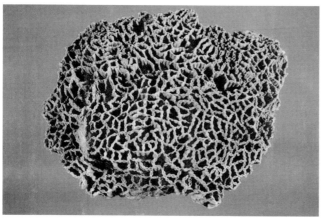

Figure 6.20   *Halysites* (Ordovician to Devonian). The specimen is about 8 cm in diameter.

ORDER HELIOLITINA (ORDOVICIAN TO MIDDLE DEVONIAN)

The Heliolitina resemble tabulates in their colonial habit, in having well developed tabulae, and in possessing calcitic skeletons. The corallites, however, are separated by areas of coenenchyme secreted buy a mass of tissue called a **coenosarc,** which in the living form was present between polyps. Twelve septa are characteristic of this order. *Heliolites* (Fig. 6.21) is a representative form with worldwide distribution in Silurian and Middle Devonian strata.

ORDER SCLERACTINIA (MIDDLE TRIASSIC TO HOLOCENE)

Scleractinid corals occur in rocks ranging in age from Triassic to Recent. The order is defined by the manner in which septa are inserted between paired mesenteries in multiples of six. This accounts for their being alternately designated as hexacorals. When scleractinid larvae settle to the seafloor and have secreted the base of the juvenile theca, the first septa, designated the **prosepta,** are secreted all at once between each of the six pairs of mesenteries. The subsequent septa, designated **metasepta,** are then inserted in simultaneous cycles of 6 second-order septa, 12 third-order septa, 24 fourth-order septa, and so on, with each septum of the new cycle centered in the interseptal spaces of the previous cycle. Thus, each successive cycle after the first doubles the number of septa, although the third and subsequent cycles may be incomplete. Typically, each successive cycle of septa is slightly shorter than those of the preceding cycle.

Because rugose corals appear before the scleractinids and have so many similarities in structure, paleontologists have traditionally perceived the scleractinids as having evolved from rugosids. Another theory gaining favor proposes that scleractinids arose from sea anemones that may have existed as far back as the Cambrian.

The polyps of living scleractinids not only reside within their calyces, but they also have an overlapping fleshy layer on the outside of the theca. This layer is capable of secreting vertical ridges and costae that are external continuations of the septa. In colonial forms, the fleshy layer of adjacent individuals may fuse and form a **coenosarc.** The coenosarc functions in the secretion of coenosteum. Presumably, a coenosarc existed in the extinct Heliolitina and produced conenchyme in a similar manner. During growth, polyps move upward in the calyx and close off underlying areas by secreting tabulae and dissepiments.

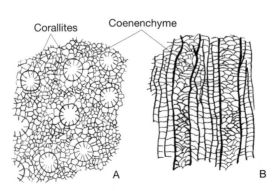

Corallites   Coenenchyme

A   B

Figure 6.21   Transverse (A) and longitudinal (B) sections of *Heliolites.* Corallites are about 3 mm in diameter.

In scleractinids, skeletal material is secreted by specialized cells called **calicoblasts**. These form the **calicoblastic layer**. The space beneath the calicoblastic layer is rich in organic compounds, and these facilitate the combination of calcium and bicarbonate ions to form calcium carbonate. The reaction is written:

$$Ca^{2+} + 2HCO_3^- \rightarrow CaCO_3 + H_2O + CO_2$$

Skeletal construction is enhanced in reef corals by the presence of symbiotic algae (zooxanthellae), which photosynthetically extract carbon dioxide from calcium bicarbonate in solution while generating oxygen. The relationship that corals have with these algae accounts for their faster growth during daylight hours when photosynthesis is possible. Corals experimentally deprived of their zooxanthellae add skeletal mass at a far lower rate than they do under natural conditions. The zooxanthellae in corals are nonmotile dinoflagellates that impart their brown or yellow-brown color to their coral hosts.

With regard to the development of septa, calicoblasts in scleractinids secrete calcium carbonate as tiny bundles of aragonite crystals that radiate from a common center. The bundles, called **sclerodermites**, are stacked vertically to form rods named **trabeculae** (Fig. 6.22). Trabeculae are also present in some rugose corals. In Scleractinia they may consist only of a single column of sclerodermites (simple trabeculae), or they may be composed of aggregates of six or more (compound trabeculae). In the formation of septa, trabeculae ar aligned in parallel or fanlike fashion, rather like the posts of a palisade wall. If the space between adjacent trabeculae is solidly filled with calcium carbonate, the septa are termed **laminar**. Perforated septa are termed **fenestrate**, and if there are cross members (**synapticulae**), the septa are called **synapticulate**. Along their free edges, septa may be smooth or toothed (**dentate**) like a saw blade.

In the classification of the Order Scleractinia, the structure and arrangement of septa, the development of coenosteum, and overall form are of primary importance.

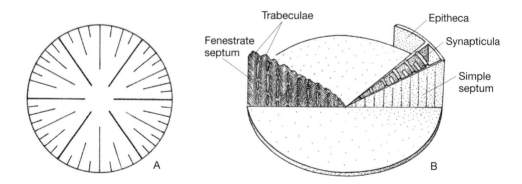

Figure 6.22   (A) The order in which septa appear in the growth of scleractinid corals. The first-formed or primary septa extend farthest inward, and subsequent septa appear sequentially in the order indicated by decreasing length of lines. Although the length of progressive cycles of septa decrease in many corals, certain groups may not follow this pattern. (B) Diagram depicting the alignment of trabeculae to form septa and trabecular cross-bars known as synapticulae.

The five suborders of scleractinid corals are the Astrocoeniina, Fungiina, Faviina, Caryophyllina, and Dendrophyllina.

Suborder Astrocoeniina (Middle Triassic to Holocene)   Most members of the Suborder Astrocoeniina are reef-building, colonial corals that construct relatively small corallites (1 to 3 mm in diameter). Although a few of the more primitive species may have septa represented only by spines, most families have laminar, dentate septa composed of up to eight trabeculae. *Acropora* (Fig. 6.23), known commonly as the staghorn coral, is an abundant and important component of reefs in the Pacific Ocean. Members of this familiar genus are known in rocks ranging from the Eocene to the Holocene.

Suborder Fungiina (Middle Triassic to Holocene)   Fungiid corals include both colonial and solitary species. All have more than eight trabeculae in each septum. The septa are linked by synapticulae, are dentate along their free margins, and are fenestrate at least in the early growth stages. *Fungia* (Fig. 6.24), whose skeleton resembles the underside of a mushroom, is a representative genus that has thrived since Miocene time. Another familiar genus is *Porites* (Eocene to Recent), species of which are among the most common of all reef corals today.

Figure 6.23   The staghorn coral *Acropora* (Holocene). Species of this genus are common in tropical reefs of the Atlantic and Indo-Pacific.

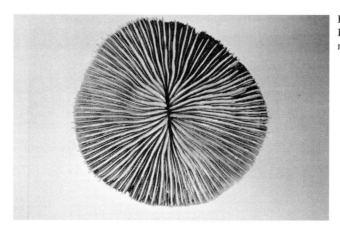

Figure 6.24 *Fungia* (Miocene to Holocene). The specimen has a maximum diameter of 11 cm.

Suborder Faviina (Middle Triassic to Holocene)   Faviinids have laminar septa with fanlike trabeculae. Septal margins are dentate. Synapticulae are rarely seen, but dissepiments are common. The suborder includes both solitary and colonial forms. *Favia* and *Platygyra* (Fig. 6.25) are representative members of this suborder. Because it reminds one of the sinuous curves (meanders) of a stream, the convoluted thecal pattern is termed **meandroid**. The so-called brain coral *Diploria* shows the meandroid pattern well.

Suborder Caryophyllina (Jurassic to Holocene)   Most caryophyllinids are solitary corals characterized by laminar septa with smooth margins. Synapticulae are lacking, and various skeletal structures are secreted on the outside of the theca by overlapping ectodermal tissue. Caryophyllinids are nonreef corals. *Parasmilia* (Cretaceous) and the Eocene "button coral" *Trochocyathus* (Fig. 6.26) are examples.

Suborder Dendrophyllina (Upper Cretaceous to Recent)   Like the caryophyllinids, members of the Suborder Dendrophyllina are mostly nonreef corals having laminar septa with smooth margins. They differ in having synapticulae, however. Also, septa fuse into distinctive branching patterns according to a

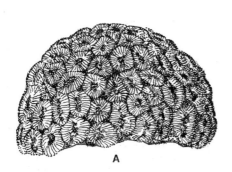

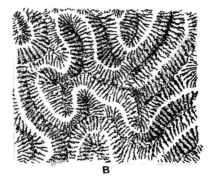

Figure 6.25   (A) Corallum of *Favia* (Cretaceous to Holocene); diameter of corallites is about 7 mm. (B) *Platygyra* (Cretaceous to Holocene), width of field 60 mm.

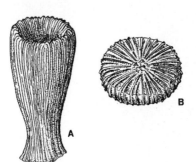

Figure 6.26   (A) *Parasmilia* (Cretaceous to Holocene), ×2. (B) *Trochocyathus* (Eocene), ×3. (Redrawn from R. C. Moore (Ed.), *Treatise on Invertebrate Paleontology.* Boulder, CO, and Lawrence, KS: Geological Society of America and University of Kansas Press, 1956.)

pattern of septal insertion termed the **portales** plan. *Endopachys* (Fig. 6.27) is an Eocene representative of this suborder.

CONULARIIDS

Conulariids are a group of problematic organisms of uncertain affinities that have some features suggesting they may be allied to the Cnidaria. Based on studies of exceptionally well-preserved fossils, Feldmann and Babcock (1986) argue that the structure of the conulariid exoskeleton is unique and warrants the group being considered a separate phylum. However, examination of the conulariid ultrastructure as revealed by electron microscopy has convinced van Iten (1991) that conulariids are related to scyphozoans.

The conulariid exoskeleton resembles a four-sided cup (Fig. 6.28). It has a chitinophosphatic composition, a pointed apex, and an expanded free end, which, in most species, could be closed off by neatly fitted flaps of exoskeletal material. The most conspicuous features of the exoskeleton are transverse, closely spaced, somewhat arcuate ridges along each face. These ridges are the external expression of calcium phosphate rods that support the otherwise delicate exoskeletal wall. Specimens are usually between 4 and 10 cm in length, but exceptionally large specimens up to 40 cm long are known.

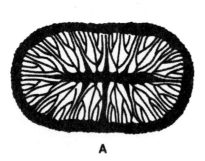

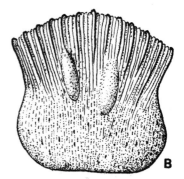

Figure 6.27   Transverse section (A) and side view (B) of *Endopachys* (Eocene to Holocene), ×1.5. (After T. W. Vaughn, *United States Geological Survey Monograph 39*, 1900.)

Figure 6.28   Excellent preservation of two specimens of *Paraconularia*, ×0.9, from the Cuyahoga Formation (Mississippian) of Ohio. The specimen at the right has been somewhat flexed. (Courtesy of R. M. Feldmann and Loren E. Babcock.)

Conulariids, like the Mississippian species *Paraconularia chesterensis*, were widely distributed and are found in diverse kinds of sedimentary rocks. This suggests a pelagic life style for at least some conulariids. A few species show the vague outlines of what may be attachment stalks, or perhaps these outlines represent marine plants used by conulariids for anchorage. Some investigators believe a pseudoplanktonic habit is indicated by the stalklike outlines. Groupings of conulariids at collection sites suggest they may have lived clustered together, perhaps to facilitate reproduction.

CORAL HABITATS

With regard to habitat, there are two categories of corals. Those that are important components of reefs are termed **hermatypic**, whereas nonreef corals like the dendrophyllinids are called **ahermatypic**. Hermatypic corals flourish at shallow depths (up to about 90 meters) where temperatures range from 25° to 29°C. By inference, fossil hermatypic corals had similar preferences. Locations of ancient coral reefs have been useful in paleogeographic studies for determining the approximate location of the paleoequator. Ahermatypic corals usually lack the symbiotic relationship with zooxanthellae that is common to their hermatypic cousins. The group includes about equal numbers of solitary and colonal forms, both of which have been successful inhabitants of the seafloor from depths of only a few meters to 6000 meters. Ahermatypic corals can tolerate temperatures as low as 1°C.

The Scleractinia that inhabit tropical seas today are not only abundant but also quite diverse. They thrive in areas having good circulation where oxygen is abundantly available to polyps. Because turbid water tends to retard photosynthesis in the symbiotic algae, and because suspended sediment tends to smother polyps, Scleractinia grow best in well-mixed clear water very near the ocean surface. As one follows the reef front to greater depth, it is often possible to see a downward increase in the diversity of scleractinids. This relationship has been used to estimate depth of water in which ancient coral-bearing sediments were deposited. In addition to their sensitivity to water depth, coral species vary according to their position along the shoreline. For example, species living on the windward side of an island differ from those on the leeward side, and corals growing in quieter areas behind the reef front are different from those that inhabit the surf zone.

Although modern reef corals have relatively few enemies, they do provide food for certain starfish and for certain fish that eat polyps. The most notorious of the starfish that consume coral polyps is *Acanthaster*. Within the last three decades, this echinoderm has severely damaged Pacific reefs. Recovery of the damaged reefs is difficult because algal growths quickly cover the dead corals and prevent coral larvae from reestablishing colonies. Certain activities of humans can also be destructive of coral reefs. Silts from farmlands and construction have destroyed fringing reefs at many localities around the world. Oil pollution produced during oil-tanker spills or when oil tanks are flushed and cleaned has been lethal to coral growths. In some underdeveloped countries, living reefs are destroyed by quarrying coral for the manufacture of cement and building stones.

Much like the Scleractinia, rugose corals were also inhabitants of shallow, warm seas. Many had a preference for areas of the seafloor covered with soft sediment. The convex undersides of solitary and some colonial forms improved stability on soft substrates. Many of the cuplike forms were anchored by roots or talons. Studies of groups of rugose corals relative to other associated invertebrates and sediment types indicate that they can be separated into platform assemblages and basinal assemblages. The former preferred shallow areas of slow deposition and limey substrate. They are found primarily in limestone. The deeper water basinal assemblages are common in shales, reflecting their tolerance for a muddy substrate. Pelagic organisms like cephalopods often occur in shales bearing the basinal assemblage of rugose corals.

Like the modern hermatypic corals, most tabulate corals seem to have preferred shallow marine areas having little turbidity. These corals were important in constucting the framework of Paleozoic reefs. Tabulates were especially active as reef-builders during the Silurian, Devonian, and Early Pennsylvanian.

CORALS THROUGH TIME

The geologic range of the entire Phylum Cnidaria is Late Proterozoic to Holocene (Fig. 6.29). Corals have a somewhat shorter geologic range, with the first tabulates making their appearance during the Ordovician. By the Silurian, tabulates were represented by abundant Favositina and Halysitina. The Syringoporina, although present in strata as old as the Ordovician, are more abundant in rocks of the Devonian, Mississippian, and Pennsylvanian.

Rugosa appear in the Ordovician, shortly after the appearance of tabulates. Initially, small solitary species of the Streptelasmatina and Columnariina predominated. These were followed by an impressive expansion of zaphrentids, cyathophyllids, and cystiphyllids during the Devonian. In the subsequent periods of the Paleozoic, rugose corals characterized by extravagant development of dissepiments and complex axial structures became common.

Neither the Tabulata nor the Rugosa survived beyond the Permian; indeed, strata representing the earliest part of the Triassic provide no known fossils corals. By the

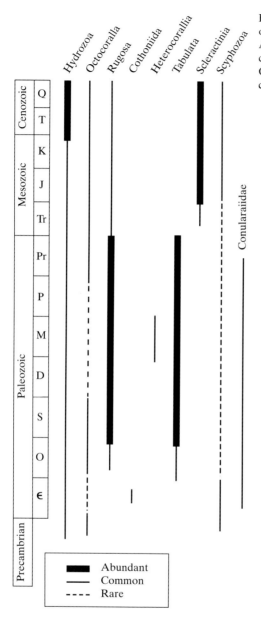

Figure 6.29   Stratigraphic ranges of important groups of cnidarians. Although their classification as cnidarians is questionable, the Conulariida are included for convenience.

middle of the Triassic, however, the Scleractinia had begun their expansion. Astrocoeniina and Fungiina had appeared and were joined by the Faviina before the close of the Triassic. By Late Jurassic time, scleractinid reef-building was underway in tropical and semitropical regions around the globe. These reefs were particularly well-developed along the Tethys Sea, which extended across the southern margin of Eurasia. Following the Jurassic, there were alternate episodes of reef expansion and contraction, with the former occurring late in the Early Cretaceous, in the Late Cretaceous, and in the Early Tertiary.

REEFS AS LANDFORMS

Hermatypic corals, in combination with encrusting invertebrates and calcareous algae, are important geologically in the construction of reef landforms. Examples of reef structures that developed around volcanic islands in the Pacific Ocean are plentiful. Some of these extend as narrow aprons around the margins of islands and are appropriately named **fringing reefs**. If there is a lagoon between the inner edge of the reef and the shoreline of the island, the landform is termed a **barrier reef**. If there is no central island at all, but merely a circle of reefs and coralline islands, the feature is termed an **atoll**. All of these reef landforms are related in origin, a fact perceived long ago by Charles Darwin during his service aboard the survey ship H. M. S. *Beagle* in 1832.

At a meeting of the Geological Society of London during the summer of 1837, Darwin proposed that volcanic islands tend to sink slowly until they become submerged. He believed the reason for their subsidence was an inability of the seafloor to support the great weight of the island. This may be true for some volcanoes, but seafloor spreading provides another mechanism for subsidence. As ocean crust moves laterally away from spreading centers along mid-oceanic ridges, they subside slowly owing to thermal contraction and isostatic adjustment at a rate of about 9 cm annually for the first million years. Thus, a volcano that formed at the crest of the mid-oceanic ridge is carried down a slight incline of the lithospheric plate until volcanoes on its surface become submerged.

Darwin understood that before there was appreciable subsidence of the island, the motile larvae of corals invaded shallow offshore areas and established the colonies that grew to form the framework of fringing reefs. As the host island subsided, the living corals built their structures upward so as to maintain their optimum habitat near the ocean surface. The result of the contemporaneous subsidence and upward growth of the coral colonies produced barrier reefs (Fig. 6.30). Eventually, the central island would sink beneath the level of the waves, leaving a circle of coralline island—the atoll. In Darwin's words, "As the land with the attached reef subsides very gradually . . . the coral-building polypi soon raise again their solid masses to the level of the water: but not so the land; every inch of which is irreclaimably gone: as the whole gradually sinks, the water gains foot by foot upon the shore until the last and highest peak is finally submerged."

Deep borings into atolls have validated Darwin's theory. A boring at Eniwetok in the Pacific encountered the basaltic summit of the volcano at a depth of 1200 meters, after first drilling through a thick cap of Eocene reef material. There is, however, an alternate possibility for atoll formation. During glacial intervals of the Pleistocene, when sea level was lower, corals constructed fringing reefs around erosionally reduced volcanic islands. During interglacial stages when sea level rose, corals built the reef upward to stay at their preferred near-surface habitat. In this way atolls were formed as a result of eustatic (worldwide) changes in sea level.

REVIEW QUESTIONS

1. What are the major differences between Porifera and Cnidaria in morphology, skeletal structures, and feeding mechanisms?

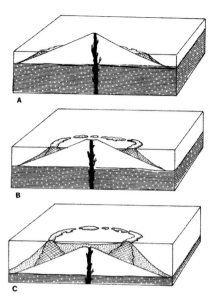

Figure 6.30   The origin of an atoll. (A) In the initial stage, a fringing reef forms around the shoreline of a volcanic island. (B) The island slowly subsides (or sea level rises eustatically) and corals build upward so as to maintain their shallow water life zone. The result is the development of a barrier reef backed by lagoons. (C) As subsidence continues, the original volcanic land area is submerged, and a circle of coralline islands (an atoll) remains.

2. Which of the cell layers in a cnidarian contain cells that function in secretion of skeletal matter, and which in the assimilation of food? What kind of cells are important in the capture of prey?

3. What is the relation between zooxanthellae and reef corals?

4. What are the three major taxonomic categories of corals, and how do they differ from one another?

5. What is the principal function of the asexually produced medusae in Hydrozoa?

6. What is the purpose or function of mesenteries, siphonoglyphs, columella, and dissepiments?

7. Why are Rugosa also referred to as Tetracoralla?

8. What is the relationship among sclerodermites, trabeculae, and septa?

9. Differentiate between hermatypic and ahermatypic corals.

10. Describe two geologic situations that might cause the formation of an atoll.

SELECTED READINGS AND REFERENCES

Clarkson, E. N. K. 1993. *Invertebrate Paleontology and Evolution.* 3rd ed. London: Allen and Unwin.

Easton, W. H. 1960. *Invertebrate Paleontology.* New York: Harper & Row.

Feldmann, R. M. & Babcock, L. E. 1986. Exceptionally preserved conulariids from Ohio: Reinterpretation of their anatomy. *National Geographic Research* 2(4):464–472.

Hill, D. 1981. Rugosa and tabulata. In Teichert, C. (Ed.), *Treatise on Invertebrate Paleontology,* Pt. F, Supplement 1. Boulder, CO, and Lawrence, KS: Geological Society of America and University of Kansas Press.

Moore, R. C. (ed.). 1956. *Treatise on Invertebrate Paleontology,* Pt. F, Coelenterata. Boulder, CO ,and Lawrence, KS: Geological Society of America and University of Kansas Press Press.

Oliver, W. A. Jr. & Coates, A. G. 1987. Phylum Cnidaria. In Boardman, R. S. (Ed.), *Fossil Invertebrates.* Palo Alto, CA: Blackwell Scientific.

van Iten, H. 1991. Anatomy, patterns of occurrence and nature of conulariid schott. *Paleontology* 34:939–954.

Wells, J. W. 1963. Coral growth and geochronometry. *Nature* 197:348–950.

CHAPTER 7

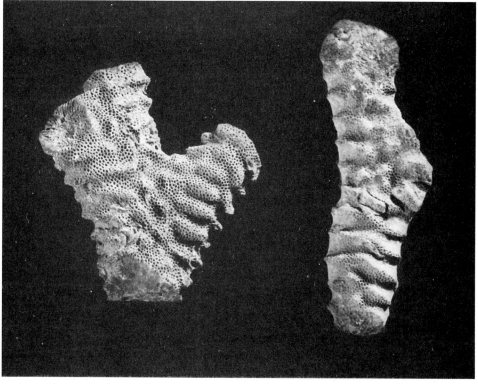

The trepostome bryozoan *Hallopora rugosa* from the Late Ordovician of Ohio. (Courtesy of Ward's Natural Science Establishment, Inc.)

Bryozoa

> *The phylum Bryozoa, popularly known by the paradoxical term "moss animals," have long been noticed by strollers along seashores, where their delicate plantlike colonies are washed ashore after every storm.*
>
> R. S. Bassler, 1953, *Treatise on Invertebrate Paleontology*

As suggested in the above epigram, the tiny shrublike organisms that wash up on beaches, along with "air ferns" that "live on air alone," are not plants at all. They are colonies of invertebrates that were given the name Bryozoa by C. G. Ehrenberg in 1831. J. O. Thompson assigned the name Polyzoa to these creatures a year earlier, but the term Bryozoa was more commonly used, particularly in North America.

The Bryozoa have been a long-ranging, prolific, and diverse phylum. Approximately 4000 living species are known, and nearly four times that number have been described as fossils. Throughout their history they have been mostly marine, colonial animals of very small size. The majority of fossil bryozoa secreted calcareous, branching, cylindrical, or matlike colonies called **zoaria**. Each individual or **zooid** (sometimes called a *polypide*) rarely exceeded a half millimeter in length, and was encased in an external skeleton termed a **zooecium**.

THE LIVING ANIMAL

Unlike coral polyps, which have a single opening that serves both for taking in food and discharging waste, bryozoa can boast a complete **U**-shaped digestive tract with a mouth at its beginning and an anus at its end (Fig. 7.1). In addition, they possess a circular fold of the body wall that encircles the mouth and bears numerous tentacles. This distinctive structure, which is also found in the Brachiopoda to be described in the next chapter, is called a **lophophore**. The presence of a lophophore in both Bryozoa and Brachiopoda suggests that they are sister phyla with a common ancestor. The tentacles of the bryozoan lophophore are ciliated hollow outgrowths of the body wall. When extended, their cilia generate currents that drive microscopic organisms and food particles into the mouth. The mouth itself can be a simple opening, or covered by a fleshy hood called the **epistome**. Bryozoa lack organs for gas exchange, circulation, and excretion. They do, however, have a nervous system consisting of a ganglion from which extends a nerve that encircles the pharynx. From the ganglion and encircling nerve ring, additional nerves extend to each of the tentacles and other parts of the body.

Perhaps the most obvious feature of the bryozoan zooecia are the apertures through which the polypides partially extend themselves during feeding. The apertures of some groups may be provided with a movable lid or **operculum**. When threatened, retractor muscles extending from the inside of the zooecium to the body of the polypide contract and pull the polypide back into the zooecium as the operculum simultaneously closes.

The living bryozoan *Bowerbankia* provides a glimpse of the soft organs of the polypide. Within the body wall there is a large cavity or **coelom** that contains the digestive tract. The mouth of the animal leads into an esophagus, stomach, intestine, and ultimately to an anus that is positioned close to the mouth just outside the circle of tentacles (Fig. 7.1). Attached to the tentacles is a sheet of tissue, the *tentacular sheath*, which fully encloses the tentacles when they are retracted, but is everted (turned inside out) when the lophophore is pushed outward during feeding. A threadlike strand called the *funiculus* extends from the base of the stomach to the wall of the zooecium.

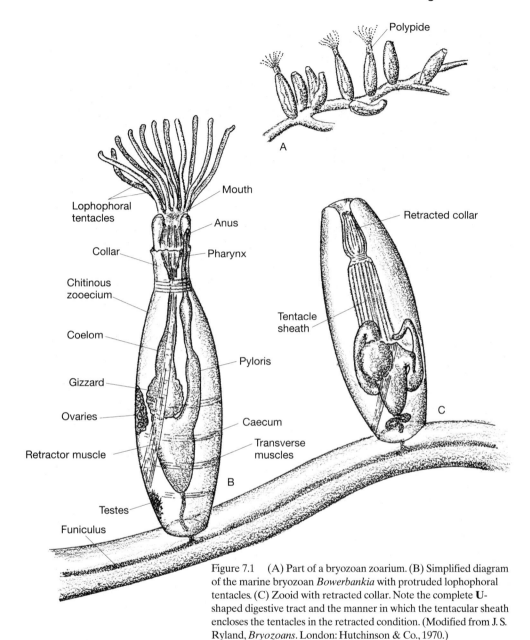

Figure 7.1   (A) Part of a bryozoan zoarium. (B) Simplified diagram of the marine bryozoan *Bowerbankia* with protruded lophophoral tentacles. (C) Zooid with retracted collar. Note the complete U-shaped digestive tract and the manner in which the tentacular sheath encloses the tentacles in the retracted condition. (Modified from J. S. Ryland, *Bryozoans*. London: Hutchinson & Co., 1970.)

One of the more fascinating aspects of bryozoan physiology is the manner in which tentacles are everted for feeding. This is accomplished by increasing the hydrostatic pressure on the coelom. Pressure is increased either by contraction of transverse muscles acting on a flexible membrane stretched across the zooecium, or on a sac that is specialized for this purpose. The precise mechanism is illustrated by cheilostome bryozoa to be described in the section on Gymnolaemata (see Fig. 7.5).

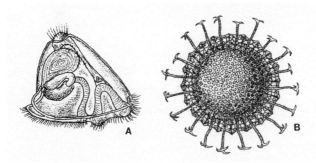

Figure 7.2   (A) Trochophore larva of the bryozoan *Membranipora*. (B) Plan view of a discoidal statoblast of a freshwater bryozoan. Both figures are greatly enlarged. (Modified from J. S. Ryland, *Bryozoans*. London: Hutchinson & Co., 1970.)

Bryozoa are hermaphroditic. Ovaries are usually located in the upper area of the zooid, and testes near the base. Eggs and sperm are discharged into the coelom. Fertilized eggs may be retained in a special brood chamber where they undergo development to the larval stage, or less commonly they may be released into the water to complete their development away from the parent colony. The ciliated, bell-shaped **trochophore** larvae (Fig. 7.2) of the bryozoan may then settle onto a suitable substrate, undergo an extensive metamorphosis, and develop into the first zooid or **ancestrula** of a new colony. By asexually budding, the ancestrula may then give rise to additional zooids, and these too bud either in succession (successive budding), or several at a time (simultaneous budding).

Freshwater bryozoa have yet another means of propagation. They produce tiny, highly resistant, discoidal bodies called **statoblasts** that contain bryozoan cells and nutritive substances. The statoblasts function rather like the gemmules of freshwater sponges. During a dry spell or as winter approaches, statoblasts form on the funiculus and then bulge out into the coelom. On death of the organism, the statoblasts are released. They are able to withstand drying and freezing, and are often picked up and transported by water currents and wind. After they have settled on a suitable substrate, the two valves of the statoblast separate, and a new zooid develops from the cells that were inside.

Bryozoa undergo metamorphosis, not only in the larval stage but also periodically as adults. Generally at intervals of a few weeks, the lophophore and digestive organs metamorphose into a structureless mass, leaving only the body walls and muscles intact. Most of the degenerated material is incorporated into a conspicuous dark ball termed a *brown body*. Shortly after the degenerative event, there is a regeneration of a new lophophore and digestive tract. The brown body may remain in the coelom or the reconstituted zooid, or it may be incorporated into the new stomach and expelled during defecation.

Colonies of certain bryozoa are known to contain more than one kind of zooid. The feeding zooids, or **autozooids**, are always the most abundant in the colony, but others are modified so as to perform functions such as defense, attachment, or reproduction. Two distinctive defensive zooids are **avicularia** and **vibracula** (Fig. 7.3). An avicularium rather resembles the head of a bird. Its "jaws" are operated by muscles that allow them to snap shut on any small animals that approach the colony. The avicularia, however, do not capture food for the colony. Rather, their function is to prevent

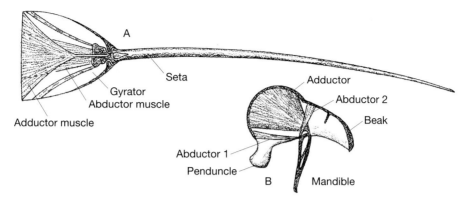

Figure 7.3   Vibraculum (A) and avicularium (B). (Modified from J. S. Ryland, *Bryozoans*, London: Hutchinson & Co., 1970.)

larvae and other small creatures from settling on the colony and interfering with its feeding activities. They also help defend the colony against predators. Vibracula are zooids having opercula that have been modfied into long whiplike bristles. Their rapid back-and-forth movement sweeps away detritus that might otherwise smother parts of the colony.

Although most bryozoa are sessile (stationary), a relatively few species are capable of movement. In some living bryozoa, vibracula located along the margins of the colony act like tiny legs in moving the colony across the substrate. Others extend sticky contractile processes into pores between sand grains, and having anchored the end of the process, pull themselves forward. A few bryozoa are able to creep along by means of circular and longitudinal muscles located on the "sole" of the colony.

THE MAJOR CATEGORIES OF BRYOZOA

It is possible to summarize the principal characteristics of the Bryozoa in a brief definition of the phylum. Bryozoa are small, lophophorate, colonial animals having a U-shaped gut and an anus outside the tentacular ring, and with a body usually covered by a chitinous or calcareous exoskeleton. Individual zooids are linked to one another by means of a cordlike funiculus or by pores in zooecial walls. Colonies arise from an ancestrula (less commonly from a statoblast) and develop branching, cylindrical, or encrusting forms. The phylum is divided into three taxonomic classes: the Gymnolaemata, Stenolaemata, and Phylactolaemata.

CLASS GYMNOLAEMATA

The Gymnolaemata are largely marine polymorphic bryozoa that normally construct skeletons of calcium carbonate. As a result, they have left a good fossil record. The lophophore of the gymnolaemates is circular. Living species lack an epistome over the mouth, and it is assumed that extinct species also lacked this feature. Zooecia are usually cuplike or boxlike in shape. Included in the Gymnolaemata are the orders Ctenostomata and Cheilostomata.

Order Ctenostomata   Ctenostomata means "comb mouths." The name is suggested by the manner in which the collar folds into pleats when retracted, to appear comblike. Zoaria are typically chitinous or membranous, with only a few known calcareous species. As a result, ctenostomids are not abundantly represented in the fossil record. Occasionally one can discern the former locations of ctenostomid colonies by their borings or the delicate traces they produce on shells of the larger invertebrates upon which they bore and encrust themselves. Ctenostomata have been entirely marine throughout their Ordovician to Holocene geologic range. *Bowerbankia* (see Fig. 7.1), a living member of the Ctenostomata, is used to study the anatomy, physiology, and habits of this group.

Order Cheilostomata   Of the two orders of gymnolaemates, the Cheilostomata are by far the most abundantly represented in the fossil record. Cheilostomes build attractive zoaria composed of regularly arranged boxlike zooecia (Fig. 7.4). Zooecial apertures are located at the distal ends of the "boxes" and may be closed by hinged opercula. Avicularia and vibracula are exclusive features of the cheilostomes. These unmineralized structures are not preserved, but their locations may be indicated by perforations or pits on the zooecial walls. Some cheilostomes also have special zooecia called *ovicells* that serve as brooding chambers for eggs. When the eggs open, larvae

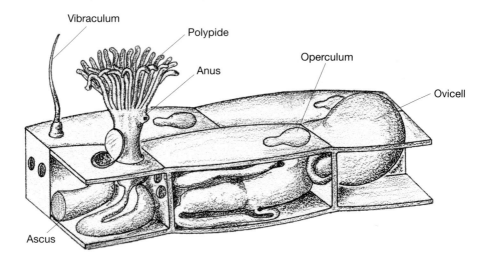

Figure 7.4   Morphology of a cheilostome bryozoan. (Modified from U. Lehman & G. Hillmer, *Fossil Invertebrates*. Cambridge: Cambridge University Press, 1980.)

leave the ovicells and are discharged through the aperture. The size, shape, and surface features of the ovicells are necessary for identification of cheilostomes.

Although branching and frondose colonies of cheilostomes are known, the more typical colony consists of a single encrusting layer of zooecia that are positioned rather like a pavement of bricks. The exposed surface of each zooecium is called the *frontal wall*. When the polypides are feeding, their lophophores are extruded through the apertures of the frontal wall. Otherwise they are safely retracted into the zooecia beneath the closed opercula.

Two methods are employed for the protrusion and retraction of the polypide in cheilostomes. In the first and simplest method, the central region of the frontal wall is a flexible membrane called the **ectocyst** (Fig. 7.5A,B)). Attached to the underside of the ectocyst are parietal muscles. When these contract, they depress the ectocyst and thereby decrease the volume of the coelom. The resulting increased hydrostatic pressure forces the lophophore out through the aperture. Conversely, when the parietal muscles relax, the ectocyst springs back because of its natural resiliency, and the polypide is forced back into the zooecium by the greater external water pressure. Muscles attached to the end of the zooid assist and are coordinated with the muscles serving the ectocyst. In many cheilostomes, a calcareous panel or **cryptocyst** lies beneath the ectocyst (see Fig. 7.5C,D). It is perforated to allow passage of the parietal muscles. There cheilostomes have developed a perforated calcareous roof above the ectocyst that is called the **pleurocyst**. In either case, the ectocyst remains the critical structure for changing the hydrostatic pressure within the coelom.

In the second method for protrusion and retraction of the lophophore, hydrostatic pressure in the coelom is increased by changing the interior volume of a flexible bag called the **ascus** (see Fig. 7.5E–H). Muscles attached to the ascus cause it to expand and thereby draw water into the bag through an orifice just behind the aperture. Space for the zooid is thereby diminished, and it is forced upward so that the lophophore protrudes. When the muscles serving the ascus relax, its volume decreases and the lophophore is retracted. Changes in the volume of the ascus are perfectly coordinated with the action of muscles attached to the polypide itself. The operculum in many species has a posterior flange, so that as it swings upward to open the aperture, the posterior flange swings downward to allow water to enter the ascus. One cannot help but marvel at the efficiency of this design.

The cheilostomes are considered by many to be the most highly developed of the bryozoa. They are also the most dominant group living today. Those cheilostomes having an ascus are placed in the Suborder Ascophorina, whereas those dependent on an ectocyst comprise the Suborder Anascina. Cheilostomes range from the Jurassic to the Holocene. Examples include *Membranipora* (Cretaceous to Holocene), *Ochetosella* (Eocene), and the living uncalcified *Bugula* (Fig. 7.6).

CLASS STENOLAEMATA

The Stenolaemata are marine bryozoa with cylindrical calcareous zooecia. Apertures are circular and are located at the terminal end of the zooecia. Unlike the Gymnolaemata, profusion of the lophophore is not dependent on such structures as an ectocyst or ascus. The Stenolaemata have a significant fossil record that extends from

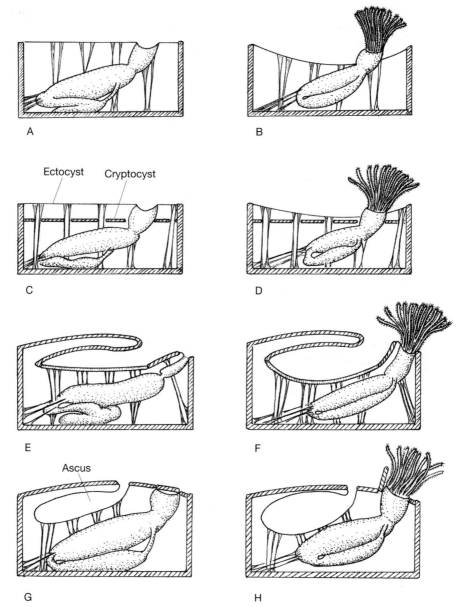

Figure 7.5   Lophophore extrusion and retraction in Cheilostomata. (A) Anascan, polypide retracted. (B) Anascan, polypide extruded. (C) Anascan with cryptocyst below ectocyst and polypide retracted. (D) Anascan with cryptocyst below ectocyst and polypide extruded. (E) Simple ascophoran plan, polypide retracted. (F) Simple ascophoran plan, polypide extruded. (G) Advanced ascophoran plan, polypide retracted. (H) Advanced ascophoran plan, polypide extruded. (Adapted from W. H. Easton, *Invertebrate Paleontology*. New York: Harper & Row, 1960.)

the Ordovician to the Holocene. The class has been variously divided into four, five, or six orders. Here the Class Stenolaemata is divided into five orders: the Tubuliporata, Cystoporata, Trepostomata, Cryptostomata, and Fenestrata.

Order Tubuliporata (Ordovician to Holocene)   As suggested by their name, the zooecia of Tubuliporata are tubular and have terminal circular apertures. Their circular apertures prompted earlier designation of this group as cyclostomata or "round mouths." Tubuliporates do not secrete horizontal partitions within their zooecial tubes, nor do they have opercula. Pores within the walls of adjacent zooecia provide contact between members of the colony. There are no indications of vibracula or avicularia.

The shapes of zoaria even within the same species of Tubuliporata may vary according to the presence of currents or wave action in their environment. As a result, this feature is of little use in classification. Fortunately, ovicells differ consistently between species and are useful in identification. The genus *Stomatopora* (Fig. 7.7) from the Ordovician is an example of the delicate encrusting type of Tubuliporata. Huskier forms are represented by *Entalophora* (Jurassic to Holocene), which builds a colony resembling a tree trunk, and *Spiropora* (Jurassic? to Tertiary), whose apertures are grouped into successive whorls (see Fig. 7.7).

Order Cystoporata (Ordovician to Triassic)   Typical members of this order (Fig. 7.8) possess long tubular zooecia separated by areas of curved structures called *cystiphragms*. The frothy appearance of cystiphragms are readily identified in species of *Fistulipora* (Silurian to Permian). In addition, the apertures of most cystoporates have an overhanging crescentic projection or **lunarium**, and the outer wall of the colony may have regularly spaced small mounds (**monticules**), and/or small bowl-

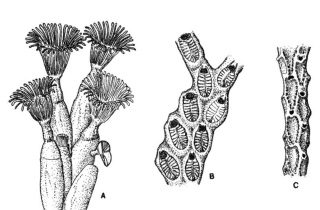

Figure 7.6   Representative cheilostomes. (A) Part of a colony of the common extant cheilostome *Bugula*, ×50. (B) *Membraniporella*, frontal structure of zooecia, ×20. (C) *Ochetosella*, ×20.

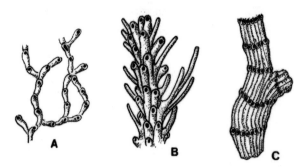

Figure 7.7 (A) The tubuliporates *Stomatopora*, ×5; (B) *Entalophora*, ×20; and (C) *Spiropora*, ×8.

shaped depressions (**maculae**). Microscopic examination reveals that both of these structures are composed of clusters of small accessory zooecia called **mesozooecia**, which form **mesopores** at the surface of the zoarium. Presumably, mesopores were occupied by specialized zooids.

Ceramopora (Ordovician to Devonian) is a well-known cystoporate that built bowl- or disk-shaped colonies. Far more distinctive in shape, however, is the stellate zoarium of the Mississippian genus *Evactinopora*.

Order Trepostomata Trepostomes built sturdy zoaria composed of tightly packed tubular zooecia. Many of the compact colonies were important components of

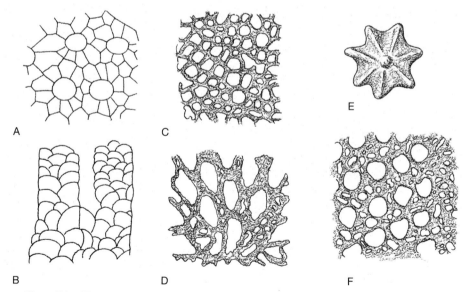

Figure 7.8 Three members of the Cystoporata. Tangential (A) and longitudinal (B) sections of *Fistulipora*, ×20. Tangential (C) and longitudinal (D) sections of *Ceramopora*, ×20. Colony (E) of *Evactinopora*, ×3, and tangential section, ×20. (Redrawn from R. S. Bassler. In R. C. Moore (Ed.), *Treatise on Invertebrate Paleontology*. Boulder, CO, and Lawrence, KS: Geological Society of America and University of Kansas Press, 1953.)

Paleozoic reefs, and this accounts for their being named the "stony bryozoa." Trepostomes also display a variety of form and numerous external and internal features that assist in their classification and identification. Zoaria may take the form of encrustations, twiglike branches, or frondlike sheets. The encrustations do not resemble the pavements formed by the boxlike zooecia of the gymnolaemates. Rather, they consist of tightly packed zooecial tubes that rise perpendicularly from the surface that is encrusted. Often encrusting zoaria grow upward into mounds that range in diameter from a few centimeters to over a meter. The frondlike zoaria may be constructed of either one (**laminar**) or two (**bifoliate**) layers.

Examination of the surfaces of trepostome zoaria with a hand magnifier is usually sufficient to see the often polygonal primary apertures, among which may be dispersed smaller aperture openings or mesopores of the correspondingly smaller mesozooecia. As in cystoporates, the mesozooecia contained specialized zooids. Clusters of mesopores form maculae and monticules on the surfaces of zoaria. Often these clusters have distinctive outlines and spacings that can be useful in identifying particular species. For example, the genus *Constellaria* (see Fig. 7.9E) is readily recognized because of the starlike pattern developed on its maculae. Trepostome maculae are believed to mark the location of excurrent chimneys that formed part of a water circulation system. The characteristics most useful for identification, however, are usually not maculae and other features seen on the surface of the zoarium, but the internal features seen in thin sections.

Thin sections of bryozoa are prepared by cutting a thin slice out of the zoaria with a diamond saw. The cut piece is polished, cemented to a glass slide, and ground to a flat surface on one side. Next, it is inverted, recemented, and ground down to a uniform thickness of about 0.025 mm. After applying a cover glass, the specimen is ready for viewing under the microscope. Both a **tangential** and **longitudinal section** are normally required (see Fig. 7.9A). The plane of the tangential section must pass just under and parallel to the surface of the zooecia (that is, at right angles to the zooecial axes.). A view of the zooecia parallel to their axes is provided by the longitudinal section. It is also necessary to prepare a **transverse section** perpendicular to the longitudinal section.

When rapid identification of large numbers of specimens is required, acetate peels may speed the work. To prepare a peel, the selected cut face of the specimen is polished with tin oxide polishing powder. The polished surface is etched for a few seconds in 5 percent formic acid or 20 percent hydrochloric acid so as to produce the necessary surface relief. Acetone is applied to the surface, and a thin piece of acetate film is immediately placed over the wet surface. The film that has been softened by the acetone records an accurate impression of the bryozoan skeletal features beneath it. When dry, the film is peeled from the specimen and mounted for study.

When one examines longitudinal or transverse sections of trepostomes, two distinct zones can be discerned. The outer or **mature region** is marked by zooecia with closely spaced diaphragms and abundant mesozooecia. This mature region encloses an inner **immature region** rather like a rind. The immature region has fewer diaphragms, a lack of mesozooecia, and the zooecial walls are thin and wrinkled. This change in the character of the zooecial tubes in trepostomes is the basis for the name of the order. The term *trepos* means "change."

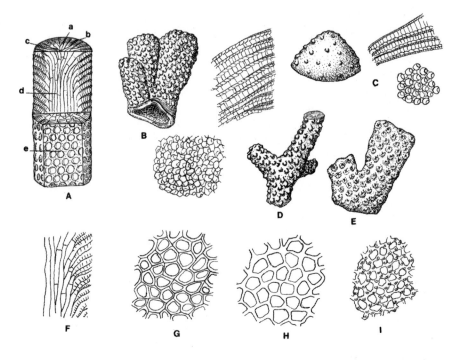

Figure 7.9   Trepostome bryozoans. (A) Branch of a trepostome to illustrate immature (a) and mature (b) regions of a zoarium, as well as the orientation of transverse (c), longitudinal (d), and tangential (e) sections. (B) Zoarium, longitudinal, and tangential sections of *Monticulopora*. (C) Zoarium, longitudinal, and tangential sections of *Prasopora*. (D) Zoarium of *Hallopora*. (E) Zoarium of *Constellaria*. (F) Longitudinal section of *Hemiphragma* showing the hemiphragms that extend only partly across zooecia. (G) Tangential section of *Batostoma* showing integrate walls. (H) Amalgamate walls in *Dekayella*. (I) Tangential section of *Stigmatella* with numerous acanthopores. (Redrawn from R. S. Bassler. In R. C. Moore (Ed.), *Treatise on Invertebrate Paleontology*. Boulder, CO, and Lawrence, KS: Geological Society of America and University of Kansas Press, 1953.)

Thin sections reveal several other distinctive zooecial features that have critical taxonomic importance in trepostomes. The diaphragms, for example, may extend completely across zooecial tubes, or they may extend only partway across the tube to form a shelflike platform called a **hemiphragm** (see Fig. 7.9F). Strongly curved partitions called **cystiphragms** may be present one above the other along one side of the zooecial tubes. Also present are longitudinal strands composed of cone-in-cone layers that have a narrow central tubule. They are called **acanthopores** (see Fig. 7.9I), and ancanthopores terminate at the surface of the zoarium as small nodes or spines. In life, these may have supported defensive structures, but there are no acanthopores in living bryozoa to support this hypothesis. The walls of trepostomes are not pierced by mural pores. In tangential sections one can ascertain if the walls of adjacent zooecia meet at a median line, or are fused to form a common wall. The former condition provides **integrate walls**, and the latter **amalgamate walls**. Tangential sections permit one to measure the diameter of zooecia, mesopores, and acanthopores, as well as the thickness of zooecial walls and whether or not diaphragms are continuous or perforated.

About 120 genera of trepostomes are known for the Paleozoic, with the greatest abundance in Ordovician rocks. Among the well-known Ordovician genera are *Monticulopora, Dekayella, Prasopora, Hemiphragma, Hallopora, Stigmatella*, and *Constellaria* (see Fig. 7.9).

Order Cryptostomata (Early Ordovician to Late Permian)

In cryptostomates, the true openings of the zooecial chambers lie hidden beneath apertures at the surface of the zoarium. The name Cryptostomata means "hidden mouth." Beneath the surface aperture there is a chamber called the **vestibule**, at the base of which one finds an incomplete diaphragm called a **hemiseptum**. (Hemisepta resemble hemiphragms but are confined to the near-surface part of the zooecial tube.) The hemiseptum marks the location of the true (hidden) zooecial aperture.

The zoaria of Cryptostomata are commonly branching or **dendroid** in form, although bifoliate fronds also occur. Maculae and monticules are well developed, and there is a distinct differentiation of mature and immature regions. *Stictopora* (Ordovician) and *Graptodictya* (Ordovician) are representative genera (Fig. 7.10).

Order Fenestrata (Ordovician to Permian)

The Fenestrata are mostly lacy, erect Bryozoa with thin meshlike surfaces that expand into funnel, fan, or spiral shapes. The zoaria have great delicacy and beauty. Close examination of the fronds reveals that they are constructed of thin vertical members called **branches**. Each branch contains two to eight longitudinal rows of apertures. Branches are connected laterally to one another by cross-bars termed **dissepiments**. Hemisepta and vestibules are often well developed. The Fenestrata take their name from the small "windows," or fenestrules, which pierce the fronds at regular intervals. *Fenestella* (Fig. 7.11) is among the best known of the Fenestrata. Species of *Fenestella* are found in rocks ranging in age from the Ordovician to the Permian. The most distinctive form, however, is

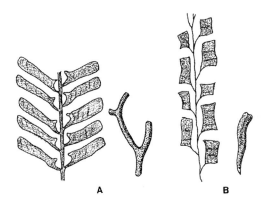

Figure 7.10   Longitudinal sections and zoaria of *Stictipora* (A) and *Graptodictya* (B). Sections ×20, zoaria ×1.5. (Modified from R. S. Bassler. In R. C. Moore (Ed.), *Treatise on Invertebrate Paleontology*. Boulder, CO, and Lawrence, KS: Geological Society of America and University of Kansas Press, 1953.)

A   B

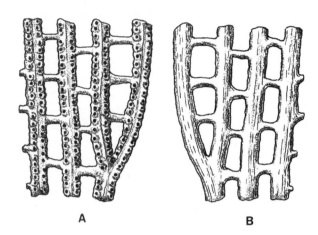

Figure 7.11 Apertural (A) and reverse (B) sides of *Fenestella*, ×20.

A **B**

Archimedes (Fig. 7.12), which has a netlike frond arranged helically around a central axis. The central axis is considered an extrazooidal skeletal structure in that it is outside the boundaries of the zooids throughout life. At one time it was thought to be formed by algae living in consort with the bryozoa of the colony. Subsequent study, however, indicated that the screw like axis along with the attached mesh were completely covered by epithelial tissue responsible for the secretion of the entire zoarium. Because of its robust character, the helical extrazooidal structure is more readily preserved than the fragile frond portion, and is more frequently seen in the field.

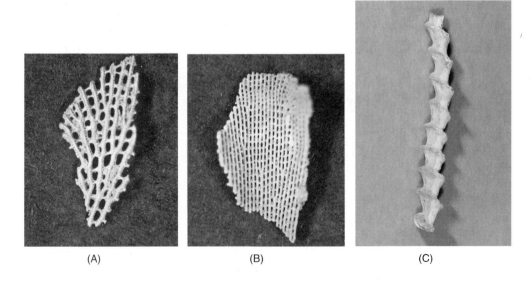

(A) (B) (C)

Figure 7.12 *Fenestella pulchella* (A), and *Fenestella variapora* (B) are two Devonian fenestellate bryozoans from the Falls of the Ohio. *Archimedes wortheni* (C) is an important Mississippian fossil.

Archimedes occurs in Mississippian limestone and calcareous shales of eastern and central North America, but in more western locations its range is extended into the Permian.

As suggested by Cowen and Rider (1972) one can make some interesting deductions about the way fenestrates lived by examining their skeletal morphology. In addition to their meshlike construction, fenestrate bryozoa have zooecial apertures located on the frontal sides of branches. In those groups having two rows of zooecia on each branch, the zooecial apertures are directed slightly toward the fenestrules. Branches tend to be streamlined and possess keels or longitudinally arranged spines on their frontal surfaces. These observations led to the hypothesis that water currents were drawn into the colony from the frontal to the reverse side (Fig. 7.13), and that the zooids were ideally positioned to harvest food particles from water currents being moved through the fenestrules by ciliary action. The streamlined shape of the branches and their accessory spines or keels served to divide water currents more or less equally into the fenestrules on either side of the branches, and also to minimize friction. The size of fenestrules was probably closely coordinated with the area covered by the expanded lophophore tentacles, thus ensuring good pumping action and little opportunity for food particles to escape through the mesh. In conical or cup-shaped colonies, the cilia-driven currents would be directed either inward or outward depending on whether the frontal surface faced inward or outward (see Fig. 7.13B). In spiral colonies

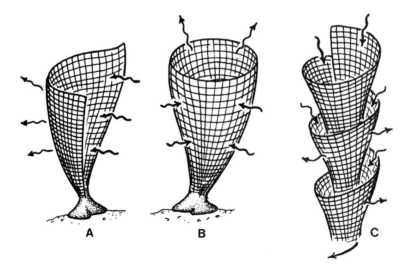

Figure 7.13   Hypothetical flow of water through the zoaria of fenestellate bryozoans. In (A), the apertural surface faces the incoming flow. In (B), water flows into the cone from the sides and the apertural surface faces to the outside of the cone. In forms with the apertural surface on the inside of the cone, water flows in from the top and passes out through the sides. In *Archimedes* (C), water is taken in at the top and at the upper surface of each deck. An axial current spirals downward to emerge at the base. (Modified from R. Cowen & J. Rider, Functional analysis of fenestellid bryozoan colonies, *Lethaia* 5:145–164, 1972.)

like *Archimedes*, a helical current may have flowed downward between "decks" and emerged near the base of the colony.

THE CLASS PHYLACTOLAEMATA

Phylactolaemata are exclusively freshwater bryozoa. The polypide of a phylactolaemate has a horseshoe-shaped aperture and a distinctive lip or **epistome** that overhangs the mouth. Because phylactolaemates lack mineralized hard parts, fossils are unknown, although the hard-shelled statoblasts of phylactolaemates are sometimes recovered from Tertiary and Quaternary sedimentary rocks. These same statoblasts account for the wide geographic distribution of living species, for they are transported across great distances in the feathers of migratory water fowl. *Plumatella* (Fig. 7.14), which often aggregates in jellylike masses, is probably the most familiar of the phylactolaemates.

BRYOZOAN HABITS AND HABITATS

Today, marine Bryozoa range from polar to tropical regions, although they are most abundant in warmer climatic zones. Their optimum temperature for larval survival is 20°C to 28°C. Polar species may undergo disintegration at the onset of winter, but tropical Bryozoa are active throughout the year. Bryozoa thrive in shallow shelf seas of normal salinity, but many species do very well in brackish water bodies as well. Their optimum living depth for marine groups is from 10 to 70 meters. Although water depth has an influence on the shape of zoaria, other factors such as currents, wave action, and the kind of substrate exert a more direct effect. Fragile forms clearly have a preference for quiet water, whereas sturdier colonies populate relatively agitated zones. Except for soft mud or shifting sands, Bryozoa will grow on any reasonably firm substrate, including the skeletons of other invertebrates and blades of marine algae. Some species show a distinct preference for particular kinds of algae. Bryozoan fossils are often abundant in Paleozoic calcareous shales and shaley limestones. They are common members of Paleozoic reef assemblages, preferring the quieter areas away from the reef fronts.

Figure 7.14   The freshwater phylactolaemate *Plumatella*. Part of a colony with lophophores extended. (Courtesy General Biological Supply House, Chicago.)

Bryozoa interact with humans in many, often unfavorable ways. When they proliferate so rapidly as to smother oyster beds, they become a problem for oyster harvesters. Colonies grow extensively on the hulls of ships and periodically must by scraped off. Commercially important fish populations can be poisoned by toxins emitted when colonies disintegrate. The gelatinous colonies of freshwater Bryozoa clog air conditioning systems and cooling towers. They are known to foul the pipes of water systems and the conduits that supply water to hydroelectric plants. The relation of Bryozoa to humans is not entirely unfavorable, however, for some species contain chemical compounds being tested for treatment of cancer (see Box 7.1).

BRYOZOA THROUGH TIME

Bryozoa are virtually unknown before the Ordovician. By the middle part of the period, however, they had become varied, complex, and abundant (Fig. 7.15). The relatively sudden advent of advanced forms suggests an earlier period of differentiation. Perhaps the ancestral forms had not developed calcareous skeletons, and are therefore not preserved in the Cambrian fossil record. In any case, Middle and Late Ordovician Bryozoa are so numerous as to constitute the bulk volume of some strata and reeflike masses called *bioherms*. All of the bryozoan taxonomic orders except the Cheilostomata are present by Late Ordovician, with trepostomes clearly the most abundant. The trepostomes decline somewhat during the Silurian and Devonian.

BOX 7.1   MEDICINE FROM BRYOZOA

Although the Bryozoa are regarded as major pests by oyster fisherman and the owners of ships with bryozoan-encrusted hulls, these diminutive creatures may also provide a direct benefit to humans by means of pharmaceuticals extracted from their tissues. Certain populations of *Bugula neritina* living along the southern coast of California produce a compound, biostatin 1, that promises to have value in combating cancer. Harvesting natural growths of the Bryozoa is possible, but would be exceptionally costly and might be ecologically harmful. Growing colonies of *Bugula* on land in tanks has also proven to be so expensive as to contribute heavily to the ultimate cost paid by the consumer for the drug. Largely for these reasons an in-the-ocean culture system was devised that would allow colonies to feed naturally while posing little or no hazard to the marine environment. In this system, *Bugula* colonies are grown on perforated plastic plates. Once the colonies are well established, tens of thousands of laboratory-grown *Bugula* larvae are settled onto the plates. The plates with their bryozoan occupants are then transported out to sea where divers fasten them to underwater attachment structures. The larvae mature and the adults multiply rapidly, producing a dense growth of bryozoa on the plates. The "crop" is then removed, the biomass chemically extracted, and the drug purified. Aside from the value of the drug itself, biochemists are keenly interested in the technology used in culturing bryozoa at sea, for in the future, other marine invertebrates may yield important biomedical products.

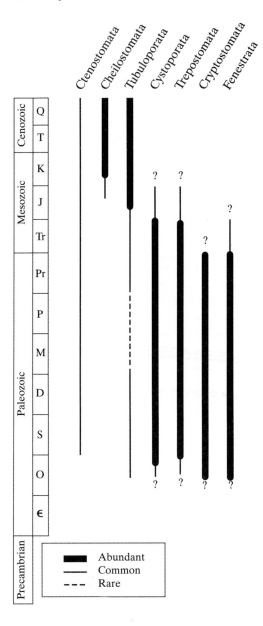

Figure 7.15   Geologic range chart for Bryozoa.

During the Devonian and Carboniferous, lacy Bryozoa and those with slender branching colonies predominate. Finally, the cheilostomes make their appearance during the Jurassic and continue in abundance to the present day.

REVIEW QUESTIONS

1. What group of extinct Cnidaria resemble bryozoa in the construction of tubular skeletons having horizontal partitions? How would you distinguish one from the other?

2. What organs present in the coelum of a bryozoan are not present in the coelenteron of a coral polyp?

3. What are the three taxonomic classes of the Bryozoa? What are the general characteristics of each?

4. Describe the hydraulic mechanism used by cheilostomes for extrusion of the lophophore.

5. What are the three thin sections needed for adequate study of trepostomes? In which of these would integrate and amalgamate walls be most readily observed? In which would mature and immature areas be most easily observed?

6. Distinguish between cystiphragms and hemiphragms in trepostomes.

7. In what taxonomic order of Bryozoa does one find zooecia with vestibules? What purpose is served by the vestibule? What skeletal structure lies at the base of the vestibule?

8. Distinguish between branches and dissepiments in fenestrate bryozoa. How are the apertural openings arranged in these bryozoa? What distinctive helical fossil forms the axial support for certain fenestrates?

SUPPLEMENTAL READINGS AND REFERENCES

Bassler, R. S. 1953. Bryozoa. In Moore, R. C. (Ed.), *Treatise on Inverterate Paleontology.* Boulder, CO, and Lawrence, KS: Geological Society of America and University of Kansas Press.

Boardman, R. S. et al. 1983. Bryozoa, Vol. 1, Pt. G. (2nd ed.). In Robison, R. A. (Ed.), *Treatise on Invertebrate Paleontology.* Boulder, CO, and Lawrence, KS: Geological Society of America and University of Kansas Press.

Boardman, R. S. & Cheetham, A. H. 1987. Phylum Bryozoa. In Boardman, R. S. (Ed.), *Fossil Invertebrates.* Palo Alto, CA: Blackwell Scientific.

Cowen, R. & Rider, J. 1972. Functional analysis of fenestellid bryozoan colonies. *Lethaia* 55:145–164.

McKinney, F. K. and Jackson, J. B. C. 1989. *Bryozoan Evolution.* Boston: Unwin Hyman.

Ryland, J. S. 1970. *Bryozoans.* London: Hutchinson & Co.

CHAPTER 8

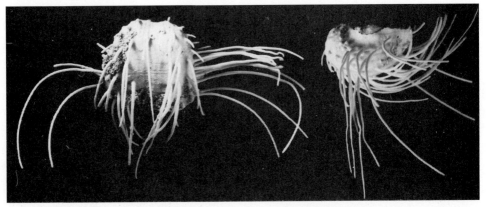

Pedicle valve(left) and lateral view(bottom) of the Permain productid brachiopod *Marginifera ornata* from the Salt Range of West Pakistan, x1. (Courtesy of R.E. Grant, U.S. Geological Survey.)

Brachiopoda

It may be useful to the student to know that a brachiopod differs from ordinary bivalves, mussels, cockles, etc., in being equal-sided and never quite equivalved; the form of each valve being symmetrical, it may be divided into two equal parts by a line drawn from the apex to the centre of the margin.

Sir Charles Lyell, *The Student's Elements of Geology*, 1882

Although brachiopods are not abundant in the ocean today, members of the phylum were once common and richly diverse. Over 30,000 fossil species have been described. For the biostratigrapher, fossil brachiopods offer the advantages of abundance, rapid change through time, identification of many species from external features alone, and generally good preservation. In addition, brachiopods have been exceptionally useful for studies of evolution and adaptation. Brachiopod skeletal features often bear the imprints of soft organs, permitting interpretation of unpreserved internal features. The study of extant species provides important constraints to such studies.

THE LIVING ANIMAL

Brachiopoda are bilaterally symmetrical animals with bivalved shells. Throughout their lengthy history, they have been marine creatures. Although a few species have been tolerant of brackish water, most preferred seawater of normal salinity in a range of bottom environments from subtidal to the abyss. The majority of brachiopods lived attached to the seafloor or to objects on the seafloor, and their often widespread distribution was dependent upon a free-swimming larval stage. Lacking the ability to pursue their food aggressively, adult brachiopods have existed as filter-feeders that gather tiny particles of food from the surrounding water. Food brought to the mouth is passed into an arched esophagus and moved by peristalsis to the stomach. The articulates have a narrow, blind intestine, but in the inarticulate brachiopods the intestine ends in an anus. Complex sensory organs and special defensive mechanisms are largely absent in brachiopods, although adjustment of the pedicle may provide limited evasive movements. Sexes are separate, although some living forms have been found to be hermaphroditic. For the majority in which sexes are distinct, egg and sperm are released from the body cavity through delicate funnel-shaped openings called *nephridia* and fertilized in surrounding waters. Once fertilized, the eggs develop into motile larvae. Some living brachiopods brood their young in the mantle cavity, in nephridia, or in the arms of the lophophore.

CHARACTERISTICS OF THE SHELL

Because, like clams, brachiopods have shells constructed of two valves, they were once considered mollusks. The resemblance to mollusks, however, is only superficial, for in brachiopods half of each valve is a mirror image of the other half. Thus, the plane of symmetry passes through the valves, not between them as in mollusks such as clams (Fig. 8.1). The gape of the valves defines the anterior margin, and the opposite end, where the valves may be hinged, is designated the posterior. Dissimilarities between the valves provide criteria for designation of the **brachial valve** and **pedicle valve**. The fleshy stalk by which brachiopods are attached to the substrate is called the **pedicle**.

In brachiopods that have a definite hinge structure, the pedicle valve is the valve from which the pedicle emerges from the shell. In some species, the pedicle valve has an opening for the pedicle that is called the **pedicle foramen**. The pedicle valve is often the larger of the two valves, contains the hinge teeth, and may display a broad linear depression or trough called the **sulcus**. The brachial valve is usually the smaller of the two valves, bears the sockets that receive the hinge teeth, and often exhibits an arch called a **fold**. The fold on one valve usually occupies a position corresponding to the sulcus on the other valve. The line along which the valves are in contact is called the

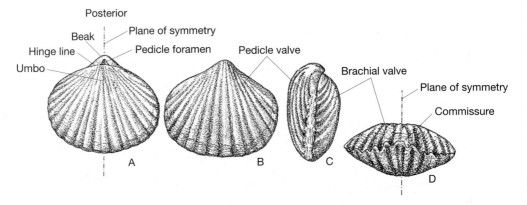

Figure 8.1   Orientation and symmetry in brachiopods, as indicated by *Fenestrirostra*, ×4. (A) Brachial (dorsal) view. (B) Pedicle (ventral) view. (C) Lateral view. (D) Anterior view.

commissure. The most posterior part of the shell represents secretion by the juvenile brachiopod, and is often rounded to form a **beak**.

If the two valves are hinged together, the hinge structures are located along the posterior separation between the valves. Not all brachiopods, however, have a definite hinge. Whether or not this feature exists is one of the criteria for dividing the phylum into its two major divisions, the Articulata and the Inarticulata. The former possess a definite hinge along which teeth and sockets articulate with one another. Inarticulates are not rigidly hinged, and hence they actually have somewhat greater freedom of valve movement.

BRACHIOPOD SOFT PARTS

In most living brachiopods (Fig. 8.2), the most obvious soft organ is the pedicle. At its proximal end, the pedicle is attached to the inner surface of the larger valve by muscles. Although the pedicle usually protrudes from the shell through the pedicle opening, in a relatively few brachiopods it emerges between the valves.

When one forces apart the two valves of a living articulate brachiopod, the most conspicuous internal organ is the **lophophore**, with its two prominent spirally coiled arms that give brachiopods their name (Greek *brachion*, arm + *podus*, foot). Extending along the arms or **brachia** of the lophophore is a ciliated groove, and adjacent to the groove is a row of tiny ciliated tentacles. The cilia sweep tiny organisms and particles of food into the mouth, which is located between the base of the two brachia. Like the valve in which it rests, the lophophore has bilateral symmetry. It may be attached by interior skeletal structures known as the **brachidia**. The presence of brachidia serve to identify the brachial valve.

The other soft-tissue components of articulate brachiopods occupy only a small space in the posterior of the shell. Included here are muscles for movement of the valves and the pedicle, a simple digestive system, excretory and reproductive organs, and components of the nervous and circulatory systems. These organs lie behind a thin

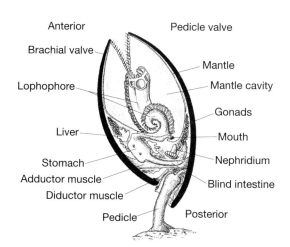

Anterior

Brachial valve

Lophophore

Liver

Stomach

Adductor muscle

Diductor muscle

Pedicle

Pedicle valve

Mantle

Mantle cavity

Gonads

Mouth

Nephridium

Blind intestine

Posterior

Figure 8.2   Components of an articulate brachiopod. (Modified from A. Williams & A. J. Rowell. In R. C. Moore (Ed.), *Treatise of Invertebrate Paleontology*, Pt. H. Boulder, CO, and Lawrence, KS: Geological Society of America and University of Kansas Press, 1965.)

body wall from which extend two **mantle** sheets. The sheets are thin, double-layered membranes that function primarily in respiration and the secretion of the shell. The lophophore, however, assists the mantle in its respiratory function.

INARTICULATE BRACHIOPODS

Members of the Class Inarticulata (Fig. 8.3) are brachiopods having valves that are usually not hinged by teeth and sockets. In addition, most inarticulates construct chitinophosphatic shells in which the shell is composed of calcium phosphate with a variable proportion of chitin (a nitrogenous polysaccharide) and protein. The shells are ornamented by weak concentric growth lines. Internal supports for the lophophore and other skeletal features common to articulate brachiopods are lacking in the inarticulates. Because inarticulates lack hingement, they require a complex system of muscles to move one valve forward or backward or side to side relative to the other valve. An equally complex system of muscle scars may be seen on the interior of well-preserved specimens.

Inarticulates include such long-ranging genera as *Lingula* and *Orbiculoidea* (Cambrian to Holocene). In these inarticulates, the pedicle emerges between the posterior margin of the two valves. There is no pedicle opening. The *Lingula* shell is elliptical in outline and the generally slightly convex valves are similar in shape. The animal lives in a vertical burrow anchored by its elongate pedicle. One might surmise that the burrow was excavated by the pedicle, but this is not the case. As indicated by direct observations of living linguloid brachiopods, the animal digs into the soft substrate anterior end first ("head first"), using the pedicle initially as a prop. Once the shell is directed downward, tunneling commences by rotary and scissorlike movements of the valves. When a suitable depth has been attained, tunneling proceeds to the side and then upward so as to produce a **U**-shaped burrow. On reaching the surface, the animal has attained its vertical feeding position.

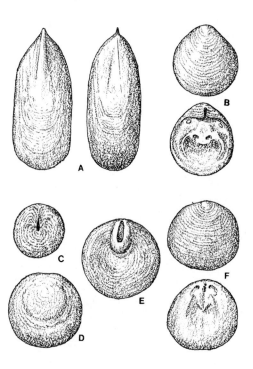

Figure 8.3   Inarticulate brachiopods. (A) External views of pedicle (left) and brachial (right) valves of *Lingula*, ×1. (B) Exterior (top) and interior of pedicle valve of *Obolus*, ×4. (C) Exterior of pedicle valve of *Orbiculoidea*, ×3. (D) Exterior of brachial valve of *Crania*, ×2. (E) Exterior of pedicle valve of *Discinisca*, ×1. (F) Exterior (top) and interior of pedicle valve of *Obolella*, ×4. (Drawings from photographs in A. Williams & A. J. Rowell. In R. C. Moore (Ed.), *Treatise on Invertebrate Paleontology*, Pt. H. Boulder, CO, and Lawrence, KS: Geological Society of America and University of Kansas Press, 1965.)

As compared to *Lingula*, the inarticulates *Discinisca* (Triassic to Holocene) and *Orbiculoidea* (Ordovician to Cretaceous) have pedicle openings and a more rounded shape. *Crania* (Ordovician to Holocene) is an unusual inarticulate in having a shell constructed of calcium carbonate. The shell is rounded and there is no pedicle. Instead, the relatively flat pedicle valve is cemented to the shells of other invertebrates, to larger brachiopods, or to objects on the seafloor. *Crania* takes its name from the pattern of sculpting on the interior of the ventral valve, which bears a vague resemblance to a human skull.

Inarticulates have been present on Earth through nearly all of the Phanerozoic. They are, however, a far smaller group than the articulates. They reached their greatest diversity during the Ordovician. By the end of the Devonian, most inarticulate families had become extinct. Those that have survived closely resemble their Paleozoic ancestors.

ARTICULATE BRACHIOPODS

The most common and geologically useful brachiopods are the articulates. As noted, these brachiopods differ from inarticulates in having a definite hingement based on the development of teeth and sockets, a shell composition that is consistently calcite, and a diagnostic musculature that can be reconstructed from muscle scars located on the inner surfaces of the valves. In addition to these features associated with the shell, differences exist in the soft anatomy. The gut of articulate brachiopods lacks an anus, and the pedicle is a muscle that originates from a posterior portion of the larvae, rather than as an outgrowth of the mantle and coelom as in inarticulates. The fossil record provides few clues as to the origin of the articulates. Most paleontologists assume they were derived from inarticulates that had evolved a definite hingement with associated musculature.

MUSCLES OF ARTICULATE BRACHIOPODS

In order to close its valves, the articulate brachiopod must bring its **adductor muscles** into play (Fig. 8.4), whereas **diductor muscles** serve to open the valves. The adductors extend from the brachial to the pedicle valve and are located anterior to the hinge line. By simple contraction of the adductor muscles, the valves are brought together. The adductor muscles are paired, and thus one finds two adductor muscle scars lying close together on either side of the median line of the pedicle valve. Often, they produce a heart-shaped muscle scar. As the adductors ascend toward the brachial valve, each divides to form two strands, so that the brachial valve bears four adductor scars. The

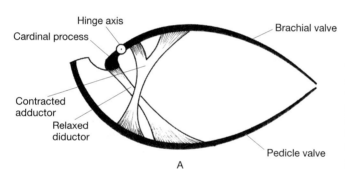

Figure 8.4   Opening (A) and closing (B) mechanism in an articulate brachiopod. The adductor muscles close the valves, whereas the diductors contract to open the valves. Note that the base of the cardinal process is posterior to the hinge axis. Thus, when the diductors contract, the shell opens by means of the lever-fulcrum principle.

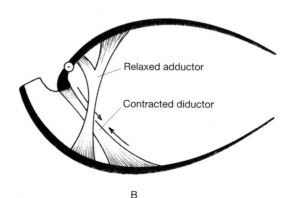

smaller posterior strand provides quick action in response to threats or annoyances, whereas the large strand reacts more slowly and is able to hold the shell firmly closed for long periods of time.

Diductor muscles, which extend in slender pairs from the posterior end of the brachial valve behind the hinge line to the floor of the pedicle valve, open the valves by contraction. Only slight contraction is required to draw the posterior end of the brachial valve downward, and this lever action causes the anterior end to rise and provide the necessary gape. The areas of attachment of the diductor muscles within the pedicle valve produce large muscle scars on either side of the more centrally located adductor scars. Diductor attachments within the brachial valve, however, produce smaller scars. The slender diductor strands are attached posteriorly to the variably shaped **cardinal process**, which not only provides a base for attachment but may also be extended ventrally to improve leverage. Cardinal processes can take the form of granules, knobs, ridges, shafts, blades, or prongs.

In addition to the adductor and diductor muscles, articulate brachiopods may also have accessory diductors to assist the principal diductors, as well as **protractor muscles** for retracting the pedicle, and **adjuster muscles** for turning the shell from side to side on the pedicle. Particularly during the Paleozoic, brachiopods evolved a rich diversity of shell form. Their many shapes and sizes produced corresponding modifications in musculature, although a constant requirement was the attachment of the diductor muscles to the brachial valve at a point posterior to the hinge axis. There were various ways in which this requirement was met (Fig. 8.5), although the simplest involved the extension of the diductor from its attachment in the pedicle valve, through the pedicle opening, to a point of attachment at the cardinal process located posterior to the hinge axis. In this plan, part of the diductor passes outside the shell, and is thus exposed except for protection afforded by its being held close to the substrate.

The pedicle opening in many Paleozoic brachiopods results from the interruption of shell growth near the center of the hinge line. It is often triangular, and hence termed a **delthyrium**. The delthyrium may remain open throughout the growth of the shell, or it may be closed to varying degrees by coverings of shell material. Usually, closure of the delthyrium occurs progressively as the pedicle gradually atrophies and is eventually pinched off entirely. The covering of the delthyrium is called a **deltidium** if it consists of a single plate, or if two separate plates extend from the sides of the delthyrium, the plates are called **deltidial plates**. These covering plates prevent predators from entering the shell.

In some articulates, the posterior area of the brachial valve is strongly curved and is arched into the ventral delthyrium (Fig. 8.5C). The diductors are thereby protectively within the shell and delthyrial covers are unnecessary. Yet another plan is seen in the flat-shaped brachiopods known as *strophomenids*. In this group, the diductors are attached to the brachial valves by a cardinal process that extends into the interior of the ventral valve (Fig. 8.5D). As illustrated in Figure 8.5, muscle attachment areas in some articulate brachiopods were raised above the interior surface of the pedicle valves. Such muscle platforms were particulary prevalent among articulates with strongly convex valves.

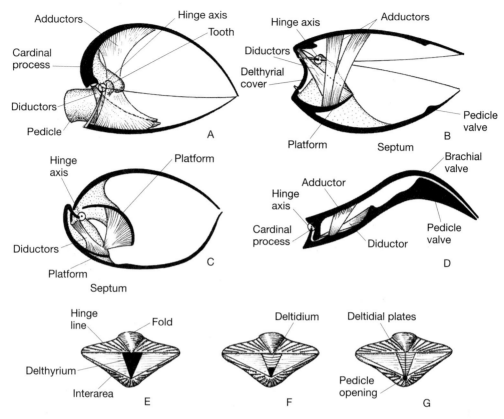

Figure 8.5   Muscle systems and delthyrial modifications in articulate brachiopods.
(A) Muscle system in *Hebertella*. In *Eastlandia* (B), there are delthyrial covers and raised
muscle platforms. (C) Musculature in *Stenoscisma*, a Permian brachiopod that also has
muscle platforms. (D) Muscle system in the saucerlike shell of the strophomenid
Strophophonelloides. (E) Posterior view of a brachiopod with no closure plates to the
delthyrium. (F) Partial closure of the delthyrium by a single plate (a deltidium). (G) Closure
of delthyrium by two plates accompanied by migration of pedicle opening to the beak.
(A–D after M. J. S. Rudwick, *Living and Fossil Brachiopods*. London: Hutchinson University
Library, 1970.)

FEATURES OF THE ARTICULATE SHELL

Shell Structure   The primary function of the brachiopod shell is protection.
This is evident from the fact that the shell is external to the soft tissue, and when the
valves are closed, internal organs are effectively closed off from possible dangers in the
external environment. The shell is secreted by ectodermal cells of the mantle, each half
of which lines the internal surface of a valve. The mantle consists of two layers of ectodermal cells separated by an internal sheet of connective tissue (Fig. 8.6). The margin
of the mantle is divided by a mantle groove so as to form two marginal lobes. The outer
lobe is responsible for the secretion of shell matter, whereas the inner lobe is sensory.

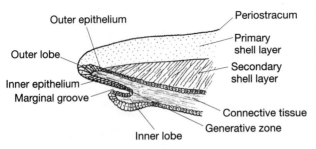

Figure 8.6 Section through the margin of the mantle of an articulate brachiopod. (Simplified from A. Williams & A. J. Rowell, *Biol. Reviews of the Cambridge Philosophical Soc.* 31, 1956.)

New cells are added to the mantle within the mantle groove and then moved, conveyor-belt fashion, toward and around the tips of the mantle lobes. As new cells approach the edge of the outer lobe, they secrete the surficial chitinous layer of the shell. Known as the **periostracum**, this layer is similar to the thin, dark covering seen on the outside of clam and mussel shells. After secreting the periostracum, ectodermal cells continue to move around the tip of the outer lobe where they secrete a layer of calcite next to the periostracum. This so-called **primary layer** (Fig. 8.7) consists of laminae of calcite arranged approximately parallel to the surface of the shell. Beneath the laminar layer is the thicker **secondary layer**, which can be modified during the life of the brachiopod. This layer is secreted by the entire outer mantle surface. It usually consists of calcite fibers that are separated by thin sheaths of protein. In some brachiopods, a termination of protein secretion may result in large fibers or prisms of calcite arranged perpendicular to the surface of the valve.

Sections cut transverse to the valve surface (Fig. 8.7) reveal that some valves are solid, whereas others are perforated by small tubes or *punctae*. A brachiopod with such perforations is said to be **punctate**. Those lacking this feature are said to be **impunctate**. The weathered valves of some brachiopods may have tiny pits that on first observation appear to be punctae. Actually, the pits of these **pseudopunctate** brachiopods are produced by differential weathering of rod-shaped elements in the shell called **taleolae**. The calcite in taleolae is less dense than that in the surrounding shell. It therefore weathers more rapidly, producing pits. Pseudopunctae may have functioned

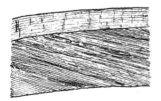 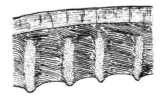 

Figure 8.7 Sections of calcareous brachiopod valves illustrating impunctate (A), pseudopunctate (B), and punctate (C) shell structures. The relatively thin outer layer is laminar in a direction parallel to the valve's exterior surface. The underlying layer is composed of inclined elements. Punctate valves are traversed by small tubes, whereas pseudopunctate valves contain solid rods of calcite.

in halting posterior propagation of fractures, thereby confining such damage to the anterior region of the shell where repairs could be made.

In life, the punctae of punctate brachiopods were occupied by slender columns of mantle epithelium called **caeca**. Studies of living brachiopods indicate that caeca function as storage localities for certain proteins and lipids. They may have also had a respiratory function. In living articulates, their presence seems to inhibit predation by boring organisms; however, this was apparently not the case in many Paleozoic punctate brachiopods whose shells are burdened with heavy infestations of encrusting and boring organisms.

Shape and Ornamentation of the Shell

The shape of a brachiopod shell is the result of the rate at which calcium carbonate is added to specific areas of the valve margin during growth. In this regard, the space between successive growth lines provides a relative indication of the rate at which shell matter is being secreted by the ectodermal cells of the mantle. One can visualize the process by imagining a valve in the shape of a disk. If new material were added at the same rate all around the periphery of such a disk, the circular shape would persist as it became increasingly larger. If, however, the rate of addition of shell matter was more rapid at one end than at the other, the bivalve would assume an oblong shape. Variations in rate of calcium carbonate secretions for the vertical components of the shell would produce convex or concave shapes, downturned and upturned areas, sulci, and folds.

Finding names to describe the varied shapes of brachiopod shells is simplified by the use of a linguistic convention in which the name of the shape of the brachial valve precedes that of the pedicle valve. Thus, there are **biconvex, plano-convex, concavo-convex**, and **convexo-concave** shells (Fig. 8.8). The term **resupinate** refers to shells that are concavo-convex near the beak, but become convexo-concave anteriorly. **Pseudoresupinate** shells have the resupinate shape in reverse (convexo-concave near the beak; concavo-convex anteriorly).

The surface sculpture of brachiopod shells is referred to as *ornamentation*. Both concentric and radial ornamentation can occur. The former include very coarse **growth lines** called **rugae** (see Fig. 8.12), overlapping frills or **laminae**, and concentrically aligned spines. Radial features can consist of ridges called **costae** (normally 1 mm or more in height) or there may be fine threadlike ridges called **costellae** (compare Fig.

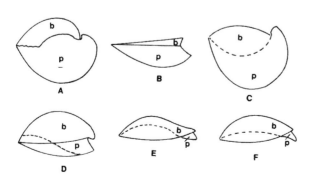

Figure 8.8   Nomenclature for designating the shape of brachiopods. In each drawing, the top valve is brachial. (A) biconvex; (B) plano-convex; (C) concavo-convex; (D) convexo-concave; (E) resupinate; and (F) pseudoresupinate. In resupinate shells, the brachial valve is mostly convex but becomes concave near the hinge with corresponding changes in the pedicle valve. Pseudoresupinate shells appear similar, but the valves are reversed.

8.12 and 8.17). Pronounced radial ridges that corrugate the entire thickness of the valve are called **plicae** (see Fig. 8.16). A netlike pattern called **reticulate** (see Fig. 8.13) results from the crisscross of costae and rugae of about equal size. Granules, tubercles, various kinds of rugosities, and spines provide additional variety to the surface sculpture of articulates. Spines are hollow, pointed structures that are particularly characteristic of the productid brachiopods. Some paleontologists regard sulci and folds as additional forms of radial ornamentation.

The Posterior Margin   The posterior margin of a brachiopod shell includes the hinge line that forms the axis of articulation between the valves. Between the hinge line and the beak, there is usually a curved or plane surface called the **interarea**. It is invariably wider on the pedicle valve than on the brachial, and may be triangular in shape (see Fig. 8.5E). The interarea is produced by growth of valve margins along the hinge line. Growth lines mark the progressive migration of the hinge during growth of the interarea. The delthyrium is a prominent feature on the interarea, and in some brachiopods, there is an equivalent of the delthyrium in the brachial valve that is called the **notothyrium**. Just as the delthyrium may be closed by deltidial plates, the notothyrium may be covered by a calcarous plate termed a **chilidium**. In brachiopods having prominent beaks, the boundaries of the interarea may be transitional and obscure. This is not the case, however, for brachiopods with long, straight hinge lines (see Fig. 8.17C,F). The shells of such brachiopods are referred to as **strophic**.

The Pedicle Valve Interior   Structures found in the interior of the pedicle valve (Fig. 8.9) provide support for the pedicle valve interarea, platforms for muscle attachment, and barriers for separation of muscle groups. As already noted, articulation in many brachiopods is dependent upon two teeth located in the pedicle valve. Two sockets in the brachial valve receive these two teeth. The teeth usually take the form of knobs of calcium carbonate that may be strengthened by a pair of vertical dental plates. These plates join the teeth to the floor of the pedicle valve. Additional support may be provided by a median septum, which extends vertically along the valve midline as a sort of interior keel. It is possible that this structure not only added strength to the valve, but also provided for the separation of muscle strands.

The pedicle valve features of some brachiopods (i.e., Pentamerida) include raised, curved, muscle attachment platforms called **spondylia**. A spondylium may serve to raise muscle attachments above the level of the mouth, and thereby facilitate feeding. Spondylia also add strength to the shell where it is needed to counter the pull of muscles. Alternatively, spondylia may be related to the type of muscles that had evolved in different groups of brachiopods. Particular groups of brachiopods have either columnar or tendonous muscles. Columnar muscles are composed of contractile fibers throughout, and hence the entire length of the muscle can contract. Tendenous muscles in brachiopods have contractile fibers only at one end, and the remainder of the muscle consists of noncontractile tendon tissue. Thus, a longer tendenous muscle may have no greater net contraction than a shorter columnar muscle. Spondylia shorten the distance to be spanned, and may have been most characteristic of brachiopods that had columnar muscles. Once tendenous muscles had evolved in certain lineages, spondylia were not needed.

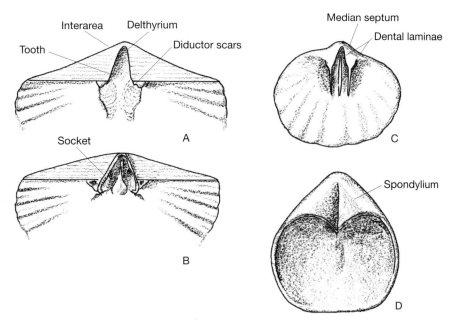

Figure 8.9   Features of brachiopod valve interiors. Pedicle (A) and brachial (B) valve interiors of *Hesperorthis*. (C) Dental plates and median septum in *Enteletes*. (D) Spondylium simplex in *Gypidula*.

A spondylium that rests on the floor of the ventral valve is called a **sessile spondylium**. One supported by a median septum is called a **spondylium simplex**, whereas if the platform is supported by two dental plates, the structure is referred to as a **spondylium duplex**.

The Brachial Valve Interior   The structures of the brachial valve interior exhibit considerable parallelism with features of the opposing pedicle valve. The interarea, for example, may be strengthened by **crural plates**, which are analogous to the dental plates of the pedicle valve. The crural plates may extend anteriorly to form a spoon-shaped **cruralium**, which is roughly similar to the spondylium of the pedicle valve. The cruralium forms a base for the attachment of the adductor muscles.

Interior features along the posterior margin of the brachial valve constitute the **cardinalia**. Here one finds the sockets that receive the teeth of the opposing valve. In many brachiopods, the space between the sockets may be covered by a plate, along which are attached the posterior ends of the diductor muscles.

Not all the internal features of the brachial valve mimic those of the pedicle valve, for only the brachial valve has the brachidia for the support and attachment of the lophophore. These may consist of crura, loops, and spiralia (Fig. 8.10). **Crura** are slender calcareous arms that extend anteriorly from the cardinal plate. In **loops**, the arms double back on themselves. **Spiralia** (see Fig. 8.10B–D) consist of paired calcareous coils whose axes extend either laterally or vertically within the shell.

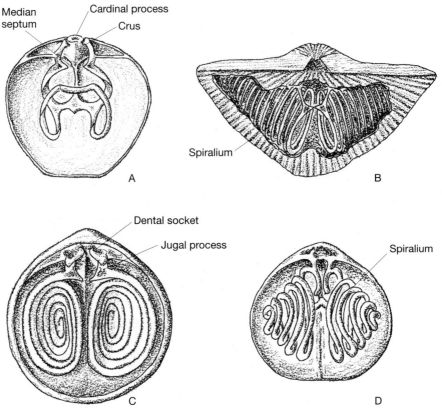

Figure 8.10   Common types of brachidia. (A) loop type. The spirally enrolled brachidia include (B) spiriferid, (C) atrypid, and (D) athyrid spiralia.

MAJOR TAXONOMIC GROUPS OF ARTICULATE BRACHIOPODS

We have seen that the Class Articulata embraces all brachiopods having definitely hinged shells composed of calcium carbonate, and usually having structures for the support of the lophophore. The taxonomic orders and suborders of articulates are distinguished largely on the basis of shell structure, the nature of the pedicle opening, and characteristics of the brachidia.

ORDER ORTHIDA (CAMBRIAN TO PERMIAN)

The orthids are relatively simple brachiopods that have unequally biconvex shells with rounded to semielliptical outlines and radial ornamentation in the form of costellae, costae, or plicae. The shells of orthids have straight hinge lines, and this is the basis for their name (*orthos* is the Greek word for "straight"). The pedicle is unobstructed. Brachidia in the Order Orthida are not developed. These brachiopods include the oldest articulates known. Many biostratigraphically important fossils species occur within

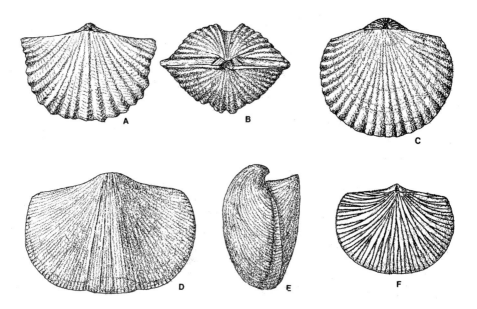

Figure 8.11   Orthida. (A, B) Brachial and posterior views of *Platystrophia*, ×1.5. (C) Brachial view of *Orthis*, ×3.6. (D, E) Pedicle and lateral views of *Hebertella*, ×1.3. (F) Brachial view of *Resserella*, ×1.5.

the Orthida. Among these are such Ordovician genera as *Orthis*, *Hebertella*, *Platystrophia*, and *Resserella* (Fig. 8.11).

ORDER STROPHOMENIDA (ORDOVICIAN TO JURASSIC)

The Strophomenida are thought to have evolved from the Orthida. They have pseudopunctate valves with long straight hinge lines. Typical strophomenids have very compressed shapes and externally concave brachial valves. This gives the shell a distinctive saucerlike form. Ornamentation normally consists of fine radial ridges or **costellae** (Fig. 8.12). A functional pedicle is lost in many species, and a hood-shaped deltidium or chilidium covers the pedicle opening.

   Among the suborders of the Strophomenida, the Strophomenidina (Ordovician to Triassic), Chonetidina (Silurian to Jurassic), the Productidina (Devonian to Permian), and Oldhamidina (Pennsylvanian to Triassic) are particularly useful in biostratigraphy. Such well-known guide fossils as *Rafinesquina* and *Strophomena* (both Ordovician) exemplify the Strophomenidina. Rows of fine, short ridges (**rugae**) provide the distinctive ornamentation (see Fig. 8.12C) of the longer-ranging *Leptaena* (Ordovician to Mississippian). The compressed form exhibited by strophomenida suggest that many preferred quiet, oxygen-poor waters. The broad, flat shell may have given the animals a maximum area for oxygen-gathering tissue. Some forms, like *Strophomena*, developed resupinate shells, perhaps for stability on particularly soft-bottom sediment. The resupinate form places the anterior margin above the surface of

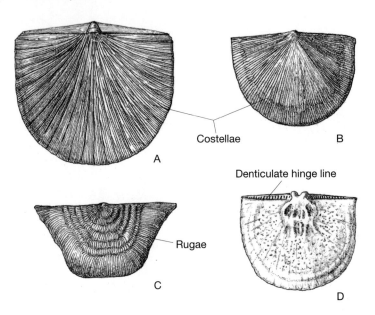

Figure 8.12   Strophomenida. (A) Brachial view of *Rafinesquina*, ×1.3. (B) Brachial view of *Strophomena*, ×1.5. (C) Pedicle view of *Leptaena*, ×1. (D) Interior of brachial valve of *Stropheodonta*, ×1. Note denticles along the hinge line of *Stropheodonta*.

the mud while at the same time the hollow on the ventral valve just anterior to the beak prevents the valve from being overturned by currents.

The Chonetidina (Fig. 8.13) are a suborder of the Strophomenida in which spines are typically present along the pedicle valve interarea and the shells are concavo-convex. *Chonetes* and *Mesolobus* (both Carboniferous) are common representatives. Although chonetid brachiopods range from the Silurian to the Permian, they occur most abundantly in Carboniferous rocks.

Most genera of the Suborder Productidina (Fig. 8.13) have concavo-convex shells with elaborate development of spines over the entire surface, not just along the hinge as in the chonetidinids. The arrangement of spines has importance in identification. Although many productinids were attached by a pedicle during the juvenile stages of their development, the delthyria and notothyria were closed off during later growth. Well-known productinids include *Echinoconchus*, which had parallel transverse bands of fine spines; *Dictyoclostus*, which developed a reticulate pattern of ornamentation on the umbo (arched area near the beak); and *Waagenoconcha*, which had spines arranged in regularly intersecting oblique rows.

Richthofenids are a bizarre group of articulates derived from the Productidina. *Richthofenia* (Fig. 8.13D), for example, had an erect ventral valve with septa sealing off the unoccupied part of the shell. This gave the brachiopod an uncanny resemblance to a solitary rugose coral. The brachial valve in *Richthofenia* had the shape of a flat operculum. Richthofenids populated warm tropical seas during the Permian.

The Productidina made their appearance in Late Devonian time, and were present as the most abundant and diverse brachiopods until the end of the Permian. Some

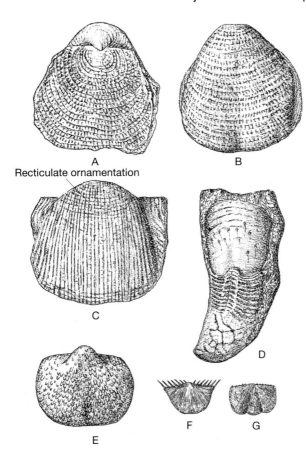

A
Recticulate ornamentation

B

C

D

E

F   G

Figure 8.13   Productidina and Chonetidina. Brachial (A) and pedicle (B) views of *Echinoproductus*, ×1.2. (C) Pedicle view of *Dictyoclostus*, ×1.2. (D) Pedicle valve of *Richthofenia*, ×1.2. The pedicle valve in *Richthofenia* grows erect like a coral, and the vacated parts of the valve have transverse septa. The brachail valve (not shown) has the shape of a flat lid. (E) Pedicle view of *Waagenoconcha*, ×1.2. (F) *Chonetes*, ×1.2. (G) *Mesolobus*, ×1.2. (*Richthofenia* redrawn from A. Williams et al. In R. C. Moore (Ed.), *Treatise on Invertebrate Paleontology*. Boulder, CO, and Lawrence, KS: Geological Society of America and University of Kansas Press, 1965.)

Mississippian productidina are noted for their enormous size. *Gigantoproductus*, a Mississippian genus, had a total width of over 37 cm.

The final suborder of the Order Strophomenida is the Oldhamidina. *Collemataria* (Carboniferous to Permian) is a representative genus known from its concavo-convex shell and spoon-shaped pedicle valve, which contains a median septum from which lateral septa extend. The odd brachial valve is featherlike, and is recessed within the pedicle valve (Fig. 8.14). There is no interarea, hinge, or pedicle opening.

ORDER PENTAMERIDA (CAMBRIAN TO DEVONIAN)

Like the Strophomenida, pentamerids are descendants of the orthids. Pentamerids have large, impunctate, strongly biconvex shells in which the beak of the larger ventral valve overhangs the opposite valve. The hinge is short and the delthyrium and notothyrium are uncovered. A distinguishing feature of the pentamerids is the well-developed spondylium in the pedicle valve. The Silurian genus *Pentamerus* (Fig. 8.15) is an easily recognized guide fossil. It is often preserved as an internal mold that clearly

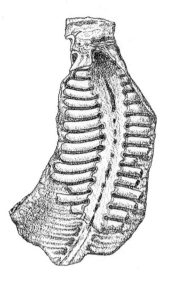

Figure 8.14 *Collemataria* viewed from the dorsal side, ×1. (After photograph in G. A. Cooper, *Smithsonian Contributions to Paleobiology* 15:689, 1974.)

displays the impressions of the spondylium. *Conchidium* (Late Ordovician to Devonian) is among the relatively few pentamerids that were ornamented with costae.

ORDER RHYNCHONELLIDA (ORDOVICIAN TO HOLOCENE)

Rhynchonellids are rostrate brachiopods, which means they have prominent beaks. Their shells are strongly convex, and many bear strong costae or plicae. It has been suggested that strong plication in these brachiopods provided the valves with greater resistance to damage, but an important additional concept is that the zigzag juncture

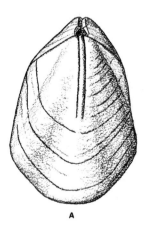
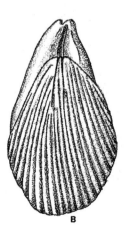

Figure 8.15 Pentamerida. *Pentamerus* (A) and *Conchidium* (B), both ×1.

A

B

caused by the plications along the anterior margin of the shell served as a sieve to prevent entry of harmful large particles into the mantle cavity. It may also have been useful in directing currents into and out of the shell. The fold on the brachial valve and sulcus on the pedicle valve are conspicuous features of many rhynchonellids. Most species have a tiny pedicle opening near the tip of the beak. In some globular forms, the beak overhangs the brachial valve to a degree that prevents more than a slight opening of the shell. Rhynchonellids were a very successful group of brachiopods, often occurring in such great abundance that they formed rhynchonellid coquinas (rock composed of fossil shell debris). They appear to have been adaptable to diverse bottom conditions, as indicated by their presence in pure and clayey limestones as well as shales. *Lepidocyclus* (Ordovician) and *Rhynchotreta* (Silurian) are widely used in biostratigraphic correlation (Fig. 8.16).

ORDER SPIRIFERIDA (ORDOVICIAN TO JURASSIC)

The spiriferids are a large and important group of articulates distinguished principally by spirally coiled brachidia. Most construct impunctate shells that are strongly biconvex and that may have either open or closed delthyria. The Order Spiriferida can be divided into four suborders: the Atrypidina (Ordovician to Jurassic), Retziacea (Silurian to Permian), Athyrididina (Ordovician to Jurassic), and Spiriferidina (Silurian to Jurassic).

Most species in the Suborder Atrypidina have vertically coiled brachidia, although in some species these structures are directed parallel to the hinge line. In those with vertically coiled spiralia, the brachial valve is hemispherical so as to accommodate the coil. In contrast, the pedicle valve is relatively flat. Hinge lines are short, and interareas are either very narrow or absent. These characteristics are readily seen in *Atrypa* (Fig. 8.17), a Silurian to Devonian genus whose brachidium consists of two

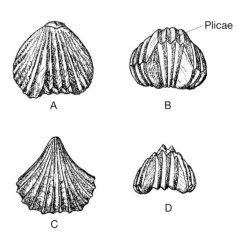

Figure 8.16   Rhynchonellida. Brachial (A) and anterior (B) views of *Lepidocyclus*. Pedicle and anterior views of *Rhynchotreta*. (All ×1.5.)

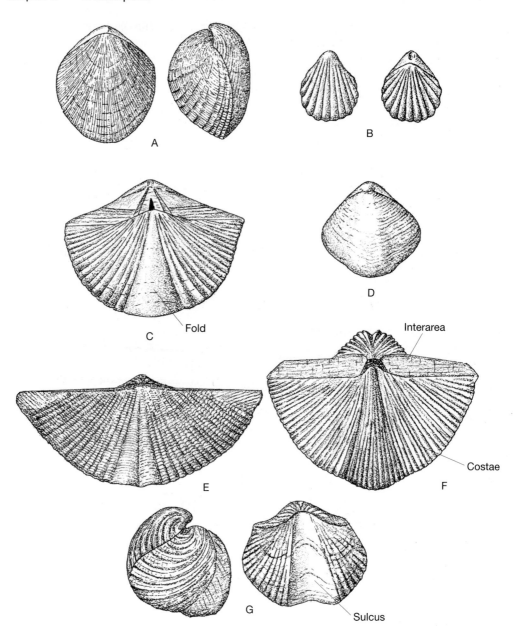

Figure 8.17   Spiriferida (A) Brachial and lateral views of *Atrypa*, ×1. (B) Pedicle and
brachial views of *Hustedia*, ×2. (C) Brachial view of *Platyrachella*, ×1.3. (D) Brachial view of
Composita, ×2. (E) Brachial view of *Mucrospirifer*, ×1.5. (F) Brachial view of *Neospirifer*, ×1.3.
(G) Brachial view of *Paraspirifer*, ×1.

spirally coiled cones with their narrow ends extending toward the middle of the strongly convex brachial valve.

Species of Retziidina have laterally directed spiralia, are rostrate, and possess strong, ridgelike costae. As a result, they somewhat resemble rhynchonellids in external appearance. *Hustedia* (see Fig. 8.17B), a Pennsylvanian to Permian form, has characteristics typical of the Suborder Retziacea.

As seen in *Composita* (see Fig. 8.17D), members of the Suborder Athyrididina have smooth shells ornamented by growth lines. Internally, there are laterally directed spiralia joined by a complex, bridgelike structure called a **jugum**. *Composita* has a broad fold and sulcus and a round foramen at the beak of the pedicle valve. It is a common brachiopod in some Carboniferous and Permian rocks.

Fossil collectors find the shells of members of the Suborder Spiriferidina attractive because of their symmetry and resemblance to the winged insignia of airline pilots. Spiralia that coil laterally to a point at either end of a straight hinge line account for their distinctive shape. In the majority of families the shell is impunctate, has a pronounced sulcus and fold, and is costate. Beneath the beak, there is commonly a distinct delthyrium, which may be partly closed by a deltidium. The trilobate shape centered on the fold and sulcus facilitated the movements of currents into and out of the shell, and thereby helped to aerate the mantle, bring in food, and expel waste. Perhaps also, the winglike form of some genera provided greater surface area in contact with the substrate so as to prevent overturning. The ample surface area may also have buoyed the shell enough to prevent its sinking into very soft sediment. Many excellent guide fossils (see Fig. 8.17) are within this suborder, including *Mucrospirifer* (Devonian), *Platyrachella* (Devonian), *Neospirifer* (Pennsylvanian and Permian), and the interestingly shaped *Paraspirifer* (Devonian).

ORDER TEREBRATULIDA (DEVONIAN TO HOLOCENE)

The majority of living articulate brachiopods belong to the Terebratulida. *Terebratulina* (Fig. 8.18) exhibits the characteristics of the order very well. The most distinctive feature of terebratulids is their loop-shaped brachidium. Unfortunately, this internal structure is rarely seen in fossil terebratulids, and tentative identifications rely

Figure 8.18   *Terebratulina septentrionalis*, coastal Maine. (Length 3.5 cm.)

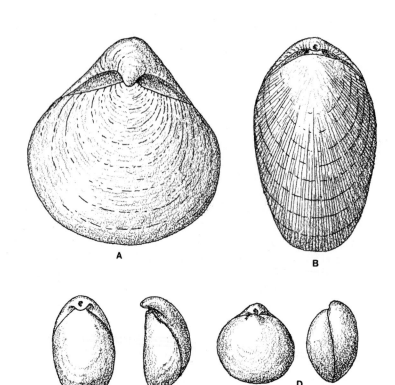

Figure 8.19   Fossil terebratulida. (A) *Stringocephalus*, brachial view, ×1. (B) Brachial view
of *Rensselaeria*, ×1. Brachial and lateral views of *Dielasma* (C) and *Kingena* (D), both ×2.

upon less specific external characteristics. The shell is biconvex, punctate, rostrate, and
more commonly smooth than costate. The pedicle opening is usually distinct and lies at
the posterior tip of the pedicle valve. In some forms the opening is closed by a system
of plates. Fossil terebratulids (Fig. 8.19) include *Stringocephalus* (Devonian),
Rensselaeria (Devonian), *Dielasma* (Mississippian to Permian), and *Kingena*
(Cretaceous).

Incertae Sedis: Thecideidina (Triassic to Holocene)   Placement of this small
group of brachiopods within the classification of articulates is uncertain. They appear
to have affinities with the Terebratulida and possibly the Spiriferida and
Strophomenida as well. Thecideidinid shells are punctate, thick, and either circular or
elongate. They rarely exceed a centimeter in diameter. A pedicle is not developed, and
stability is provided by cementation of the pedicle valve to the substrate. *Thecidea* (Fig.
8.20), a Cretaceous genus, serves as a representative of this problematic group of inarticulates.

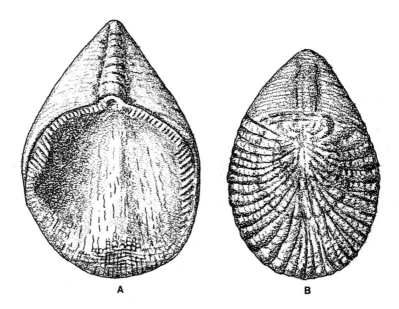

Figure 8.20   Interior of pedicle valve (A) and exterior of brachial valve (B) of
Thecidea, ×5. (Redrawn from A. Williams & A. J. Rowell. In R. C. Moore (Ed.), *Treatise
on Invertebrate Paleontology*, Pt. H. Boulder, CO, and Lawrence, KS: Geological
Society of America and University of Kansas Press.)

BRACHIOPOD PALEOECOLOGY

Brachiopods are marine, benthic, mostly sessile and epifaunal, filter-feeding invertebrates whose environmental range extends from the sublittoral zone to the abyssal
depths of the ocean. Except for some inarticulates that have a burrowing habit and live
in intertidal zones and brackish water, most living articulates prefer marine waters of
normal salinity. They are most abundant in the cooler waters of temperate zones where
they occur in sporadic clusters. This is not the case for fossil brachiopods, for many of
these proliferated in tropical and semitropical seas. Sensitivity to light is another characteristic of living forms, something that does not appear to have been essential in
ancient forms.

Like many of their living descendants, most articulate brachiopods were able to
tolerate considerable turbidity. Sand and silt particles that found their way into the
shell were ejected by currents generated by the cilia of the lophophore. Rapid snap-
ping of the shell in some groups served to dislodge particles that had settled within the
shell.

Today as in the past, animals that feed on brachiopods include starfish, crustaceans, polychaete worms, gastropods, cephalopods, and fish. Brachiopods have also
been afflicted by a variety of parasites. Their shells are often encrusted with other
invertebrates, many of which serve as camouflage.

Some brachiopods had skeletal features that appear to be associated with particular kinds of substrates. Productids such as the Permian genus *Waagenoconcha* utilized

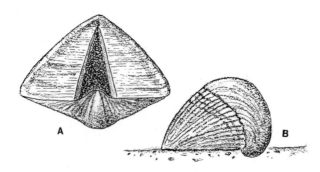

Figure 8.21 The broad interarea of certain spiriferid brachiopods may have served as a stable support on the substrate, as suggested by M. J. S. Rudwick. (A) Posterior view of *Syringothyris* with its broad, flat, triangular interarea. (B) Lateral view of the same genus indicating its probable living position.

spines to stabilize their position in soft sediment. Long slender spines along the cardinal margin of some chonetidinids and on the valve surfaces of some orthids functioned as attachments to sponges and corals. By means of these attachments, the brachiopods were supported above the seafloor. Certain other productids lived with most of their pedicle valve buried, and with a thin layer of sediment covering the brachial valve as camouflage. Many Devonian and Carboniferous spiriferids rested on their particularly broad interareas with valves oriented vertically (Fig. 8.21).

Not all brachiopods are restriced to a purely sessile mode of life. Some attached themselves to drifting objects and thereby achieved a measure of passive mobility. Some, like the flat-shelled strophomenids, may even have been able to swim above the seafloor for short distances by rapid movement of their valves. One might argue, however, that such "valve-flapping" would not be possible unless the brachiopods had the essential striated muscles. A lesser motility, however, may have been possible in forms that used the pedicle and associated muscles to pull the shell across short distances of the ocean floor. At least one living species of terebratulid displays this behavior.

In retrospect, it is evident that Paleozoic brachiopods were richly endowed with adaptations for a variety of habitats and living strategies. Yet, with all their adaptive versatility, they declined severely after the Paleozoic. Perhaps these mostly sedentary suspension feeders became too vulnerable to predation or were replaced by more motile and aggressive members of the Bivalvia.

GEOLOGIC HISTORY OF THE BRACHIOPODA

Brachiopods are the oldest known lophophorate animals. Because Lower Cambrian rocks contain large, complex, and abundant brachiopods, there is a strong probability that the phylum appeared earlier in the Late Proterozoic. The majority of earliest Cambrian brachiopods are inarticulates with phosphatic shells. Substitution of calcium carbonate for calcium phosphate followed, and by Middle Cambrian time, brachiopods with calcareous shells were abundant. Predominant among these early calcareous articulates were members of the Order Orthida (Fig. 8.22).

During the Ordovician, articulates assumed numerical superiority over the inarticulates. The already well-established orthids were joined by members of the Suborder Strophomenidina and the Order Rhynchonellida. The Order Pentamerida, and Suborders Atrypidina and Athyrididina also made their appearance before the end of the Ordovician.

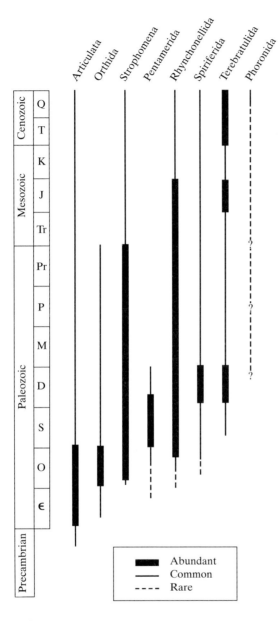

Figure 8.22   Geologic ranges of Brachiopoda (Phoronida are not brachiopods, but are lophophorate animals included on the chart for the reader's convenience).

    The Silurian lasted only about 30 million years. The relatively short duration of the period may account for the conservative nature of brachiopod evolution during the Silurian. The previously successful orthids declined, while the Athyrididina increased in abundance and diversity. The most distinctive group of Silurian brachiopods were the pentamerids, which are so abundant in some strata that they constitute pentamerid coquinas.

    Devonian marine strata often contain an abundance of the Athyridian and Spiriferidina. Rhynchonellids, already important in the Silurian, continued their expansion, as did more advanced groups of strophomenids. Some of the latter lost their dependency on pedicles and cemented their pedicle valves to hard substrates of various kinds. Certain Devonian Rhynchonellida developed tubular spines that encased thin columns of mantle tissue. These structures may have functioned as sensory antennae, allowing the animal to detect changing conditions of water temperature, chemistry, or possibly even the approach of predators. Punctate shells, which had first appeared among orthids during the Ordovician, became increasingly abundant during the Devonian. The end of the Devonian, however, was a time of crisis for brachiopods. All of the impunctate orthids became extinct, along with the pentamerids and entire families of strophomenids and atrypidinids. It was an episode of mass extinction that affected all of the other phyla of marine invertebrates as well.

The decline in brachiopod diversity that marked the close of the Devonian continued into the Carboniferous. Strongly costate spiriferids persisted, as did chonetidinids. The most characteristic Carboniferous brachiopods, however, were the large, strongly plano- to concavo-convex, spinose members of the Suborder Productidina.

The most abundant Permian brachiopods are also productinids, although rhynchonellids and spiriferids are also common. The most distinctive of the Permian articulates, however, were cone-shaped forms such as *Richthofenia* (see Fig. 8.13D). The end of the Permian was another time of crisis for brachiopods. Along with many other marine invertebrates, entire families became the victims of mass extinction. Although spiriferids lingered on into the Jurassic, Mesozoic brachiopod faunas were dominated by terebratulids and rhynchonellids, both of which continue to the present time. In addition, inarticulates such as *Lingula* have survived with little change since Cambrian time.

PHORONIDA

The Phoronida are a phylum of lophophorate animals considered a sister taxa to the Brachiopoda. The lophophore structure and derivation of the coelomic cavities within the lophophore are similar in phoronoids, brachiopods, and bryozoa. Living phoronids are slender, wormlike creatures. The body is surmounted by the lophophore and is enclosed in a tubular chitnous sheath that the animal covers with particles of sediment. Extant species live in all regions of the ocean, often in dense aggregations. They dwell in burrows and borings with only their lophophores projecting above the substrate.

Phoronids lack a durable skeleton, and therefore have a sparse fossil record. They have been reported in rocks as old as Devonian. Phoronid borings are occasionally found on Cretaceous and Tertiary mollusk shells. *Phoronis* (Fig. 8.23) and *Phoronopsis* are living examples. It has been suggested that the trace ichnofossils *Scolithus*, which consist of vertical tubular shafts excavated into lower Paleozoic sandstones, are actually the burrows of ancient Phoronida.

REVIEW QUESTIONS

1. Until about 100 years ago, brachiopods were classified as bivalve mollusks. In what ways are they fundamentally different from Bivalvia?
2. What are the most abundant and diversified of extant Brachiopoda?

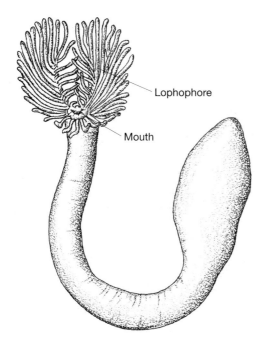

Figure 8.23   *Phoronis architecta* shown separated from its chitinous sheath.

Lophophore

Mouth

3. What phyla, other than the Brachiopoda, possess a lophophore?
4. What is the purpose or function of the following:

a. lophophore	d. mantle	g. delthyrium
b. diductor muscles	e. brachidia	h. deltidium
c. adductor muscles	f. cardinal process	i. spondylium

5. How do inarticulate brachiopods differ from articulate brachiopods?
6. Describe the method used by linguloid articulates in producing the burrow in which they live.
7. What skeletal structures seen on the interior of the brachial valve of an articulate brachiopod do not occur in the pedicle valve?
8. How does the spiralia of *Atrypa* differ from that of *Mucrospirifer*?
9. What terms best describe the shell shapes of *Platystrophia, Echinoconchus*, and *Rafinesquina*?
10. What was the probable living position of adult productids like *Waagenoconcha*? What purpose was served by the extreme spinosity of such productids?
11. During what geologic periods were the following brachiopod groups particularly abundant?

a. Orthida	d. Pentamerida
b. Strophomenida	e. Spiriferida
c. Productidina	f. Terebratulida

SUPPLEMENTAL READINGS AND REFERENCES

Boardman, R. S., Cheetham, A. H. & Rowell, A. J. (Eds.). 1987. *Fossil Invertebrates.* Palo Alto, CA: Blackwell Scientific.

Hyman, L. H. 1959. *The Invertebrates.* New York: McGraw-Hill.

Moore, R. C. (Ed.). 1965. *Treatise on Invertebrate Paleontology*, Pt. H, Brachiopoda. Boulder, CO, and Lawrence, KS: Geological Society of America and University of Kansas Press.

Rudwick, M. J. S. 1970. *Living and Fossil Brachiopods.* London: Hutchinson & Co.

C H A P T E R   9

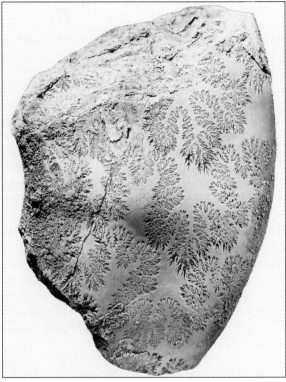

Ammonitic sutures on the outer whorl of the Cretaceous cephalopod
Tragodesmoceras socorroense from the Mancos Shale (height 16cm).
(courtesy of W.A. Cobban, U.S. Geological Survey)

Mollusca

This is the most varied, dominant, and successful of the primarily aquatic phyla, both at the present time and throughout most of the fossil record.

G. G. Simpson, *The Meaning of Evolution*, 1949

If the success of a phylum is determined by the number of species and individuals, then the mollusks are an enormously successful group. Over 50,000 living species and 35,000 fossil species have been described. The fossil record for mollusks extends back to the earliest Cambrian. Although they have always been most abundant in the ocean, many members of the phylum have made the transition to freshwater and even land. Whereas most mollusks possess shells, many slugs and all octopods do not. In size, mollusks range from microscopic clams and snails to the largest of all living invertebrates, the 16-meter-long giant squid.

As presently classified, the Phylum Mollusca includes primitive, limpetlike monoplacophorans, curiously plated amphineurans (chitons), scaphopods (tusk shells), bivalves (formerly called pelecypods), diverse gastropods, the extinct rostroconchians of the Paleozoic, and those most complex of all mollusks, the cephalopods. The extraordinary variety of mollusks might cause one to question the inclusion of all these animals in a single phylum. Yet even with their apparent differences, the diverse classes of mollusks are but variations of a single basic morphologic plan. There are also developmental similarities in that all have a trochophore larval stage resembling that of annelid worms. These observations not only link the various molluscan classes to one another, but indicate their phylogenetic proximity to the Phylum Annelida.

BASIC MOLLUSCAN ATTRIBUTES

Nearly everyone has some familiarity with members of the Phylum Mollusca, whether it be through clam chowder, pearl jewelry, aquarium snails, garden slugs, or the cuttlebone in the canary cage. Typically, Mollusca are unsegmented, bilaterally symmetrical animals having a muscular foot that may be enlarged at the front to form a head containing a mouth, special sense organs, some internal concentration of nerve tissue, and sometimes, tentacles. A scraping organ called the **radula** (Fig. 9.1) is present on the floor of the mouth in all mollusks except the bivalves. The radula is a chitinous ribbon bearing longitudinal rows of chitinous teeth. It is supported by a movable tonguelike organ called the odontophore. Alternate contractions of muscles move the tooth-bearing ribbon back and forth to provide a rasping action that loosens food particles from the substrate and conveys those particles to the gut.

Mollusks possess a heavy fold of tissue called the **mantle** that covers the visceral mass and that usually contains glands that function in secretion of the shell. The mantle generally overhangs the visceral mass, forming a mantle cavity, which often contains gills.

Other molluscan features include a circulatory system with heart, a respiratory system with complex featherlike gills, a digestive system with mouth, buccal cavity,

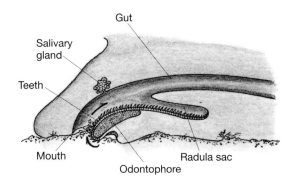

Gut

Salivary
gland

Teeth

Mouth

Odontophore

Radula sac

Figure 9.1   One type of
gastropod radula depicted during
forward retraction. In this position,
food is rasped from the substrate
by radular teeth and transported
back into the esophagous.
Movement is then quickly reversed
to return the rasp to its forward
position so that the operation can
be repeated. The result is a rapid to
and fro movement. Alternate
contractions of antagonistic
muscles move the radular ribbon
back and forth. The structure of the
radula differs among gastropod
taxa, often reflecting a particular
mode of food aquisition, a manner
of excavating the substrate, or the
method used to bore into the shells
of prey.

esophagus, stomach, intestine and anus; an open circulatory system in which the heart pumps blood into an aorta that branches into other vessels, an efficient excretory system, and in most cases a calcareous shell that provides protection. Thus, mollusks have all the organ systems expected of complex animals. As expected, features of the shell are of prime importance for the study of ancient mollusks.

CLASSIFICATION OF MOLLUSKS

The anatomy, embryology, and comparative biochemistry of the Phylum Mollusca lend themselves to a classification that includes three subphyla and eight classes. The classification, however, includes a few groups that have only a scant fossil record, and one class that has no fossil members. Those mollusks of lesser paleontologic importance will be described only briefly.

    Subphylum Amphineura
        (Mollusks with radula, and wide separation of mouth and anus at opposite
            ends of the body. Early Cambrian to Holocene.)
        Class Aplacophora
        Class Polyplacophora
    Subphylum Cyrtosoma
        (Mollusks with radula and conical, univalved, often spiraled shells. Anus
            located close to mouth, and gut usually twisted or U-shaped. Early
            Cambrian to Holocene.)
        Class Monoplacophora
        Class Gastropoda
        Class Cephalopoda
    Subphylum Diasoma
        (Mollusks with bivalved, univalved, or pseudobivalved shells that are often
            open at both ends. Early Cambrian to Holocene.)

Class Bivalvia
Class Scaphopoda
Class Rostroconchia

CLASS APLACOPHORA

The Aplacophora are a class of the Subphylum Amphineura. A radula is present. The mouth and anus are located at opposite ends of the body. Of the two amphineuran classes, the Aplacophora are not known as fossils. Living aplacophorans are tiny creatures, usually less than about 4 mm in length. Their mantle borders are rolled inward ventrally, which gives the animal a superficial resemblance to a worm. Where the mantle margins meet on the ventral surface, there is a longitudinal groove containing one or more ridges. This feature is probably homologous to the foot of other mollusks (Fig. 9.2A).

Aplacophorans lack a shell, but they derive some protection from an enclosing cuticle that bears small aragonitic spicules. Members of the class are exclusively marine. They live at all depths in the sea, from intertidal to the deepest abyss. Mostly, they burrow in sediment or live on the skeletons of larger invertebrates.

CLASS POLYPLACOPHORA

Because they include the more frequently seen and readily recognized chitons (Fig. 9.2B), the Polyplacophora are the more familiar of the two classes of the Subphylum Amphineura. Chitons are sometimes called "armadillo snails" in reference to the shell, which in most species consists of eight articulating calcareous plates. The plates are

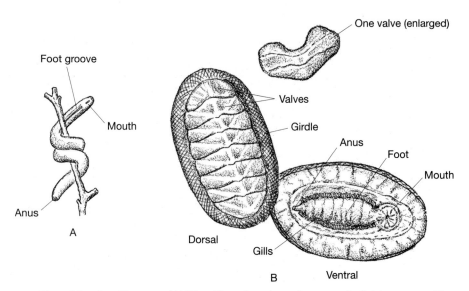

Figure 9.2   Amphineurans. (A) Wormlike aplacophoran known as the Solenogastres, ×40. (B) The common Atlantic coast chiton, ×0.5.

slightly overlapping in living forms, but such imbrication is tenuous or lacking in Paleozoic polyplacophorans. The body is ovoid in shape and may range in length from 3 to 12 cm. (One giant, *Cryptochiton stelleri*, grows to an adult length of 40 cm.) The head is indistinct in chitons and there are no cephalic eyes or tentacles. Chitons have a thick mantle and gills aligned within two mantle grooves. Most of the underside of a chiton is occupied by the broad muscular foot. The mouth is near the anterior end of the ventral surface and the anus is at the opposite end. Chitons are able to cling tenaciously to a rock and can be dislodged only with difficulty. This ability is clearly an adaptation for living in the nearshore zone of high wave energy. Chitons are herbivores. They move slowly across the surfaces of rocks along the seashore, using their radulae to browse on algae and other microvegetation.

CLASS MONOPLACOPHORA

Prior to 1957, this primitive class of mollusks was known only from a few early Paleozoic fossils. In that year, however, a research vessel cruising off the coast of Costa Rica dredged up a small, limpetlike animal that was later to be identified as a living monoplacophoran. It was given the name *Neopilina* (Fig. 9.3). The creature had multiple gills, which, along with serial muscles and nephridial (kidney) units, were arranged

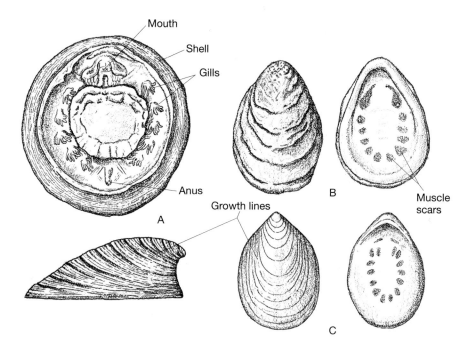

Figure 9.3   (A) Ventral and lateral views of *Neopilina*, ×4. (B) Dorsal and ventral views of *Tryblidium*, ×2. (C) Dorsal and ventral views of *Pilina*, ×1.2. (Drawing of *Neopilina* after H. Lemche & K. G. Wingstrand, *Galathea Report*. Copenhagen: Zoologisk Museum, 1959. Other drawings after J. B. Knight et al. In R. C. Moore (Ed.), *Treastise on Invertebrate Paleontology*, Pt. I. Boulder, CO, and Lawrence, KS: Geological Society of America and University of Kansas Press, 1969.)

in series along the length of the body. This serial arrangement of features in *Neopilina* is observed in segmented animals, and is regarded by some biologists as further evidence that mollusks evolved from annelid worms. Segmentation, however, is not characteristic of other mollusks. As a result, some biologists believe the serial repetition of gills, muscles, and nephridia in *Neopilina* is not comparable to that in annelids, and that these features may have evolved secondarily.

All monoplacophorans have small, depressed, cap-shaped shells. Two Silurian representatives are *Tryblidium* and *Pilina* from which *Neopilina* takes its name.

CLASS GASTROPODA

The Gastropoda are the largest class of mollusks. Nearly 35,000 living and 15,000 fossil species have been recognized. The class includes an extraordinary variety of snails, whelks, limpets, slugs, abalones, drills, pteropods, and nudibranchs. Many of these creatures are marine, but others have adapted to lake, stream, and land environments.

THE LIVING ANIMAL

From observing snails, most of us are aware that gastropods typically have a head with mouth, eyes, and tentacles, as well as a prominent ventrally flattened foot (Fig. 9.4). In most species, undulatory motion of the foot provides locomotion. Inside the shell, a fleshy mantle surrounds the viscera. Within the mouth is a radula that functions as a movable rasp, but may also serve as a grater, cutter, harpoon, and conveyor. The radula in gastropods is an elaborate organ that is so diverse in its morphology as to be valuable in systematics and determining the habits of gastropods. Over a quarter of a million individual teeth may be present in the gastropod radula. Food processed by the radula is passed from the mouth into an esophagus, and thence to the stomach. A tubular intestine extends from the stomach and opens into the mantle cavity. Respiration in aquatic gastropods is accomplished by gills, and in the terrestrial snails by a richly vascularized mantle roof that is capable of providing for gas exchange. The gills in aquatic snails usually have filaments arranged on either side of a central axis, rather like a double comb. Such a gill structure is termed **bipectinate**. In other gastropods, the comblike structure is developed only on one side and hence called **monopectinate**.

Excretion is accomplished by a tubular, variously infolded kidney. The nervous system includes a cerebral ganglia, a sensory epithelium or **osphradium**, eyes, organs for testing the turbidity of water enroute to the gills, and olfactory tissues (commonly

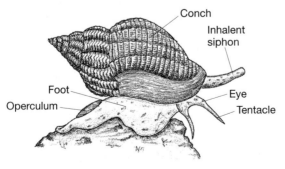

Conch
Inhalent siphon
Foot
Operculum
Eye
Tentacle

Figure 9.4 The shallow marine, carnivorous gastropod *Buccinum*. (Conch length 12 cm.)

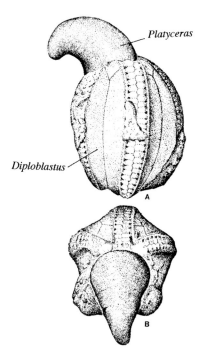

Figure 9.5   The gastropod *Platyceras* sits snugly atop the anal opening of the blastoid *Diploblastus*. It is a commensal relationship in which *Platyceras* is copraphagous. The gastropod is 4 mm in length. (A) Lateral view; (B) oral view. Specimens were recovered from the Mississippian St. Louis Limestone.

located in the tentacles). In some species, sexes are separate, and there may be slight sexual dimorphism of the shell. Other species are hermaphroditic.

Nearly every type of feeding habit can be found among gastropods. There are gastropod carnivores, herbivores, omnivores, deposit feeders, scavengers, suspension feeders, and parasites. There are also feces-ingesting (coprophagous) gastropods that attach themselves to the exhalent or anal openings of other invertebrates (Fig. 9.5). Some gather their food as planktonic or nektonic (swimming) organisms, others are benthic, either living at the surface (epifaunal gastropods) or living within the sediment (infaunal gastropods).

Perhaps the most distinctive feature of gastropods is the way in which their bodies twist during early developmental stages (Fig. 9.6). The process is termed **torsion**. During torsion, the body mass, as viewed dorsally, is twisted 180° counterclockwise relative to the head and foot. As a result, the nervous system is twisted into a figure eight, and the gut and nerve cords are looped. Typically, some of the organs on the inside of the curve are lost during the process. Torsion is a unique feature of gastropoda. Even in those forms that have achieved a degree of bilateral symmetry, torsion can be detected in their developmental stages. Embryologically, it is the evolutionary novelty or shared derived character that defines the class.

The reason for torsion in gastropods is not well understood. It is clear that torsion evolved from an earlier bilateral condition. This is indicated by the fact that gastropods begin their development as bilaterally symmetrical organisms, and subsequently experience the differential growth that results in torsion. One of several hypotheses, termed the "larval retraction hypothesis," suggests that torsion provided protection for the larvae. When gastropod eggs hatch, they develop into trochophore

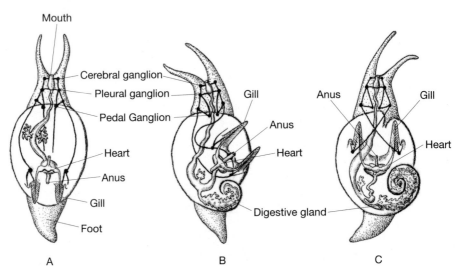

Figure 9.6   Diagram of the effects of torsion as observed in the marine gastropod *Buccinum*. (Torsion occurs in the embryonic stages, but is shown here in adults.)

larvae. The **trochophore** is top-shaped and encircled by a band of locomotor cilia. Below the band of cilia is the mouth, which leads to a gut and thence to the anus (Fig. 9.7). As the trochophore develops, it is transformed into a structure called a **veliger**. When this occurs, a delicate shell is secreted and the area under the gut is expanded to form a bilobed, ciliated swimming organ called a **velum**. Torsion probably occurred suddenly prior to the secretion of the shell as a result of a mutation that caused unequal timing in the development of retractor muscles. Contraction of early formed retractor muscles, having no counterparts, might have produced the twisting required for torsion. Once established, the twisted condition was imposed on the adult. Torsion may have benefited the organism by allowing retraction of the head and velum

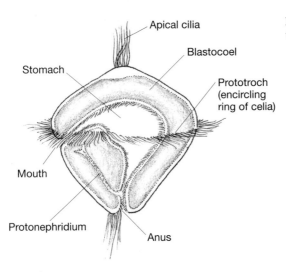

Figure 9.7   Major features of the trochophore larva.

into the shell before the foot, and by providing for better utilization of the incoming water stream.

THE SHELL

Although most slugs and nudibranchs do not construct shells, the majority of gastropods build a shell or conch that provides support for the visceral mass and space for retraction of the head and foot (the cephalopedal mass). Typically, the shell is a planispiral or helicoid (trochospiral) tube of regularly increasing diameter. The **apex** of the shell contains the first formed and smallest **whorl**, and each whorl consists of a complete 360° turn of the tube (see Fig. 9.8). The successively larger whorls are coiled around a central axis. Where the inner margins of the whorls become intergrown, a solid axial rod called the **columella** is formed. The surface of contact between whorls is expressed by a line called a **suture**. The last and largest whorl is termed the **body whorl**. It terminates at the opening or **aperture** from which the cephalopedal mass of the living animal may protrude. In some forms the aperture may be closed by a lid or **operculum**, which is attached to the foot. Withdrawal of the foot into the conch serves to position the operculum over the aperture.

As one examines collections of gastropod shells, it becomes evident that they exhibit two basic plans of coiling. Where the axis of coiling is horizontal, **patelliform** and **planispiral** conchs are produced (Fig. 9.8C,D). The latter may be discoidal or globular. The second and more frequently seen method involves helical coiling around a vertical axis (Fig. 9.8A,B). It has been suggested that helically coiled shells had the advantage of greater compactness over planispiral forms, and thus offered greater resistance to crushing. Most gastropods coil to the right (dextrally). The direction of coiling in most helically coiled conchs is determined by holding the conch with its apex directed upward and its aperture facing you. If the aperture is on your right, the conch is coiled dextrally; if on the left, it is sinistrally coiled.

Gastropod shells display a variety of ornamental features such as ridges, grooves, bumps, spines, and other configurations of aragonitic calcium carbonate arranged in more or less ordered fashion. This ornamentation or sculpture is useful in gastropod classification and identification. Even smooth shells possess growth lines that record incremental secretion of calcium carbonate along the edge of the mantle. Where transverse ridges or ribs cross spiral ridges of equal magnitude, the ornamentation is termed **reticulate**. Often the intersections of these two groups of ridges are surmounted by nodes, in which case the pattern is called **nodose-reticulate**.

Although torsion in gastropods primarily affected soft parts, this twisting also produced certain features in the conch itself. As we have seen, torsion brought the mantle cavity, which in the pre-torsion state was posterior, to the anterior end of the animal. Thus, the exhalent currents containing waste flowed directly onto the head of the gastropod. During the evolution of the Opisthobranchia, this unfastidious arrangement was nullified by detorsion. This reversed the original situation, although organs lost as a result of torsion in ancestral species were never regained. Other gastropods evolved various kinds of slits, reentrants, holes, and tubes that helped to ventilate the mantle cavity and provide separation of waste-bearing exhalent currents from inhalent

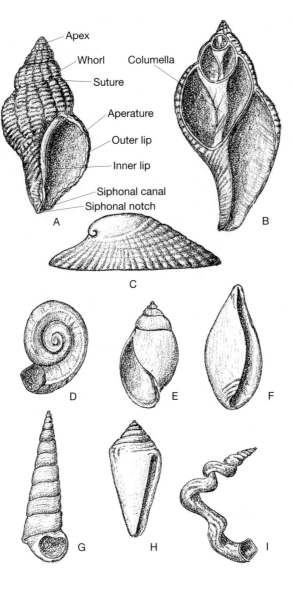

Figure 9.8   Variety in gastropod shells. (A) Descriptive terms for gastropod shells as exhibited by the oyster drill *Urosalpinx*, ×0.5. (B) *Urosalpinx* with portion of shell cut away to expose columella. (C) *Symmetrocapulus*, a Jurassic patelliform gastropod, ×2. (D) Nearly planispiral, discoidal shell characteristic of *Straparolus*, ×0.5. (E) Sinistrally coiled shell of *Physa*, ×0.5. (F) Convolute shell of *Acteonella*, ×2. (G) Turretted shell of *Turritella*, ×1. (H) Biconical shell of *Conus*, ×1. (I) *Vermetus*, which coils in an irregular manner in its adult stage, ×1.

ones. In many of the small cap-shaped gastropods, the tubelike exhalent siphon protrudes through a dorsal hole called a **trema** (plural, *tremata*), so that waste is discharged in a posterior direction away from the head. More commonly, waste is carried out through a cleft or **siphonal notch** in the aperture. As the animal grows, successive crescent-shaped increments of calcium carbonate are added to the perimeter of the notch so as to produce a trackway of former notch positions that is called a **selenizone** (Fig. 9.9A).

Another commonly seen device for separating exhalent and inhalent currents in gastropods is the **siphonal canal**, which partially protects the tubelike inhalent siphon

Figure 9.9   Archaeogastropoda. (A) *Mikadotrochus*, ×1. (B) *Maclurites*, ×0.8. (C)
Platyceras, ×0.8. (D) Left side and apertural views of *Bellerophon*, ×2. (E) *Micromphalus*, ×2.
(F) *Nacella*, ×1. (G) *Murchisonia*, ×2.5. (H) *Pleurotomaria*, ×1. (Redrawn from J. B. Knight et
al. In R. C. Moore (Ed.), *Treatise on Invertebrate Paleontology*, Pt. I. Boulder, CO, and
Lawrence, KS: Geological Society of America and University of Kansas Press, 1969.)

(see Fig. 9.4). The canal usually takes the form of a narrow fold in the conch that
extends downward from the base of the aperture (see Fig. 9.8A). Typically, the siphonal
canal is a feature of gastropods that lack selenizones or tremata.

MAJOR CATEGORIES OF GASTROPODS

Subclass Prosobranchiata   Members of the Subclass Prosobranchiata include
marine, freshwater, and terrestrial forms in which torsion is complete. As a result of the
torsion, nerve tracts are crossed, and the mantle cavity with its contained organs are
brought forward. One or two gills are located within the mantle cavity. Most forms
have a conch, and the conch is usually equipped with an operculum. In most species,

sexes are separate. The subclass is divided into three taxonomic orders: the Archaeogastropoda, Mesogastropoda, and Neogastropoda.

Order Archaeogastropoda (Early Cambrian to Holocene)   The Archaeo-gastropoda are herbivorous prosobranchs with two nephridia, two auricles, and bipectinate gill structure. Usually there are two such gills, but in some archaeogas-tropods only one persists. Shells may be planispiral or helical, although the latter are more common. In archaeogastropods, the aperture is not notched for the reception of an inhalent siphon. Also, gills are of the primitive type, consisting of flattened filaments projecting to either side of an axial support. The orders Mesogastropoda and Neogastropoda are believed to have been derived independently from the Archaeogastropoda. Archaeogastropods achieved their highest levels of diversity dur-ing the Devonian and Triassic. Representative genera (see Fig. 9.9) include *Bellerophon* (Ordovician to Triassic), *Maclurites* (Ordovician to Devonian), *Micromphalus* (Devonian to Mississippian), *Pleurotomaria* (Jurassic to Cretaceous), *Murchisonia* (Ordovician to Cretaceous), *Nacella* (Eocene to Holocene), and *Platyceras* (Silurian to Permian). The Order also includes the familiar abalone, *Haliotis*.

Order Mesogastropoda (Ordovician to Holocene)   Mesogastropods are prosobranchs with a mostly helical shell. Many are characterized by a definite notch for the inhalent siphon. Mesogastropods have only a single, monopectinate gill on the left side. The heart has only a single auricle, and only one nephridium remains. Sexes are separate, and reproduction usually involves internal fertilization. The classification of living mesogastropods is based largely on the characteristics of the radula, which consists of seven large teeth in a transverse row. Most mesogastropods are marine, but fewer numbers of freshwater and terrestrial forms are also recognized. Since their first appearance in the Early Ordovician, mesogastropods have shown an almost continu-ous increase in diversity, which appears to have been unaffected by mass extinctions that deciminated other benthic invertebrates. Among the more familiar genera (Fig. 9.10) are *Loxonema* (Ordovician to Holocene), *Viviparus* (Pennsylvanian to Holocene), *Turritella* (Oligocene to Holocene), *Crepidula* (Miocene to Holocene), and *Natica* (Paleocene to Holocene).

Order Neogastropoda (Cretaceous to Holocene)   Neogastropods develop distinct siphonal canals that are often remarkably elongate. Like the mesogas-tropods, they possess a single monopectinate gill, one auricle, and a single nephrid-ium. They differ in having a radula with three rather than seven teeth in a transverse row. Included here is the attractive, predatory *Murex* (Miocene to Holocene) shown in Figure 9.11, as well as the whelk *Buccinum*, (Oligocene to Holocene) (see Fig. 9.4), and the venomous *Conus* (Cretaceous to Holocene, depicted in Figure 9.8H).

Subclass Opisthobranchia (Mississippian to Holocene)   Members of the Subclass Opisthobranchia have one gill, one auricle, and one nephridium. Detorsion is characteristic of this group, and the mantle cavity with its contained organs is located on the animal's right side. As a consequence of detorsion, there has been a trend

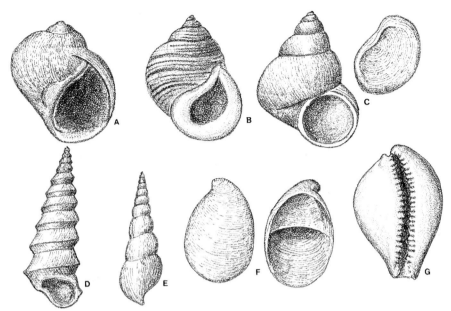

Figure 9.10   Mesogastropoda. (A) *Natica*. (B) *Littorina*. (C) Side view and operculum of *Viviparus*. (D) *Turritella*. (E) *Loxonema*. (F) *Crepidula* showing internal platform for attachment of muscles. (G) *Cypraea*, a genus in which spires are totally covered by an expanded lip. Magnification ×1.5. (Redrawn from J. B. Knight et al. In R. C. Moore (Ed.), *Treatise on Invertebrate Paleontology*, Pt. I. Boulder, CO, and Lawrence, KS: Geological Society of America and Universtiy of Kansas Press, 1969.)

toward reduction or loss of the shell and attainment of secondary bilateral symmetry. Species of opisthobranchiates, nearly all marine, number over a thousand. Included in this subclass are the brilliantly colored nudibranchs or "sea hares," and the planktonic pteropods in which the foot has been transformed into a pair of fins. The tiny shells of pteropods accumulate in vast quantities on certain parts of the ocean floor, comprising

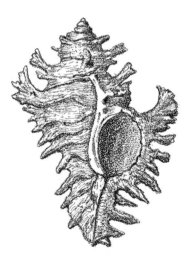

Figure 9.11   *Murex*, a predatory marine snail of the order Neogastropoda (shown one-third actual size).

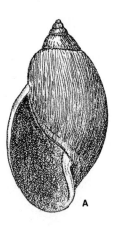

Figure 9.12   Opisthobranchia.
(A) *Acteonina*, ×1. (B) *Acteonella*, ×1.

what is termed "pteropod ooze." Fossil opisthobranchiates (Fig. 9.12) include *Acteonina* (Pennsylvanian to Jurassic), and *Acteonella* (Triassic to Late Cretaceous).

Subclass Pulmonata (Pennsylvanian to Holocene)   Biologists consider members of the Subclass Pulmonata (Fig. 9.13) to be the most advanced of all gastropods. The subclass includes not only land snails and slugs but also several genera that have secondarily returned to an aquatic environment. The pulmonates have no gills. Instead, part of the mantle cavity has been vascularized to function as a lung. In pulmonates that have become secondarily aquatic, this structure continues to serve in the exchange of gases, but with improved absorptive capacity resulting from infolding of the mantle wall. Some water dwellers continue to breathe air by periodically rising to the surface. Most pulmonates secrete a shell, although opercula for the shell are not developed. Fewer genera have either lost the shell or as in the land slug *Limax* have a tiny shell

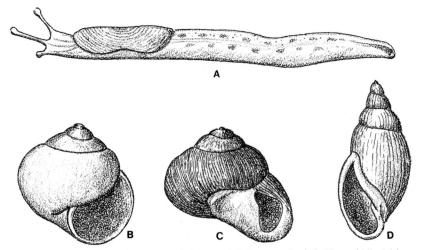

Figure 9.13   Pulmonata. (A) *Limax*. (B) *Helix*. (C) *Dawsonella*. (D) *Physa*. (All ×1.2.) (Redrawn from H. W. Shimer & R. R. Shrock, *North American Index Fossils*. New York: John Wiley & Sons, 1944.)

embedded in the mantle. Included in the Pulmonata are *Helix* (Pliocene to Holocene), *Dawsonella* (one of a diverse Pennsylvanian fauna of pulmonates), and *Physa* (Jurassic to Holocene) (Fig. 9.13).

GASTROPOD HABITS AND HABITATS

Gastropods are such a diverse group that it is difficult to generalize about their habits and environmental preferences. They are found in the ocean at all depths, on land from sea level to 6000 meters, and in all types of freshwater bodies. Gastropods include burrowers, crawlers, swimmers, floaters, and some that are sessile. Most of the extant marine gastropods show a preference for well-lighted, shallow areas of the continental shelves. By means of waves of muscular contractions that sweep along the foot, gastropods are able to move slowly across the surfaces of rocks and coral. Very small marine gastropods may travel by coordinated movement of cilia on the sole of the foot. Others, like *Natica*, are adapted for burrowing on soft, sandy substrates. The grip of some living gastropods on rock surfaces is often extraordinary. *Haliotis*, the large California abalone, can adhere to a hard substrate so tenaciously that it is dangerous to try to pry the animal loose with the hand lest one's fingers be held fast beneath the shell. Its broad shell offers minimal resistance to waves and currents when the animal is clamped against rocks. In similar fashion, other gastropods with limpetlike shells clamp securely to substrates as they graze. The adaptation protects these animals from heavy wave action in the intertidal and shallow subtidal waters they prefer. Gastropods with trochiform and turbinelike shells also tend to be epifaunal on hard substrates. Forms with conical shells like *Conus*, or with turret-shaped shells (*Turritella*), typically live partially buried in sediment.

In addition to benthic crawlers and burrowers, some gastropods live a pelagic existence. The most notable of these are the "sea butterflies" or pteropods of the Subclass Opisthobranchia and heteropods of the Subclass Mesogastropoda. Other species are pseudoplanktonic, floating on plant debris or on rafts of bubbles they have formed themselves.

Terrestrial gastropods lay down a mucous trail over which they move. They often seek shelter beneath logs and stones and are more active at night when dangers of desiccation are minimized.

The majority of gastropods are herbivorous, but carnivores and forms that live on organic debris are common as well. In general, the most active gastropods are carnivores. Predators like *Natica* and *Murex* drill holes in the shells of invertebrate prey and feed on the soft tissue inside. Their borings have been identified as trace fossils. Other carnivores feed on worms, sea urchins, and even fish. Many use an extendable proboscis to ingest their victims. The nudibranchs are predators on sea anemones, hydroids, corals, sponges, bryozoans, and fish eggs.

GASTROPODS THROUGH TIME

The earliest undisputed gastropods are Late Cambrian slit-bearing members of the Archaeogastropoda. The small shelly fossil *Aldanella* (Fig. 9.14) from Early Cambrian strata has been interpreted as a gastropod by some paleontologists, but as a worm by others. By Late Cambrian time, the class had begun a steady increase in diversity, which has continued down to the present. Archaeogastropods continue in abundance from the Ordovician through the Pennsylvanian (Fig. 9.15). Mesogastropods made steady progress during the Mesozoic and then increased significantly during the Cenozoic. Neogastropods are nearly confined to Cretaceous and younger strata. Opisthobranchs are recorded from the Mississippian to the Holocene, with the

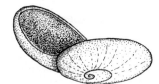

Figure 9.14 *Aldanella*, a tiny mollusk from the Tommotian stage of the Lower Cambrian. (Diameter of shell 1.5 mm.)

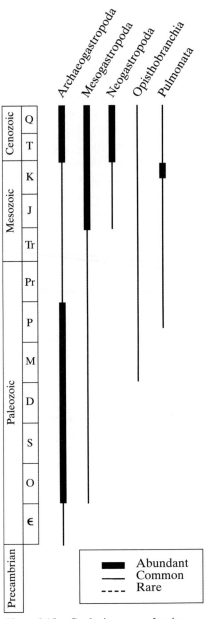

Figure 9.15 Geologic ranges of major groups of Gastropoda.

pteropods present since Eocene time. Suspected pteropods from older epochs are probably the remains of hyoliths. The pulmonates appear during the Pennsylvanian, but their most dramatic expansion came during the Cretaceous and continues to the present day. Pulmonates are useful in dating and correlating Pleistocene deposits.

CLASS CEPHALOPODA

The chambered nautilus (Fig. 9.16), cuttlefish, squid, and octopod are among the more familiar members of the Class Cephalopoda. The class includes some of the most complex of all mollusks. Cephalopods have a prominent head from which one or two circles of prehensile tentacles project. The head also bears a pair of highly efficient eyes, and a mouth equipped with parrotlike beaks and a radula (Fig. 9.17). The head and tentacles are located at the anterior end of the animal, and the shell or conch, if present, is carried posteriorly. Locomotion is accomplished by the expulsion of water through a muscular organ called the **hyponome**. In squid and cuttlefish, water is drawn into the mantle cavity by relaxation of mantle muscles. When the cavity is filled, these muscles contract, which increases pressure within the mantle cavity and forces water out through the hyponome. The cephalopod is jet-propelled in the opposite direction at a velocity controlled by the force with which the water is expelled. Streamlining also influences speed in swimming, as is apparent in the more rapid movement of the squid

Figure 9.16   The living nautiloid cephalopod *Nautilus*. (Courtesy of W. Bruce Saunders.)

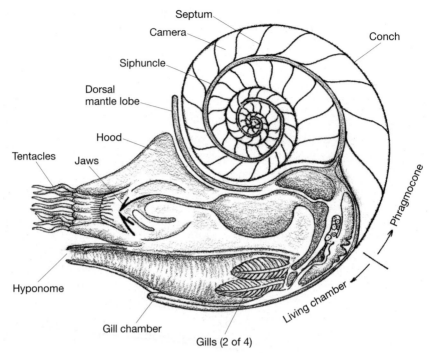

Figure 9.17   Diagrammatic section through the plane of coiling of *Nautilus*.

as compared to the chambered nautilus. Nautiloids also differ from squids in that the ejection of water through the hyponome results from retraction of the body into the conch with simultaneous contraction of hyponome muscles, rather than by mantle muscles alone.

The sexes are separate in cephalopods, and males and females exhibit sexual dimorphism. In the extinct ammonite cephalopods, the males were smaller, commonly retained juvenile ornamentation, and developed distinctive lateral projections on opposite sides of the aperture. These projections, termed **lappets**, helped the male position himself during copulation. Conchs of female ammonites lacked lappets and were usually larger.

In fossil cephalopods, it is important to recognize dimorphism in order to prevent error in the designation of species. Fortunately, there are tests that help one to avoid such errors. Because sexually dimorphic species do not appear until the individual has reached sexual maturity, the early or juvenile whorls of a conch should be the same in both males and females. This would establish the two sexes as members of the same species. One would also expect to find both morphologic forms of the dimorphic species in about a 1:1 ratio in rocks of the same age.

Fertilization of eggs in many cephalopods is preceded by a complex courtship ritual, and occurs as a result of head-on copulation. The male uses special erectable tentacles to insert a ball of sperm into the mantle cavity of the female. The eggs are then fertilized in the oviduct located within the mantle cavity or in a special sperm-containing structure deposited by the male in a receptacle around the mouth of the female. The embryo begins its development within the yolk-rich eggs. Aided by openings produced in the egg capsule by secretions from a hatching gland, tiny juvenile cephalopods emerge. There is no larval stage.

It is of at least passing interest that certain cephalopods have the distinction of having attained the largest size of any invertebrate group. The winner of the award for large size would be given to the giant squid, which may attain a length of 20 meters if tentacles are included in the measurement. Although Hollywood adventure films of several decades ago were fond of depicting giant octopods, actually these creatures seldom exceed a body diameter of more than about a third of a meter. Their tentacles, however, may extend to five times that measurement.

THE CEPHALOPOD SHELL

A completely developed external shell is common in fossil cephalopods, but among living groups this occurs only in the chambered nautilus. In cuttlefish and squid, the shell is internal and greatly reduced. Octopods lack a shell altogether. The female *Argonauta*, an octopod known more commonly as the "paper nautilus," secretes a paper-thin planispiral shell that is used as a brood chamber. The animal uses its arms to form the shell and the mantle sac as a base to shape the shell. Thus, the shell is an adult acquisition and not equivalent to the conchs of other cephalopods.

Upon seeing a cephalopod with a coiled external shell for the first time, novice fossil collectors might mistake the shell for that of a gastropod. Cephalopod shells, however, are divided into chambers (**camerae**) by partitions called **septa** and are most commonly planispiral. In contrast, the shells of gastropods are unchambered and usually helicoid. In those cephalopods having an external shell, the chambers are connected by a porous tube called the **siphuncle**, which joins the chambers to the animal. Although coiled shells are the predominant form throughout cephalopod evolution, straight, curved, uncoiled, and even trochoid forms are represented in the fossil record.

The **aperture** is the readily apparent opening at the anterior end of the shell, whereas the **apex** or first-formed part of the shell is posterior. At the mid-ventral margin of the aperture, there may be a reentrant known as the **hyponomic sinus** that accommodates the hyponome. The large, open, anterior space between the last-formed septum and the margin of the aperture is the **living chamber**. The chambered portion posterior to the living chamber is called the **phragmocone** (see Fig. 9.17). During growth, chambers change in shape, and these changes have taxonomic significance. For this reason, drawings of the cross-sections of whorls are often provided with descriptions of species as an aid to identification.

Except for the *siphuncle*, which extends backward through the conch, the cephalopod occupies only the living chamber. The siphuncle contains blood vessels, nerves, and mantle tissue, and it plays a key role in adjusting the gas and liquid content of chambers in the phragmocone. The siphuncle, which passes through the septa by way of perforations called **septal foramina** (Fig. 9.18), may be encased in a porous, spiculate, calcareous, or chitinocalcareous sheath called the **ectosiphuncle**. Two components of the ectosiphuncle are short tubelike extensions of the septal wall termed

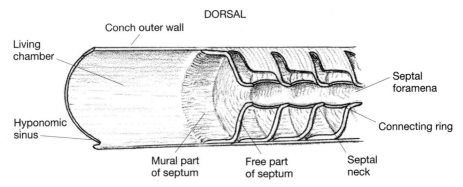

Figure 9.18   Cut-away view of an orthoconic fossil nautiloid shell showing internal features.

septal necks, and **connecting rings** that span the space between adjacent septa or septal necks.

Protection is a major function of the cephalopod shell. Living nautilids enhance the protection offered by the shell with a tough **hood** above the head. The hood closes the apertural opening as the animal withdraws completely into its living chamber. In addition to protection, the conch, along with the siphuncle, is important in providing neutral buoyancy in the ocean. The manner in which buoyancy adjustments are accomplished has been carefully studied in *Nautilus*. This interesting animal makes itself more buoyant through the use of gas-filled chambers in the shell. Salty liquid that filled the chambers shortly after the secretion of septal walls is drawn off by osmosis through the siphuncle. As the animal grows, its overall density increases because of the added growth of soft tissue and shell matter. There may also be a temporary gain in total density owing to heavy feeding. In either case, *Nautilus* is able to compensate for the increases and maintain neutral buoyancy by slowly removing small quantities of liquid from its chambers. The system appears to work very well down to a depth of about 240 meters.

Conversely, should the animal experience a loss of density through food deprivation or some other cause, the loss can be compensated by adding liquid to its shell chambers. These processes, however, are not instantaneous. It requires as long as 4 days to compensate for a gain or loss of only 5 grams. Buoyancy adjustments in *Nautilus* rather resemble those in submarines, albeit much slower. The equivalent of camerae in submarines are ballast tanks. When these tanks are empty and filled with air, the submarine's total density is less than that of water, and it will rise. It will descend when the tanks are filled. Between these two extremes, submarines cruise with neutral buoyancy resulting from partially filled ballast tanks.

In cephalopods with planispiral shells, the heavier body with head and soft organs lies below the lighter phragmocone with its partially air-filled camerae. This orientation places the center of gravity below the center of buoyancy and gives the animal vertical stability. However, not all shell-bearing cephalopods were coiled. Many early groups had long, straight, conical shells. If some adjustment were not made in such shells, the center of gravity would lie near the anterior end, and the center of buoyancy would be toward the apex. This would result in the animal being suspended in vertical orientation with head and tentacles hanging downward. Such a position might be suitable for a benthic animal feeding on other benthic organisms, but would be an unlikely orientation for an active swimmer. Evolutionary processes worked toward a solution in several ways. One solution, already mentioned, was the evolution of the coiled shell. Evidence that strong selective pressure for hydrostatic stabilty favored this solution can be seen in lineages of cephalopods that began their evolution with straight-shelled ancestors, progressed through forms with increasingly curved shells, and subsequently to tightly coiled planispiral shells.

In some groups of cephalopods that retained the straight conch, horizontal stability was provided by the weight of calcium carbonate secretions added to the posterior part of the conch (Fig. 9.19) These **siphuncular** and **cameral deposits** were thickest near the apex and along the ventral areas of the chambers. The effect of the deposits was to shift the center of buoyancy to a central position just above the center of gravity. A similar effect was achieved in some orthoconic (straight-shelled) cephalopods by crowding of septa.

Other straight-shelled Early Paleozoic cephalopods found another way to achieve a horizontal position while suspended in water. They evolved chambers on the dorsal part of the conch so that the heavier mass of soft tissue occupied a ventral position beneath the partially air-filled chambers (see Fig. 9.19C). Others modified this arrangement further by casting away the older apical part of the phragmocone and

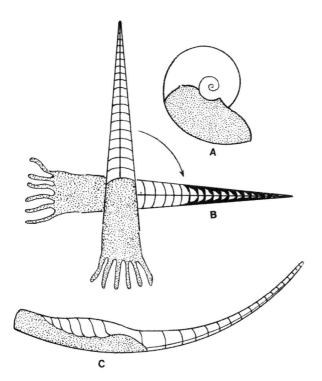

Figure 9.19   Three solutions for maintaining horizontal living orientation in cephalopods. One design, seen in *Nautilus* (A), employs planispiral coiling to place the animal's center of gravity below the center of buoyancy. In many cephalopods with straight (orthoceraconic) shells, horizontal stability was achieved by secreting counter-balancing deposits on posterior chamber walls and around the siphuncle (B). In *Ascoceras* (C), air-filled chambers are developed on the dorsal portion of the shell. On reaching maturity, the arcuate rear part of the shell is cast away, leaving only the living chamber. (Concept for *Nautilus* from R. H. Flower, Nautiloids of the Paleozoic, *Geological Society of America Memoir* 67(2):829–852, 1957.)

sealing off the opening with deposits of aragonite. The result was a conch with a peculiar bulky appearance. Apparently, many of these unstreamlined cephalopods abandoned active pelagic life for a benthic existence.

The two major subdivisions of shell-bearing cephalopods are the orders Nautiloidea and Ammonoidea. Among the former, shells are either exogastric or endogastric. The hyponomic sinus is found on the convex side of the shell in **exogastric** cephalopods. Thus, members of this group could coil planispirally without danger of the penultimate whorl impinging on the hyponome and interfering with its function. The hyponomic sinus of **endogastric** shells lies on the concave side of the shell. If tightly coiled, the penultimate whorl would interfere with the hyponome. It is probably for this reason that the shells of endogastric cephalopods are either straight or only slightly curved.

Cambrian and Ordovician cephalopods had mostly slightly curved, truncated, or straight shells, but coiled forms appeared later in the Paleozoic. Those termed **orthocones** had straight shells (Fig. 9.20), whereas **cyrtoceracones** had arcuate shells. **Gyroceracones** progressed through one or two volutions (spiral turns) but were open at the center. The first conchs to have whorls in complete contact are **tarphyceracones**. Further compactness in coiling produced **nautilicones**. If early whorls are still visible even with slight overlap of outer whorls over inner ones, such conchs are further designated as **evolute**. Where the overlap is sufficient to cover much or all of earlier whorls, the shell is said to be **involute**. In addition, one finds shells developed in the form of a helicoid spiral (**trochoceracones**), blunt shells with truncated phragmocones (**brevicones**), and shells that are planispirally coiled in their early stages and have subsequently grown straight.

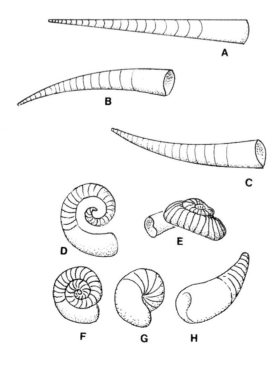

Figure 9.20   Variations in shapes of nautiloid cephalopod shells: (A) orthoceracone; (B) endogastric cyrtoceracone; (C) exogastric cyrtoceracone; (D) gyroceracone; (E) trochoceracone; (F) tarphyceracone; (G) nautilicone (involute); and (H) brevicone.

Conchs of cephalopods may be ornamented with color bands, ribbing, keels, and nodes. *Nautilus* has wavy brown bands on its upper surface. Because of these color bands, predators swimming above *Nautilus* would find the animal more difficult to see. The underside of *Nautilus* is white, which blends with the daylight whiteness of surface water when viewed by predators from below. Nodes and ribs on ancient cephalopods apparently functioned to add strength to the shell, and keels may have had a hydrodynamic function, somewhat like the dorsal fin on a fish.

SEPTA AND SUTURES

Of all the features of cephalopod shells, few are more important in phylogenetic studies than **septa** (see Figs. 9.17 and 9.18) and the sutures associated with septa. A **suture** is a line formed at the juncture of a septum and the interior of the conch wall. If the borders of the septa are relatively straight, the contact of that straight edge with the inside wall of the conch will be similarly straight. Conversely, if the septal margin is sinuous, or perhaps sinuous with secondary fluting, the suture will have a similar configuration. Because suture lines are features of the shell's interior, they are not visible on the outside of a well-preserved shell. As a result, it is often necessary to peel away external layers on part of the conch wall in order to expose the suture pattern. Alternatively, suture patterns may be evident on internal molds or on cephalopods that have lost the outer layers of the wall as a result of postdepositional decomposition. With experience, one can identify the suture pattern even when only one or two sutures from one of the chambers are exposed.

The simplest sutures occur in nautiloid cephalopods like *Nautilus* (see Figs. 9.16 and 9.17). These are only slightly undulatory. The second group of cephalopods, known as *ammonoids*, developed more elaborate sutures. Most ammonoid shells were composed of the calcium carbonate mineral aragonite. Because aragonite is chemically less stable than calcite, the conch walls of ammonites are often partially decomposed so that sutures are visible on the surface of the fossil. Internal molds exhibiting sutures are another frequently encountered type of preservation in ammonoids. Tracings of the variously waved, wrinkled, or zigzag ammonoid sutures are essential for identification. One way that these may be made is to cover part of a whorl tightly with plastic wrap and trace along the suture line with a suitable pen. The **sutural diagram** (Fig. 9.21) made in this way can then be transferred to drawing paper. The parts of the suture that are convex toward the aperture are called **saddles**. **Lobes** curve convexly toward the apex of the conch. Saddles and lobes alternate along the sides of the whorl until they disappear beneath the **umbilical seam**, formed where the two whorls are in contact. That part of the suture that is exposed is called the **external suture**, whereas the part beneath the preceding whorl is the **internal suture**. On the sutural diagram, the direction toward the aperture is indicated by an arrow extending along the imaginary midventral line.

Comparison of sutural diagrams from the entire range of ammonoids indicates sutures can be divided into four groups (see Fig. 9.21). **Agoniatitic sutures** have only a few simple undivided saddles and lobes. In these sutures there is always a narrow midventral lobe and broad lateral lobe. **Goniatitic sutures**, named after the Mississippian genus *Goniatites*, consist of simple undulatory or zigzag saddles and lobes.

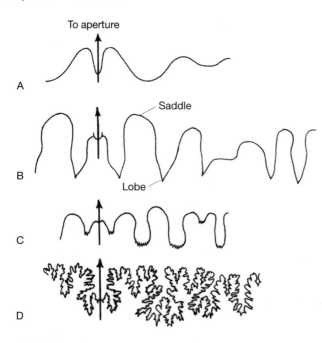

To aperture

A

Saddle

B

Lobe

C

D

Figure 9.21 Suture patterns in (A) agoniatitic, (B) goniatitic, (C) ceratitic, and (D) ammonitic cephalopods. Arrows point toward the aperture.

Modification of this goniatitic pattern so that lobes are serrated or frilled provides **ceratitic sutures**. In the still more complex **ammonitic sutures**, both saddles and lobes are intricately serrated (see chapter opener photograph).

Ammonoids with goniatitic sutures lived mostly during the late Paleozoic. Those with ceratitic sutures are mostly Triassic, and ammonitic ammonoids are found primarily in Jurassic and Cretaceous rocks. One must, however, use other criteria along with suture patterns to substantiate identifications, for overlap and repetition of suture types have occurred several times in ammonoid evolution. As a result, one finds goniatitic and ceratitic sutures that would normally be characteristic of cephalopods from Late Paleozoic and Triassic rocks, respectively, in rocks of Cretaceous age. To understand better the evolution of these convergent patterns, it is often helpful to examine suture patterns in specimens from the earliest to the latest chambers, for changes during growth (ontogenetic changes) may reflect phylogenetic relationships.

One can hardly examine suture patterns in ammonoids without speculating about the reason for their increasingly complex patterns. Many hypotheses have been proposed, including a suggestion that the convolutions provided secure attachment for soft parts. Those questioning that hypothesis argue that convolutions for the attachment of soft tissue would be more logically located over the center of a septum, because tissue along the sharply angled edge would already be secure. Another hypothesis proposes that the convolutions provided places for the attachment of muscles needed to pull the body backward against air trapped in front of the last septum. This action would compress the air and affect buoyancy. The wrinkles and undulations, however, do not resemble muscle attachments seen in nautilids.

A third hypothesis suggests that convolutions of septal margins strengthened the shell and thereby reduced the hazard of conch implosion resulting from hydrostatic pressure. Septa strengthened a cephalopod conch in much the same way as do the bulkheads of submarines. It has been observed that the depth at which implosion will occur in *Nautilus* is about 750 to 800 meters, and that shell failure occurs as septa are sheared away from the shell wall near the suture. Septal fluting lengthens the line of support along the interior of the conch wall. This greatly reduces the amount of external compressive force that each increment of the septal edge needs to bear. If the convoluted septa are also closely spaced, unsupported wall area is minimized. Two characteristics of ammonoid shells favor this "support hypothesis."

First, the outer wall in ammonoids is relatively thin. It is, in fact, distinctly thinner than the walls of nautiloid conchs. Perhaps convoluted septal margins in some ammonoids compensated for their having thinner walls. Nautiloids, with their thicker walls, lack septal fluting. Second, the outer wall of some ammonitic ammonoids has been reported as not increasing in thickness toward the aperture at a rate equivalent to the increase in chamber volume. Presumably, to compensate, the thin-walled anterior chambers are supported by septa that are not only complexly fluted, but also more closely spaced. One further notices that the cross-sections of ammonitic ammonoid whorls deviate greatly from hemispherical shape, which would have the greatest strength. A hemispherical whorl shape is characteristic of most members of the Nautiloidea.

The hypothesis that septal folding and fluting provides strength against implosion was introduced by William Buckland in 1836. Recent studies by W. Bruce Saunders, however, casts an element of doubt on Buckland's venerable hypothesis. Saunders studied 49 Paleozoic ammonoids and found no inverse relationship between shell thickness and increased sutural complexity. Possibly, septal fluting in ammonoids was related to retention and transport of chamber liquids so as to facilitate changes in buoyancy, and was not directly related to shell strength.

MAJOR CATEGORIES OF CEPHALOPODS

Debate about the taxonomic importance of various morphologic features and uncertainties caused by evolutionary convergence among families make the classification of cephalopods difficult. A classification that is useful at a preliminary level divides the Class Cephalopoda into six subclasses on the basis of conch shape, suture configuration, and the size and location of the siphuncle. The subclasses are the Nautiloidea, Endoceratoidea, Actinoceratoidea, Bactritoidea, Ammonoidea, and Coleoidea.

Nautiloidea (Cambrian to Recent) Cephalopods of the Subclass Nautiloidea possess orthoconic or coiled external conchs, typically straight or only slightly sinuous sutures, and siphuncles of moderate size. In early nautiloids, the siphuncle is located near the ventral margin, but assumes a central or subcentral position in later forms. **Septal necks** (funnel-like, backward extensions of the septa along the siphuncle) are cylindrical and directed toward the apex. In fossil nautiloids, prominent **connecting rings** link each septal neck with the previous one. Many early straight-shelled nautiloids secreted thick cameral deposits in posterior chambers.

The Nautiloidea were the first cephalopods. Following their appearance in the Late Cambrian, they expanded rapidly, so that by the Middle Ordovician all orders of the subclass were present except the Nautilida, which appeared in the Early Devonian.

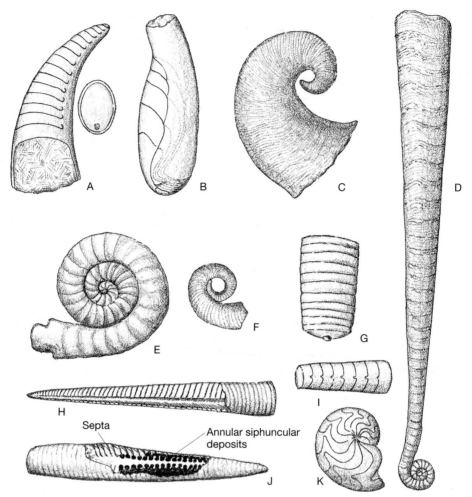

Figure 9.22 Some nautiloid cephalopods. (A) *Plectronoceras*,×2. (B) *Ascoceras*,×0.5.
(C) *Phragmoceras*,×0.5. (D) *Lituites*,×0.5. (E) *Ophioceras*,×2.7. (F) *Trochoceras*,×0.2. (G)
Michelinoceras,×0.7. (H) *Endoceras*,×0.2. (I) *Bactrites*,×1. (J) *Actinoceras*,×0.3. (K) *Aturia*,×0.3.

The only surviving genus of the Nautiloidea is *Nautilus*, of which only a few species
remain.

In perusing the fossil record of nautiloids, it is apparent that the oldest groups
had straight or gently arcuate shells, and that geologically younger forms often had
coiled shells. The oldest nautiloid known, for example, is the slightly arcuate
Plectronoceras (Fig. 9.22A), found only in three localities in northeastern China. Rapid
evolution during the Ordovician and Silurian produced an abundance of genera with
orthoconic shells like *Michelinoceras* (Fig. 9.22G) as well as forms that shed the early
portion of their phragmocones, such as *Ascoceras*. Exogastric, arcuate forms with variously restricted apertures were also abundant at that time. *Phragmoceras* (Fig. 9.22C)
was an enrolled endogastric nautiloid that became abundant during the Silurian.

Some examples of nautiloids that were initially coiled but then grew straight are *Lituites* and *Ophioceras*, both of the Devonian. In addition, there are strongly coiled nautiloids that first appear in Ordovician rocks, but become somewhat more numerous near the Silurian–Devonian boundary. These include *Trochoceras* (Devonian), *Eutrephoceras* (Cretaceous to Eocene), *Aturia* (Paleocene to Miocene), and of course the living chambered *Nautilus*. Some of the extinct nautiloids display sutures with undulatory patterns that resemble goniatitic sutures. Other shell features, however, confirm their nautiloid status. *Aturia* (see Fig. 9.22K), for example, displays such a sutural pattern and lived long after the last goniatitic ammonoid had become extinct.

Endoceratoidea (Ordovician to Silurian) Endoceratoids are medium to large cephalopods with mostly straight (**orthoconic**) shells. Fewer genera have slightly curved (**cyrtoconic**) shells. Members of this subclass possess large, subcentral (ventral) siphuncles. In many, the siphuncle occupies as much as one-fourth of the conch diameter. Septal necks are cylindrical, connecting rings distinct, and there is usually a series of invaginated cones called **endocones** that extend down the rear portions of the siphuncular tube. Cameral deposits do not occur in this group. Many Ordovician endoceratoids were truly giants, reaching lengths in excess of 9 meters. *Endoceras* (Fig. 9.22H) is a representative Ordovician member of the Endoceratoidea.

Actinoceratoidea (Ordovician to Carboniferous) Most members of this subclass have straight conchs, although a few genera have cyrtoconic shells. As was the case for endoceratoids, the siphuncles of actinoceratoids are very large. Cameral deposits are usually present, and the siphuncular tube bulges outward between septa. The siphuncle also contains radiating canals termed **endosiphuncular canals**. *Actinoceras* (see Fig. 9.22J) and *Armenoceras*, both obtained from Ordovician and Silurian sections, are representative genera.

Bactritoidea (Devonian to Triassic) The bactritoids are an interesting group having some features suggesting phylogenetic affinities to ammonoids, and others that are clearly characteristic of nautiloids. Indeed, they may have evolved from orthoceraconic nautiloids and then produced the ancestral stock for the ammonoids. Most bactritoids are small, have orthoconic or only slightly arcuate shells, and have inflated initial chambers (**protoconchs**). The siphuncle is small and located ventrally. Sutures are straight except for a distinct mid-ventral lobe. Neither siphuncular nor cameral deposits occur in members of this subclass. *Bactrites* (see Fig. 9.22I), for which the subclass was named, is a representative genus.

Ammonoidea (Devonian to Cretaceous) In terms of their value to biostratigraphy, the Subclass Ammonoidea surpasses all other suborders of cephalopods. They are particularly valuable for correlating Mesozoic strata, where their global distribution, rapid rate of evolution, and abundance have permitted delineation of biostratigraphic zones representing increments of geologic time as small as a million years. It is of some interest that the presence of ammonoid fossils in the rocks of Europe did not go unnoticed by our prehistoric ancestors. Fossil ammonoids have been unearthed among the artifacts of cave dwellers. The name *ammonoid* is taken from Ammon, an ancient Egyptian god who regarded the ram as a divine creature. Ammon was portrayed by a ram's head bearing spirally coiled horns that resemble the planispiral evolute shells of many ammonoids.

It is unfortunate that a few species of ammonoids did not survive beyond the Cretaceous, for without living representatives, inferences about soft parts of ammonoids are necessarily based on skeletal features and comparison with the living nonammonoid, *Nautilus*. Although misinterpretation can result from comparison of an extinct group of one subclass to living forms of another, most paleontologists believe that the soft parts of ammonoids were at least generally similar to those of nautiloids. With regard to skeletal parts, however, there are many features that are distinctively ammonoid. The radula is different in ammonoids, and, of course, the sutures are more complicated. Furthermore, the sutural pattern is of fundamental importance in the identification of ammonoids, whereas conch shape and siphuncular features have greater importance in nautiloids. Unlike the thick, complex siphuncles of nautiloids, the siphuncles in ammonoids are usually thin, simple, devoid of heavy siphuncular deposits, and are located near the ventral margin of the shell. (An exception occurs in the ammonoid order Clymeniida, which has septal necks and a siphuncle located along the dorsal margin of the conch.)

Other ammonoid features that help one to distinguish them from nautiloids are septa that are usually convex toward the aperture (not concave as in nautiloids), the presence in some groups of extensions of the two sides of the aperture called **lappets**, and a greater development of ornamental features such as ribs, knobs, tubercles, and keels. Some of these ammonoid features may have served to strengthen the shell, to obscure conch outlines for camouflage, or to provide resistance to pitch and roll.

The ammonoid siphuncle begins its development as a small bulb of tissue called the **caecum**, which is developed at the apertural end of the young cephalopod's bulbous protoconch. Subsequent growth of the conch is in the exogastric mode. Coiling, with some significant exceptions, is primarily planispiral. This contrasts to the many orthoconic and cyrtoconic shells seen in Paleozoic nautiloids.

Among the exceptions to the planispiral shell in ammonoids are species that begin to coil planispirally during their initial growth period, and then change to a straight mode of shell growth during most of their adult life. Other ammonoids developed helical shells that were either loosely or tightly coiled. Perhaps the most irregular pattern of growth occurred in *Nipponites*, an ammonoid that coiled in a succession of U-turns in three dimensions producing a shell aptly described as a tangle.

Occasionally, fossil collections from strata suspected of containing ammonoids include calcareous or horny plates that were originally interpreted as ammonoid opercula. Some were a paired set of plates, and these have been named **aptychi**. Others consisted of a single plate, and these are termed **anaptychi** (Fig. 9.23). Study of aptychi and

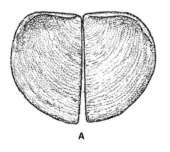

Figure 9.23   Ammonoid aptychus (A) and anaptychus (B).

A   B

anaptychi by Ulrich Lehman indicates that these structures are not opercula at all. Anaptychi are lower jaws of ammonoids that close upon upper jaws composed of the same material. The anaptychi were V-shaped in life, but have been flattened during fossilization, with the upper jaw element crushed on top. Aptychi also appear to have been part of ammonoid jaws. They apparently grew as a pair of calcareous plates on the external surface of the lower jaw. It is of interest that the anaptychi are the older of the two structures. They are known from rocks of Late Devonian to Early Jurassic age. The shorter-lived aptychi first occur in the Lower Jurassic, and disappear by the end of the Cretaceous. Often aptychi and anaptychi are found in strata that are devoid of ammonoid shells. The reason may be that they are more resistant to dissolution than are the aragonite conchs of the ammonoids with which they were associated in life.

The diversity and abundance of ammonoids (Fig. 9.24A-J) make the task of selecting a limited number of representatives difficult. The rich fossil record can only be suggested by describing a few of the more familiar genera from some of the better-known orders. Among the ammonoids with primarily goniatitic sutures (the so-called goniatites) are the orders Anarcestida, Clymeniida, and Goniatitida. The Anarcestida, represented by *Agoniatites*, lived only during the Devonian, and they are considered the group of ammonoids from which all later ammonoids sprang. The Clymeniida are unique among ammonoids in that their siphuncles have shifted from a ventral position in juvenile chambers to a distinctly dorsal position in adult chambers. This indicates that the clymeniids evolved from earlier ammonoids rather than from a branch of the nautiloids. *Clymenia* is representative of this Devonian order. The Goniatitida are characterized by an increase in the number of sutural lobes, and the lobes are distinctively angular. *Tornoceras* (Devonian) is a representative member of the Goniatitida. Ceratitic ammonoids can be classified into the orders Prolecanitida and Certatitida. The former are compressed "ceratites" that arose in Late Devonian and disappeared by Triassic time. *Prolecanites* (Mississippian) is representative (Fig. 9.24D). Ceratitida are the more progressive of the two orders of ceratitic ammonoids. Members have well-developed ceratitic sutures on all lobes, and shells are often ornamented by heavy transverse ribs that presumably provided additional strength to the shell. *Meekoceras* (Triassic) is representative of the Ceratitida.

Ammonoids loosely called "ammonites" include Triassic to Cretaceous cephalopods that are grouped within four "orders": the Phylloceratida, Lytoceratida, Ancyloceratida, and Ammonitida. The sutures of members of these orders are fluted on both saddles and lobes. Most are involute, but a few lineages became secondarily uncoiled. The characteristic feature of the Phylloceratida are their leaflike (phylloid) sutures, as seen in *Phylloceras* (Triassic to Cretaceous). The shells of phylloceratids are usually small and either smooth or only weakly ornamented.

Lytoceratine ammonites show a tendency to become evolute, as in *Scaphites* (Cretaceous), or to become uncoiled, as in *Baculites* (Cretaceous). In the lytoceratines, minor sutural elements of lobes and saddles tend to be more angular than in the phylloceratids.

The Ancyloceratida include the aberrant Cretaceous heteromorphs mentioned earlier. Although some of their shell forms seem awkward to us, the ancyloceratids were a successful group that would not have been able to evolve their seemingly bizarre shapes if such configurations were not in some way beneficial. Many of these

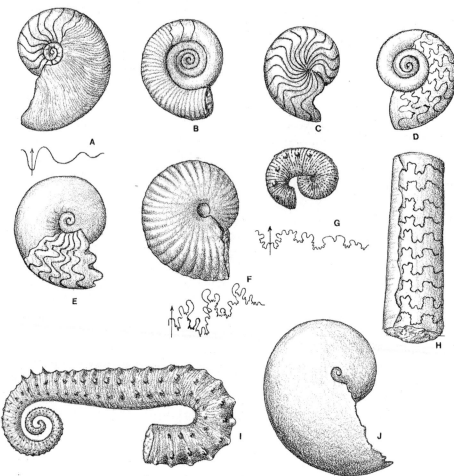

Figure 9.24   Ammonoids. (A) *Agoniatites* (with suture pattern), ×0.35. (B) *Clymenia*, ×0.35. (C) *Tornoceras*, ×0.7. (D) *Prolecanites*, ×1.2. (E) *Meekoceras*, ×0.7. (F) *Phylloceras* (with suture pattern), ×0.7. (G) *Scaphites* (with suture pattern), ×0.8. (H) *Baculites*, ×0.4. (I) *Ancyloceras*, ×0.12. (J) *Plancenticeras*, ×0.7. (Redrawn from W. J. Arkell et al. In R. C. Moore (Ed.), *Treatise on Invertebrate Paleontology*, Pt. L, Mollusca 4. Boulder, CO, and Lawrence, KS: Geological Society of America and University of Kansas Press, 1957.)

seemingly bizarre forms were probably adaptations for providing buoyancy control and stability under particular feeding and swimming strategies. Some ancyloceratids may have lived near the surface feeding on plankton. Others had benthic habits, and crawled on the oean floor or cruised just above it.

Ancyloceras (Jurassic to Cretaceous) is an example of the Ancyloceratida (Fig. 9.24I). In *Ancyloceras* the shell begins as an open spiral in which whorls are not in contact with one another. The shell then becomes straight, and finally turns 180° to form a living chamber that is directed back toward the apex.

The most important of the Subclass Ammonoidea is the Ammonitida. Included here are thousands of species of ammonoids having complex sutures and usually involute shells. Many bear heavy ornamentation of nodes, transverse ribs, spines, and keels, whereas others are sleek and discoidal like the Cretaceous ammonite *Placenticeras* (Fig. 9.24J). Other common genera are *Cardioceras* and *Harpoceras* from the Jurassic, and *Mortoniceras* and *Acanthoceras* from the Cretaceous.

Coleoidea (Mississippian to Holocene)   Except for the few surviving species of *Nautilus*, all of today's cephalopods are members of the Coleoidea. In coleoids, the shell is either internal or absent. At least a few of the tentacles in these cephalopods bear suckers, the marks of which are sometimes seen on the skins of whales. Also, all members of the subclass have one pair of gills and one pair of nephridia. The coleoids include the extinct belemnites as well as squid, cuttlefish, and octopods. Of these coleoids, belemnites find the most use in biostratigraphy. Their remains are found in rocks as old as Mississippian, but they are particularly common in Jurassic and Cretaceous rocks. Ancient people living in areas where Mesozoic rocks were exposed at the surface noticed belemnite fossils, and in at least one instance they interpreted them as the pointed tips of thunderbolts hurled from the sky by the gods.

The belemnite skeleton is a structure that is enclosed within a squidlike body, rather like the cuttlebone of the living cuttlefish. This internal shell has some similarity to the conch of an orthoconic nautiloid. At the posterior end is a forwardly directed dorsal extension of the shell termed a **proostracum** (Fig. 9.25). Beneath the proostracum and extending backward toward the apex is a chambered section that, as in cephalopods with external shells, is called the phragmocone. A siphuncle extends along the ventral margin of the **phragmocone**. The final major component of the skeleton is the solid, cigar-shaped **guard**. The guard is constructed of concentric layers of radiating calcite crystals.

The structure of the belemnite internal shell suggests that it is an evolutionary derivative of an orthoconic cephalopod resembling *Bactrites* (see Fig. 9.22I). Evolution of belemnites apparently involved a reinforcement of the bactritoid phragmocone by the addition of a guard, and subsequent envelopment of the entire shell in soft tissue. With the lighter phragmocone centrally located and separating the heavier head at one end from the dense guard at the other, horizontal orientation in water was facilitated. The oldest-known belemnite is *Eobelemnites* from Mississippian rocks of Utah and Oklahoma. *Pachyteuthis* from the Jurassic, and the Cretaceous genus *Belemnitella* are well-established guide fossils.

Squid are coleoids that first appeared in the Jurassic and are familiar fast swimmers in today's seas. They have ten arms, two of which are longer and more complex. The body is elongate and the internal shell is reduced to a flattened, unchambered plate. With an efficient jet propulsion system and ever-active fin, squids can dart rapidly, turn suddenly, or cruise along in a relatively sedate manner. Competition with

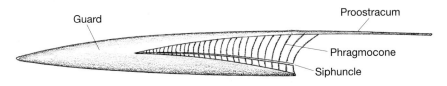

Figure 9.25   Simplified diagram of a belemnite showing the proostracum, phragmocone, and guard.

fish may have provided the selective pressures that led to the evolution of adaptations for speed and aquatic agility. The "flying squid" can even jet itself out of the water for distances of over 4 meters. *Loligo* is perhaps the most familiar living squid, but the most awesome is *Architeuthis*, a giant of the North Atlantic that achieves a total body length of 16 meters, including tentacles that are over 6 meters long. *Architeuthis* provides an occasional meal for sperm whales, although the reverse may also occur.

Like the squids, cuttlefish have ten arms. The body is relatively broad and has lateral fins. Internally, the cuttlebone retains a somewhat obscure chambered portion. The internal shell of the cuttlefish is sometimes placed in the cages of pet birds, presumably to satisfy their need for calcium.

Cuttlefish first appear in rocks of Late Jurassic age. They are represented today by *Sepia*, which is famous for its ability to eject black ink and thereby irritate or confuse predators. (Some octopods and squids also have this capability.) Cuttlefish are versatile swimmers, but not as fast as the more streamlined squid. When not darting rapidly, they can be observed moving gracefully along, propelled by the undulatory motion of paired lateral fins. The spaces between their delicate septa contain water and gas. By regulating the ratio of one to the other, they can vary their buoyancy. Mostly, cuttlefish dwell near the seafloor in shallow areas, although the deep-water form *Spirula* lives at depths of about 1000 meters.

The coleoid trend in shell reduction culminates in the octopods, most of which have no shell at all. Exceptions are the extinct genus *Palaeoctopus*, which retained the mere vestige of a shell, and the living dimorphic octopod *Argonauta*. As described earlier, *Argonauta* is known for the paper-thin unchambered shell secreted by the female for use as a brood pouch (or sometimes to carry about the small male).

Unlike other coleoids, octopods have eight arms, all of which bear suckers. The arms extend from a somewhat globular body. Margins of the mantle are fused along the top and sides of the head. In bottom-dwelling octopods, the arms, with their adhesive suckers, are used to pull the animal along or to allow it to cling tightly to objects on the seafloor. The arms are also well provided with chemoreceptor and tactile cells.

People have always found octopods intriguing animals. When octopods are part of the dinner menu, they can be a gourmet's delight, and fictional adventure stories often ascribe sinister and aggressive behavior to octopods. Actually, most octopods are timid animals that prefer to avoid contact with humans in scuba gear. Many have benthic habits, living in holes and crevices where they wait for a passing meal, or moving about in a graceful half-crawl, half-swim in search of food. Many are venomous, and are able to paralyze their prey. In this regard, the blue-ringed octopod of the Indo-Pacific has caused human fatalities. Whereas the more familiar octopods are benthic, there are also a number of pelagic families. These have webbed, umbrellalike arms. Pulsating movements of the arms are a means of swimming, but jet propulsion is also employed.

CEPHALOPODS THROUGH TIME

Cephalopods make their earliest appearance during the Late Cambrian with forms like *Plectronoceras*. These small cephalopods with arcuate conchs and marginally located siphuncles established the base for rapid diversification of the larger

Nautiloidea, Endoceratoidea, and Actinoceratoidea during the Ordovician (Fig. 9.26). The bactritoids do not exhibit marked diversification during the Paleozoic, but are nevertheless important as the probable ancestors of ammonites and belemnites. As for the Nautiloidea, all but one order disappeared by the end of the Triassic. The survivors are members of the Order Nautilida, of which only *Nautilus* remains today.

Expansion of the Ammonoidea began in the Early Devonian, and was punctuated by a series of crises throughout the remainder of the period. During the Carboniferous and Permian, however, ammonoids experienced a steady increase in diversity and abundance. This interval of relative stability ended with the terminal Permian episode of mass extinction. Survivors of the crisis expanded again during the Triassic, as ammonoids diversified into the largest number of families in their entire history. Over 80 families of ammonoids with ceratitic sutures lived during the Triassic, which consequently has been called "The Age of Ceratites." The success of the ceratitic ammonoids ended, however, when they were wiped out near the end of the Triassic. The Phylloceratida, however, survived this Late Triassic crisis. Thereafter, and until the end of the Cretaceous, ammonoids with complex ammonitic sutures predominated. In terms of abundance, diversity, and geographic distribution, the ammonites were extraordinarily successful. That success ended near the conclusion of the Cretaceous when ammonites vanished along with numerous other families of marine invertebrates as well as many marine and terrestrial vertebrates.

The earliest members of the Coleoidea were the belemnites. They made their debut in the Late Mississippian and expanded during the Jurassic and Cretaceous. By early in the Tertiary, these cephalopods had become extinct. Surviving coleoids are the squids, cuttlefish, and octopods. Of these three groups, the squids and cuttlefish appear in the Jurassic and today are the most numerous of the subclass. Octopods made their appearance at about the same time that the ammonoids disappeared late in the Cretaceous.

CEPHALOPOD HABITS AND HABITATS

Inferences about the habits of extinct cephalopods are necessarily based on conch form, nature of enclosing sediment, and comparisons with living species of *Nautilus*. However, *Nautilus* is obviously different from the arcuate or orthoconic nautiloids of the Paleozoic. The smaller and lighter of these early orthoconic cephalopods were probably good swimmers with habits somewhat like those of present-day squids. The larger and heavier forms were not as active, and probably lived on the seafloor. Their fossil remains are found in association with gastropods, corals, and brachiopods that inhabited nearshore, shallow environments.

The coiled ammonoids would appear to have habits more closely resembling those of the planispirally coiled *Nautilus* than the straight or arcuate Paleozoic nautiloids. We know that *Nautilus* lives today in the moderately shallow water around coral reefs and islands of the South Pacific. The distribution of living forms, however, is far more restricted than the distribution of the shells of *Nautilus*, for their empty shells drift for great distances and are ultimately deposited at locations different from those in which they lived. Thus, when one finds a fossil cephalopod shell, it may not indicate that the animal lived in the area of deposition. The presence of shells of juvenile forms

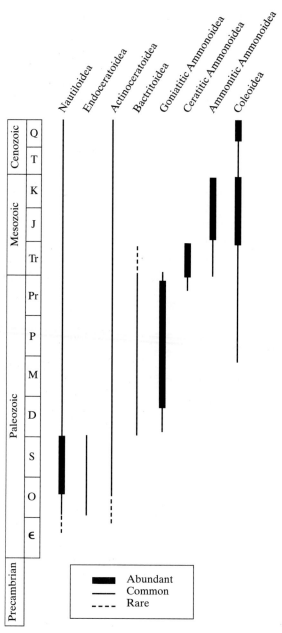

Figure 9.26   Geologic ranges of major groups of cephalopods.

along with those of adults is often strong evidence that the fossils were indeed buried near their actual habitat.

Because they are such a splendidly varied group, it is difficult to generalize about ammonoid habits. It has been suggested that ammonoids with thick and robust conchs were more prevalent in shallower waters, whereas smoother, bulkier species predominated in the deeper shelf areas. A few rather extraordinary discoveries of stomach contents indicate that at least some ammonoids fed on smaller mollusks, foraminifers, small crustaceans, and echinoderms. Based on the characteristics of their jaws, many simply shoveled food into their mouths as they coasted along just above the seafloor. As they fed upon smaller creatures, they were in turn often the victims of predatory marine reptiles, fish, and large crustaceans. Evidence of predation by crabs is suggested by the manner in which the living chambers of some ammonoid conchs are cut open, as if scissors (crab claws?) had been used. The giant marine lizard *Mosasaurus* was apparently also a major predator. One large specimen of *Placenticeras*, taken from Cretaceous rocks of South Dakota, bears tooth punctures precisely matching the battery of pointed teeth in *Mosasaurus horridus*.

CLASS BIVALVIA (PELECYPODA)

Members of the Class Bivalvia include such familiar animals as clams, scallops, and oysters, all of which have been a significant part of the human food supply for millennia. Bivalves have laterally compressed bodies and a shell composed of two valves hinged along the dorsal margin. They lack the distinct head that is characteristic of the gastropoda and cephalopoda. The terms *pelecypoda* and *lamellibranchia* have also been used for this class. The former term means "hatchet foot'" and refers to the shape of the foot in many bivalves, whereas the latter term refers to their layered gills. The gills are unusually large in bivalves, and in most species they are as important in food gathering as in gas exchange. The Bivalvia are primarily marine invertebrates, but some have successfully adapted to freshwater environments. They range in size from tiny clams no bigger than a pinhead, to the huge clam *Tridacnas*, which grows to a diameter in excess of 2 meters and weighs several hundred pounds. The Bivalvia range from the Cambrian to the present day.

THE LIVING ANIMAL

In most bivalves, the two valves are similar (**equivalved**), but some groups such as the oysters have one valve shaped differently from the other (i.e., **inequivalved**). In either case, the valves are attached to one another along the dorsal margin by an elastic proteinaceous band called the **ligament**. Immediately posterior to the beak is a depressed or flat area called the **escutcheon**, located just posterior to the beak. There may also be a heart-shaped area called the **lunule** just in front of the beak (Fig. 9.27).

The conventional way to orient the shell of a typical bivalve is to hold the shell between the hands with the **umbo** (area of strongest convexity) directed forward and the hingement at the top. This establishes the right and left valves, as well as dorsal and ventral. Height is measured dorsoventrally, and thickness from right to left directly through both valves. (This method of orientation is suitable for clams like *Mercenaria*, but is not suitable for many bivalves having attached shells or beaks that are directed posteriorly.)

Unlike brachiopods, which have teeth on one valve and sockets in the other, teeth and sockets occur in both valves in Bivalvia. The mechanisms for opening and closing the valves also differ from that in brachiopods. In bivalves, movement of the

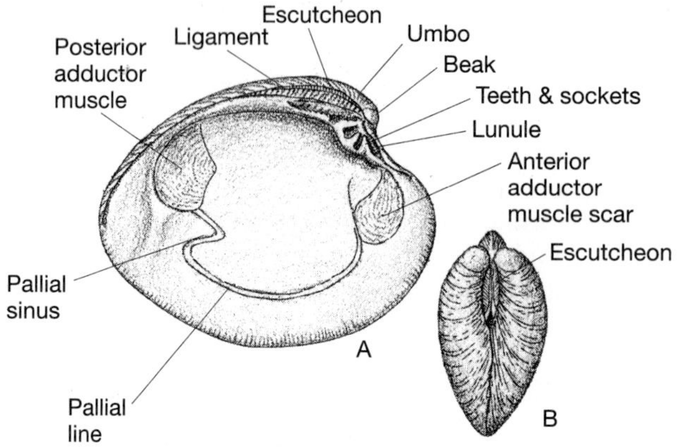

Figure 9.27   Interior of the left valve of *Mercenaria* (A), and dorsal view showing the escutcheon (B).

valves requires the antagonistic action of both the ligament and adductor muscles. The valves are held open by the elasticity of the ligament. The ligament may be either external or lie internally within a ligamental pit, or it may be supported by a special pad or **chondrophore**. Valves are closed by contraction of one or more adductor muscles. Energy must be expended to keep the valves closed. On death, the adductors relax, and the valves gape. This accounts for the infrequency with which fossils are found with both valves intact.

The scars of muscles left on the interior of valves are important criteria for the classification and identification of Bivalvia. Where the scars for two adductor muscles occur, the bivalve is termed **dimyarian** (see Fig.9.27). Further, if the two scars are about the same size, the bivalve is **isomyarian**, in contrast to the **anisomyarian** condition in which the anterior scar is distinctly smaller. Most mussels have anisomyarian muscle scars. Scallops and oysters get along well with only one adductor, and are therefore termed **monomyarian**. In some bivalves, one may also find two smaller muscle scars, typically on the dorsal side of the adductors. These are the attachment scars for muscles used in retracting the foot.

The visceral mass (Fig. 9.28) lies partially or completely enclosed within the two hinged valves. The **mantle** extends downward (ventrally) on either side as two mantle lobes. A principal function of the mantle is the secretion of the valves, although it also has sensory and muscular functions. The space between the mantle lobes comprises the **mantle cavity**. It is the most capacious mantle cavity of any class of mollusks. Within the mantle cavity lie the viscera, the gills, and the foot. Gills in bivalves are covered with cilia, which beat rhythmically so as to draw water into the mantle cavity and across the layered folds of the gills. Food particles brought to the gills by ciliary action are conveyed in a stream of mucous to the ventral margins of the gills, then forward to ciliated, leaflike fleshy appendages (palps) on either side of the mouth, and thence to the mouth and digestive tract. The gills have the remarkable ability to distinguish between food particles and mere grains of sediment. The latter are dropped from the gill surface and expelled posteriorly. The bivalve method of feeding and respiration requires a constantly renewing supply of water. To ensure an ample flow, the mantle edge may be modified into tubes or **siphons** that are able to draw water into the gills

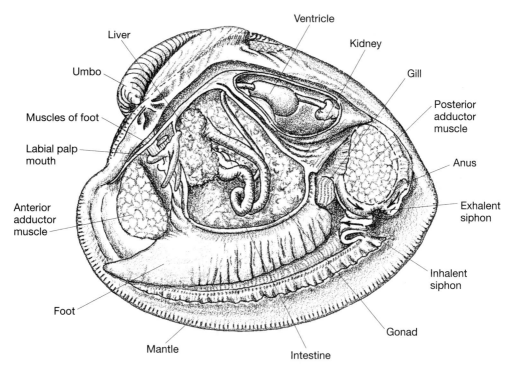

Figure 9.28   Internal organs of *Mercenaria*, revealed by removal of the left valve, mantle, and gills.

and expel it. Typically, the two tubes are located posteriorly and are designated the **incurrent** and **excurrent siphons**.

The kind of gill structure among bivalves is of great importance in classification. In the **protobranch** gill, there is a simple leaflike structure (Fig. 9.29) that is suspended within the mantle cavity. Somewhat more advanced is the **filibranch** gill, which consists of a double row of closely spaced filamentous laminae that hang downward in the mantle cavity but then bend upward. Cilia on these doubled filaments produce the currents that cross the gills. In the **eulamellibranch** gill, the laminae of the gills are joined by cross connections so as to form enclosed spaces. The **septibranchial** gill is the most advanced, and it consists of a horizontal partition that divides the mantle cavity into lower and upper parts.

Bivalvia soft parts also include a circulatory system with heart and blood vessels, paired kidneys, reproductive organs, and a nervous system with as many as four pairs of ganglia and, in some species, eyes located along the mantle margin. The sexes in bivalves are separate, although some species of scallops, oysters, and freshwater clams are hermaphroditic. In freshwater clams with separate sexes, sperm shed by the male enter the female through the inhalent opening and are transported to the eggs. Myriad tiny, bivalved, free-swimming larvae emerge from the fertilized eggs. They may live in the fins and gills of fish until they have matured sufficiently to drop off and assume an

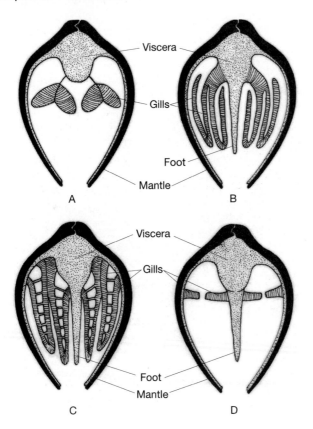

Figure 9.29   Transverse diagrammatic sections of the four types of gill structures in Bivalvia. (A) protobranch type consisting of two double-leafed gills; (B) filibranchiate type; (C) eulamellibranchiate type in which the leaves of each gill plate are interconnected; and (D) septibranchiate type in which gills are perforated partitions.

independent life. Marine clams may shed eggs and sperm directly into the water where fertilization leads to the development of free-swimming larvae.

The muscular foot of the bivalve is used for crawling, anchoring, digging, and even leaping. In many forms, the foot is extended into the sediment and anchored by causing the free end to swell. The swelling is produced by infusing blood into a cavity in the foot. Once anchored, longitudinal muscles in the foot contract, shortening the foot and drawing the animal forward. In bivalves that prefer a permanent anchorage, a gland at the ventral end of the foot secretes tensile byssal threads (tough threads or fibers) by which the animal tethers itself to a fixed object. Alternately, attachment may consist simply of cementing one valve to a hard object on the seafloor.

Exterior and interior structures and markings of the shell are important criteria for the identification of bivalves. The exterior of valves may be marked by concentric growth lines, growth lamellae, or ridges of varying degree of prominence. Radial sculpting might consist of costae and finer markings called **lirae**. Spines and nodes can further ornament the valves. In most instances, exterior markings are related to shell growth, strength, or to facilitate a particular habit, such as burrowing.

Markings on the insides of valves are generally associated with attachment of tissue and internal organs. Using *Mercenaria* as an example, one can quickly identify the two rounded scars of the adductor muscles. The next most prominent feature is a semi-circular line extending around each valve a short distance from the margin of the shell. This **pallial line** marks the line of attachment of the fixed lobe of the mantle to the shell. The pallial line in bivalves having retractable siphons will be bent inward just below the posterior adductor scar. The indentation is called the **pallial sinus** and indicates the location of a pocket into which the siphon can be withdrawn when danger threatens. The pallial sinus provides a clue to the size of the siphon in extinct species, and this may also be indicative of bivalve habits.

In addition to such features as muscle scars, pallial markings, and valve shape, identification of members of the Class Bivalvia requires recognition of shell microstructure and dentition. Dentition in Bivalvia is far more varied and complex than in articulate brachiopods, as will be evident in the brief descriptions of important categories of dentition given below.

1. **Dysodont**. As seen in *Mytilus* (Fig. 9.30A), dysodont teeth consist of very small, simple "cusps" aligned along the hinge of a narrow area.

2. **Taxodont**. The taxodont pattern consists of similar, parallel teeth and sockets that lie oblique or at right angles to the margin of the hinge, as seen in *Arca* (Fig. 9.30B).

3. **Heterodont**. This is a radial, nonuniform pattern of teeth and sockets with a large cardinal group directly beneath the beak, and adjacent smaller anterior and posterior lateral teeth. Typically, there are three cardinal teeth in the right valve and two in the left. *Cerastoderma* (Fig. 9.30C) provides an example. Most Tertiary and Holocene bivalves have heterodont dentition.

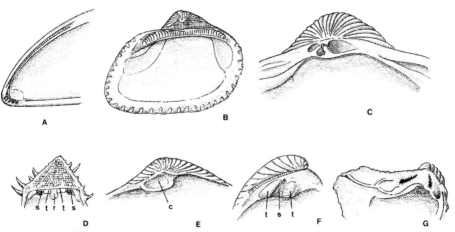

Figure 9.30   Types of dentition in bivalves. (A) Dysodont, as seen in the right valve of *Mytilus*. (B) Taxodont, right valve, *Arca*. (C) Heterodont, right valve, *Cerastoderma*. (D) Isodont, right valve, *Spondylus*. (E) Desmodont, with chondrophore, *Mya*. (F) Pachydont, left valve, *Chama*. (G) Schizodont, as seen in *Trigonia*. (Drawings not to same scale; *c* chondrophore, *r* resilifer, *s* socket, *t* tooth.)

4. **Isodont**. As seen in *Spondylus* (Fig. 9.30D), isodont dentition consists of two teeth and two sockets located symmetrically on either side of the ligament pit.

5. **Desmodont**. In desmodont dentition (*Mya* in Fig. 9.30E), the hinge teeth are reduced or absent. Thin ridges aligned along the hinge provide stability, and the ligament may be attached to inwardly projecting chondrophores.

6. **Pachydont**. Typically, pachydont dentition consists of one to three large, blunt, non-symmetrical teeth that are inserted into correspondingly shaped sockets. Pachydont teeth are seen in the aberrant rudistids (Fig. 9.30F) of the Jurassic and Cretaceous.

7. **Schizodont**. Schizodont dentition, as typically exhibited in *Trigonia* (Fig. 9.30G), consists of two large, diverging teeth in the right valve and a **Y**-shaped tooth in the left valve that fits between the two large teeth of the right valve. The sides of the teeth may be fluted.

The shell of a bivalve has an outer **periostracum** composed of proteinaceous matter that may project outward as hairlike fibers. The periostracum provides a protective covering for two or more layers of calcium carbonate (Fig. 9.31) The latter may have the mineral form of either aragonite or calcite, and may be laid down as tiny stacked prisms or minute laths, thin tablets arranged in sheets, discrete lenses, or in various imbricating or cross-layered patterns. These patterns provide important criteria for the classification of bivalves. Perhaps the most familiar pattern is that of thin sheets called **nacre** that cover the inner surface of the valves. Because of its iridescent luster, the nacreous inner layer is sometimes called the **mother-of-pearl** layer. Occasionally, foreign particles may become lodged between the mantle and mother-of-pearl layer. These particles then serve as nuclei around which concentric layers of nacreous calcium carbonate are secreted so as to form a pearl.

All of the above components of the bivalve shell are secreted by the mantle. The periostracum is laid down first by cells located along the inner edge of the mantle's outer fold. Marginal cells of the outer fold then secrete the initial underlying layer of calcium carbonate. Subsequent additions are then secreted across the entire surface of the mantle, rather than just along its leading edge.

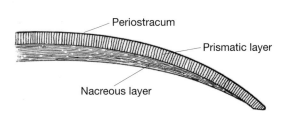

Periostracum

Prismatic layer

Nacreous layer

Figure 9.31   Transverse section of one valve showing the three fundamental shell layers in a bivalve: periostacum, prismatic layer, and the laminated or nacreous layer. Note that the nacreous layer thickens toward the older part of the valve.

MAJOR CATEGORIES OF BIVALVIA

The classification of bivalves is based on the structure of the gills, muscles (and muscle scars), features of the ligament, shell microstructure, dentition, and shell shape. Unfortunately, gill structure can rarely be inferred in fossils. As a result, paleontologists rely heavily on dentition, muscle scars, shell structure, and shell shape. The classification used here follows that in the *Treatise of Invertebrate Paleontology*. In that work, Bivalvia are grouped into six subclasses: the Palaeotaxodonta, Cryptodonta, Pteriomorpha, Palaeoheterodonta, Heterodonta, and Anomalodesmata.

Subclass Palaeotaxodonta (Cambrian to Holocene)   Taxodont dentition and isomyarian musculature are the principal characteristics of this subclass. Valves are equal and have a nacreous or cross-laminar microstructure. Palaeotaxodonts possess a single pair of bipectinate leaflike gills (protobranchiate gills). There is only one order within this subclass, namely the Nuculoida. Representatives are *Nuculana* and *Yoldia* (Fig. 9.32A, B)

Subclass Cryptodonta (Ordovician to Holocene)   Cryptodonts are small, thin-shelled, aragonitic pelecypods with dysodont dentition. As revealed in living species, the gill strucure is protobranchiate. *Cardiola* (Fig. 9.32C), a fossil helpful in recognizing Silurian strata, is representative of this subclass.

Subclass Pteriomorpha (Ordovician to Holocene)   This subclass is composed mostly of bivalves that live tethered to the seafloor by byssal threads or they simply cement themselves to hard objects. They are a diverse group, as indicated by the distinctive appearances of each of the three orders within the subclass Pteriomorpha.

Order Arcoida.   The arcoids have taxodont dentition, straight hinge lines, isomyarian musculature, and **filibranchiate gills** (gills in which filaments are separate and held together only by tufts of specialized cilia). They are equivalved with circular to trapezoidal shells. Examples are *Arca* (Fig. 9.32E) and *Glycymeris* (Fig. 9.32F).

Order Mytiloida.   Mytiloids have dysodont dentition, a well-developed byssus (organ bearing byssal threads), and both prismatic and nacreous shell layers. The gills are filibranchiate in some groups and in others the gill filaments are joined together by continuous sheets of tissue to form a gill structure termed **eulamellibranch**. *Mytilus* (Fig. 9.32D), the edible mussel, is a familiar representative.

Order Pterioida.   The musculature in the pterioids is either anisomyarian or monomyarian, and the gill structure filibranchiate or eulamellibranchiate. This order includes oysters such as *Ostrea* (Fig. 9.32K), scallops like the familiar *Pecten* (Fig. 9.32H), and *Inoceramus* (Fig. 9.32G). The disassociated calcite prisms of *Inoceramus* are so common in some well cuttings (rock chips brought to the surface when drilling for oil) that geologists use them as evidence that drilling has penetrated Late Cretaceous marine strata. Other well-known pterioid guide fossils are *Exogyra* (Fig. 9.32I) and *Gryphaea* (Fig. 9.32J).

Figure 9.32   Some common fossil members of the Bivalvia. (A) *Nuculana*, ×2. (B) *Yoldia*, ×1. (C) *Cardiola*, ×1. (D) *Mytilus*, ×0.5. (E) *Arca*, ×1. (F) *Glycymeris*, ×1. (G) *Inoceramus*, ×0.5. (H) *Pecten*, ×0.3. (I) *Exogyra*, ×1. (J) *Gryphaea*, ×0.7. (K) *Ostrea*, ×0.5. (L) *Unio*, ×0.5. (M) *Modiomorpha*, ×1. (N) *Trigonia*, interior of left valve, ×0.3. *Nuculana* and *Yoldia* are palaeotaxodonts. *Cardiola* is a cryptodont, *Mytilus* is a mytiloid, *Arca* and *Glycymeris* are arcoids. *Inoceramus, Pecten, Exogyra, Gryphaea*, and *Ostrea* are pteriomorphs. *Unio* is a unionoid, *Modiomorpha* a modiomorph, and *Trigonia* a trigonoid.

Subclass Palaeoheterodonta (Cambrian to Holocene)   Most members of this subclass have two equal valves, heterodont hingement, and isomyarian musculature. There are only a few hinge teeth. When present, elongate lateral teeth are not separated by large cardinal teeth. The subclass includes the orders Unionoida, Modiomorphoida, and Trigonoida.

Order Unionoida.   The unionoids are freshwater bivalves with heterodont dentition. Representatives are *Unio* (Fig. 9.32L) and *Corbicula Unio*, the common freshwater mussel, is a suspension feeder that burrows into the sediment of streams and lakes. It has nearly worldwide distribution.

Order Modiomorphoida.   As seen in *Modiomorpha* (Fig. 9.32M), the modiomorphoids are equivalved palaeoheterodonts with ovoid valves and differentiated radial teeth originating at the beaks. Early members of this order appear in rocks of Cambrian age, and they are thought to be the ancestors of many later orders of the Class Bivalvia.

Order Trigonoida.   Trigonoids take their name from the trigonal outline of their shells. In addition, they possess prominent schizodont dentition consisting of teeth and sockets that diverge beneath the beaks, as is seen in *Trigonia* (Fig. 9.32N). Most trigonoids have strongly sculpted shells, and the ornamentation of the anterior part of valves differs from that of the remainder of the shell.

Subclass Heterodonta (Ordovician to Holocene)   The Heterodonta are phylogenetically the youngest subclass of the Class Bivalvia. They are also the most numerous and diverse of living bivalves. The subclass is composed of groups with equal valves and dentition that characteristically includes a few large cardinal teeth that lie beneath the ligament. These distinctly larger teeth are separated by a toothless space from elongate lateral teeth. Musculature is of the dimyarian type, and the ligament is external. Living heterodonts have eulamellibranchiate gills. The Heterodonta are divided into three orders, namely the Veneroida, Myoida, and Hippuritoida.

Order Veneroida.   Veneroids are a diverse group and have adapted to a variety of habitats. They appear to be vigorously evolving at the present time. Veneroids are recognized by their heterodont dentition, isomyarian musculature, and usually equivalved shells. Included within the order is the edible marine clam ("quahog") *Mercenaria* (see Figs. 9.27 and 9.28), as well as *Lucina*, the cockles *Cardium* and *Venericardia* (Fig. 9.33A), *Cerastoderma* (Fig. 9.33B), *Astarte*; the giant clam *Tridacna*; and *Solen*, the razor clam (Fig. 9.33C).

Order Myoida.   Most myoids are burrowers with thin, inequivalved shells (Fig. 9.34) and well-developed siphons. One or two cardinal teeth are usually present in the degenerate hingement. Included here are *Mya* (the soft-shell clam), *Pholas* (the boring clam), *Teredo* (the wood-boring shipworm), and *Corbula*.

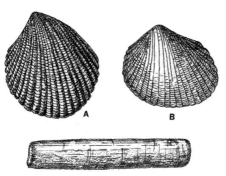

Figure 9.33   Left valves of (A) *Venericardia*, ×0.8; (B) *Cerastoderma*, ×1; and (C) *Solen*, ×1. (After E. L. Cox et al. In R. C. Moore (Ed.), *Treatise on Invertebrate Paleontology*, Pt. N, Mollusca 6. Boulder, CO, and Lawrence, KS: Geological Society of America and University of Kansas Press, 1969.)

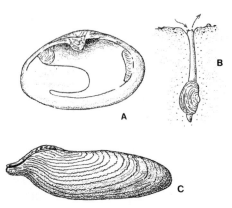

Figure 9.34   Myoids. (A) *Mya*, ×0.5. (B) *Mya* in burrow with siphon extended. (C) *Pholas*, ×0.5.

Order Hippuritoida.   The often bizarre shapes of hippuritoids (Fig. 9.35) remind one of both corals and richthofenid brachiopods. These are the fossil rudists (robust coral-like Bivalvia). They have thickened specialized teeth called *pachydont*, are inequivalved, and cement one valve to a hard substrate. Representatives are *Diceras* and *Hippurites*.

Subclass Anomalodesmata (Ordovician to Holocene)   This relatively small subclass consists of burrowing forms having eulamellibranchiate gills, thin aragonitic

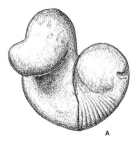

Figure 9.35   Hippuritoids. (A) *Diceras*, exterior, ×0.7. (B) *Hippurites*, side view, ×1.5. (After E. L. Cox et al. In R. C. Moore (Ed.), *Treatise on Invertebrate Paleontology*, Pt. N, Mollusca 6. Boulder, CO, and Lawrence, KS: Geological Society of America and University of Kansas Press, 1969).

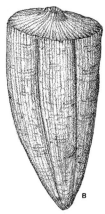

Figure 9.36   *Pleuromya*, right valve, ×1.

shells, usually isomyarian musculature, and hingement by means of a prominent internal ligament. The hinge line is thick and incurved. *Pleuromya* (Fig. 9.36) is an anomalodesmatid known from strata of Triassic to Cretaceous age.

BIVALVES THROUGH TIME

The fossil record for the Bivalvia begins in the Cambrian (Fig. 9.37). Among the earliest to appear is *Pojetaia*, from the Lower Cambrian of Australia; *Fordilla*, which occurs in Lower and Middle Cambrian rocks of eastern North America, Denmark, and Siberia; and the Middle Cambrian genus *Tuarangia*, from New Zealand. Certain features of *Pojetaia* suggest these tiny bivalves can be considered the oldest members of the Subclass Palaeotaxodonta.

Cambrian bivalves were notably small, rarely exceeding a centimeter in length. They were not abundant, and they existed as minor components of the total invertebrate fauna. Their numbers and variety improved during the Ordovician when new species of burrowers, byssally attached species, and epifaunal groups thrived. The burrowers, however, were probably somewhat limited, in that they had not yet evolved true siphons, such as characterize later bivalve burrowers.

From the Late Ordovician to the present, bivalves have been a significant presence in benthic marine faunas. They have not been restricted to the seas, however, for many groups also successfully invaded freshwater and brackish-water habitats. Freshwater bivalves are frequently found in nonmarine Devonian sedimentary rocks, whereas brackish-water forms are abundant in some Carboniferous shales and siltstones deposited in estuaries and lagoons. It is interesting that many fossil bivalves found in rocks older than Carboniferous are preserved as external molds and internal molds known as steinkerns. Perhaps their shells were composed of aragonite rather than the more durable calcite.

The Triassic Period was a turning point in bivalve history. Many competing groups of bivalves as well as brachiopods had declined or had become extinct. The stage was set for the expansion of many new groups of bivalves. The process was enhanced by the evolution in the Permian of bivalves with true siphons. Efficiently functioning siphons permitted bivalves to burrow deeply, bore into the substrate more effectively, and to exploit more fully the harsh intertidal environments around the world.

Bivalves continue in great abundance during the Jurassic and Cretaceous periods. Groups particularly useful in biostratigraphy for rocks of these ages are the trigonoids, inoceramids, pectinids, oysters, arcoids, pteriomorphs, and heterodonts.

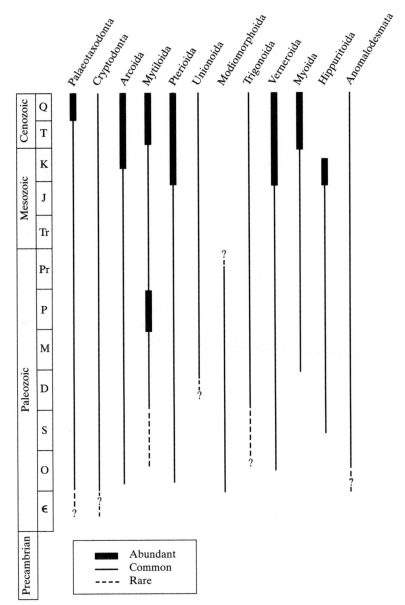

Figure 9.37 Geologic ranges of major groups of Bivalvia

Neither inoceramids nor rudists, however, survived to the end of the Cretaceous. Thus, they were not victims of the terminal Cretaceous biological crisis. Species of *Ostrea*, *Gryphaea*, and *Exogyra* not only provide valuable stratigraphic markers, but constitute a major part of the bulk volume of some Jurassic and Cretaceous formations. The aberrant, reef-building rudists came on the scene during the Jurassic and expanded during the Cretaceous. They were unable, however, to survive beyond that period. The terminal Cretaceous crisis devastated many other bivalve orders as well, but a few were able to carry on, albeit in diminished numbers. Some of these survivors were the stock for a major expansion of bivalves during the Tertiary. As is clearly apparent from the billions of shells of bivalves that litter modern beaches around the world, the final episode of bivalve success continues today.

BIVALVE HABITS AND HABITATS

Throughout their long geologic history, bivalves have been primarily aquatic, benthic animals. Mostly, they have lived in the ocean, but as noted above, freshwater species have been recorded in the fossil record since the Devonian. There are even a few hardy species of bivalves that make their living on land among moist leaves and humus.

In the marine realm, bivalves have adapted to a variety of bathymetric levels, from the intertidal zone down to the abyss. Their optimum life zone, however, is from low tide to about 200 meters. In general, pelecypods living in shallow areas where they are buffeted by waves and currents tend to build larger, more robust, and more heavily ornamented shells. Deep water and mud-dwelling species have thinner, more delicate, and smoother shells. A more important constraint on the shape or the appearance of the shell is the precise manner in which any given species lived. Many dwell on the surface of the seafloor as epifaunal animals. Some of these are free-living swimmers, such as some scallops. Others are attached by cementation (as in oysters) or byssal threads (as in mytiloids). Others are infaunal and burrow deeply into bottom sediment with siphons extended through the sediment to the surface of the seafloor. The infaunal group also includes wood borers such as *Teredo*, and species that simply dwell in pre-existing cavities and crevices.

With regard to nutrition, most bivalves are filter feeders that consume suspended organic matter and minute plankton. An examination of the stomach contents of living bivalves reveals a diet of tiny larvae, eggs, foraminifers, diatoms, and radiolaria. A fewer number of bivalves are deposit feeders, living on organic particles gleaned from soft sediment. The giant Indo-Pacific clam *Tridacna* lives with symbiotic algae in its mantle tissue. The algae provide the clam with an accessory source of nutrition. Species that live in the darkness below the photic zone appear to do very well by feeding on worms, crabs, and carrion.

CLASS ROSTROCONCHIA

The Rostrochonchia are a class of mollusks that have no living representatives. Fortunately, they are only a minor class and of limited biostratigraphic use. Only about 35 genera are known. They are restricted to the Paleozoic, appearing first in Lower Cambrian rocks, and absent in rocks younger than the Permian. Many paleontologists regard the Rostroconchia as the molluscan class from which the Bivalvia sprang.

Rostroconchia (Fig. 9.38) were benthic marine invertebrates, some of which appear to have been suspension feeders and others deposit feeders. Most species lived in shallow burrows. Superficially, adult rostroconchids resemble bivalves, and indeed they are sometimes described as "pseudobivalved" or "bivalvelike." Their shell layers, however, are secreted continuously across the dorsal margin. Thus, although they may seem to have two valves, the skeleton is a one-piece shell, and there is no ligament or

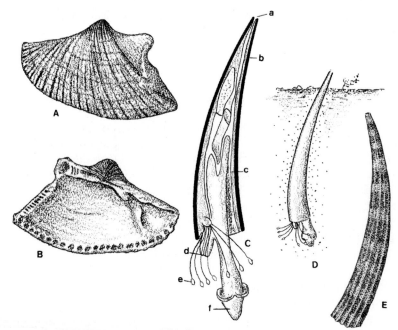

Figure 9.38   Rostroconchia and Scaphopoda. Left valve exterior (A) and interior (B) of the rostroconch *Conocardium*. (C) General scaphopod structure based largely on *Dentalium*. (D) A scaphopod in feeding position within sediment of the substrate. (E) *Dentalium*. (*a*, shell aperture; *b*, shell; *c*, mantle; *d*, mouth; *e*, tentacles; *f*, foot.)

plane of union that in bivalves would separate the two valves. Nor are muscle scars present. Rostroconchia also differ from bivalves in their early development. They begin postlarval development as tiny univalves with a single center of shell growth. Subsequently, they achieve their "bivalve appearance" by secretion of two plates called **pseudovalves**. Probably the best known of the Rostroconchia is *Conocardium* (Ordovician to Permian), which at one time was considered to be a bivalve (Fig. 9.38A,B).

CLASS SCAPHOPODA

Because of their cylindrical, curved, and tapered form, members of the Class Scaphopoda have been dubbed "tusk-shells." The shell, which averages only 3 to 6 cm in length, is open at both ends. The larger opening at the anterior end bears the mouth and ciliated tentacles, whereas the smaller end is posterior and contains the anus. Water is drawn in and out of the posterior end to assist in respiration and the expulsion of waste and gametes.

Compared to such mollusks as gastropods or cephalopods, these animals have a somewhat simpler organization. The head is ill-defined, there are no eyes, and there are either no gills or gills that are severely reduced. Respiration is accomplished by gas exchange over the mantle surface. A short radula located in the mouth assists in feeding.

Scaphopods are, and have always been, marine infaunal creatures. They live in soft sediment of the continental shelf and slope where they lie buried, head downward, with their curved shells positioned so that the posterior end projects above the surface of the sediment. The threadlike tentacles are used in capturing food, which consists of small organisms such as foraminifera and other protoctistans.

Although interesting from a biological point of view, scaphopods have not been particularly useful as guide fossils. They made their appearance in the Ordovician as probable descendants of rostroconchids and have survived modestly down to the present. *Dentalium* (Fig. 9.38) is a fairly representative scaphopod.

REVIEW QUESTIONS

1. What are the distinguishing characteristics of the Phylum Mollusca? How do the molluscan classes differ from one another?

2. Among Gastropoda, what is the function of the following:
   a. mantle   b. radula   c. siphonal notch   d. opercula

3. What features of chitons are adaptations for living on hard surfaces in high-energy intertidal zones?

4. Select the most appropriate match from the right-hand column, for entries in the left-hand column.

   _____*Helix*   A. Archaeogastropoda
   _____*Crepidula*   B. Mesogastropoda
   _____*Bellerophon*   C. Neogastropoda
   _____*Murex*   D. Opisthobranchia
   _____*Loxonema*   E. Pulmonata
   _____*Maclurites*
   _____*Acteonina*

5. What characteristics may be useful in distinguishing between the shells of male and female ammonoid cephalopods?

6. Describe two methods in which orthoconic cephalopods were able to maintain approximately horizontal balance in swimming. Describe the relation between center of gravity and center of buoyancy in planispirally coiled cephalopods.

7. Distinguish between the following:
   a. orthocones and cyrtoceracones   d. saddles and lobes
   b. evolute and involute shells   e. ceratitic and ammonitic sutures
   c. sutures and septa   f. aptychi and anaptychi

8. Advance several arguments in support of the hypothesis that septal folding and fluting in ammonoids were adaptations that provided strength to the shell in order to prevent breakage from hydrostatic pressure.

9. Explain the mechanism for valve opening and closing among clams like *Mercenaria*.

10. Distinguish between the following:
    a. monomyarian and dimyarian valves
    b. filibranch and eulamellibranch gills
    c. taxodont and heterodont dentition

SUPPLEMENTAL READINGS AND REFERENCES

Broadhead, T. W. (Ed.) with Bottjer, D. J., Hickman, C. S. & Ward, P. D. (organizers). 1985. *Mollusks: Notes for a Short Course.* Knoxville: University of Tennessee, Dept. of Geology, Studies in Geology No. 13, sponsored by The Paleontological Society.

Jenkins, M. M. 1972. *The Curious Mollusks.* New York: Holiday House.

Landman, N. H., Tanabe, K. & Davis, R. A. 1996. *Ammonoid Paleobiology.* New York: Plenum Press.

Lehman, U. 1981. *The Ammonites: Their Life and Their World.* Cambridge: Cambridge University Press.

Moore, R. C. (Ed.). 1957, 1964, 1969. *Treatise on Invertebrate Paleontology:* The Mollusca volumes, Pts. I (gastropods), K (nautiloids), L (ammonoids), and N (bivalves). Boulder, CO, and Lawrence, KS: Geological Society of America and University of Kansas Press.

Pojeta, J. Jr., Runnegar, B., Peel, J. S. & Mackenzie, G. Jr. 1987. Phylum Mollusca. In R. S. Boardman (Ed.), *Fossil Invertebrates.* Palo Alto, CA: Blackwell Scientific.

Saunders, W. B. 1988. *Nautilus: The Biology and Paleobiology of a Living Fossil.* New York: Plenum Press.

Saunders, W. B. 1995. The ammonoid suture problem: Relationships between shell and septum thickness and sutural complexity in Paleozoic ammonoids. *Paleobiology* 21(3):343–355.

Stanley, S. M. 1970. Relation of shell form to life habits of the Bivalvia (Mollusca). *Geological Society of America Memoir 125.* Boulder, CO: Geological Society of America.

Ward, P. D. 1982. *Nautilus*: Have shell will float. *Natural History* 91: 64–69.

Yonge, C. M. & Thompson, T. E. 1976. *Living Marine Mollusks.* London: Wm. Collins & Sons.

Enrolled specimen of the Flexicaymene from the Ordovician of Ohio. (Width across cephalon about 2 cm.)

Arthropoda

It is difficult to overestimate the importance of arthropods in the history of life. In terms of diversity and abundance, modern arthropods embody the ultimate success of the metazoa.

Donald G. Mikulic, 1990

With an estimated three quarters of a million species formally described, and possibly many millions yet to be recorded, the arthropods are indeed a mighty multitude. They have successfully invaded nearly every conceivable habitat, and they have an impressive geologic history that extends back nearly 550 million years and continues with remarkable vitality to the present day. The Cambrian Chengjiang and Burgess Shale faunas dramatically depict the remarkable diversity of arthropods even at the beginning of the Paleozoic. Among the Burgess Shale arthropods, *Marella* was clearly the most abundant. Second in abundance was *Canadaspis*, an arthropod regarded as the earliest crustacean. Because of their rapid evolution, arthropods such as trilobites and ostracodes have great value in biostratigraphic correlation. In addition, many species are superb indicators of ancient environmental conditions.

We all have some familiarity with arthropods. We have been annoyed by insects, enjoyed a meal of crustaceans (shrimp, lobster, crab, or crayfish), been startled by arachnids (spiders, scorpions, mites), or marveled at the armored symmetry of merostomes (horseshoe crabs). Insects, of course, are the most frequently encountered arthropods, for they comprise nearly 99 percent of living species. The least frequently seen arthropods are myriapods (centipedes, millipedes) and pycnogonids (sea spiders).

Among the most important characteristics of arthropods are paired, jointed appendages (arthropod means "jointed foot"). These appendages are used not only for walking and swimming, but also are modified to form mouth parts and accessory reproductive structures for transferring sperm. Another distinguishing feature of arthropods is their chitinous, jointed exoskeleton. The exoskeleton not only provides support and protection but it also serves as a sturdy framework for the attachment of muscles. These are affixed to the underside or inside of the exoskeletal and flex or relax in order to produce movements of the body and appendages. Joints and articular membranes between the segments facilitate movement generated by the action of muscles. The body of an arthropod is segmented, like that of an annelid worm. However, in some groups the segments are lost or fused together. Another characteristic of some arthropods is that the external segments may not coincide with the fundamental internal segmentation. Segments may become fused into groups called **tagmata**, such as head, thorax, and abdomen, each of which performs distinctly different functions. Arthropods also have an open circulatory system with a dorsal heart, and a nervous system that includes well-developed sense organs.

Growth in arthropods is accomplished by molting. During molting, the hard outer layer of the exoskeleton is discarded. At the same time, a softer inner layer is resorbed and the materials reused as the underlying epidermal cells secrete a new covering. That covering is composed of a mixture of chitin (a resistant organic material) and protein. The protein varies in composition to permit the development of skeletal tissue of greater hardness in some areas and yet provide flexibility in others. In such arthropods as trilobites, crustaceans, and ostracods, the exoskeleton may be strengthened further by impregnation with calcium carbonate, which greatly enhances preservation.

ARTHROPOD ORIGINS AND CLASSIFICATION

The origin of the arthropods is best understood in the context of the embryonic development of bilaterally symmetrical animals having a true **coelom**, or body cavity formed within and lined by mesoderm. Such animals form two branches of the phylogenetic tree. One branch includes the arthropods, as well as mollusks and annelids. These animals comprise the **protostomes**. The other branch is composed of **deuterostomes**, represented by echinoderms, hemichordates, and chordates. The differences between protostomes and deuterostomes is seen in their embryonic development. In the early embryonic development of protostomes, a group of cells move inward to form an opening (the **blastopore**) that develops into a mouth, whereas in deuterostomes the opening develops into an anus and the mouth forms at a later stage. Differences also occur in the pattern of embryonic cell division and the manner in which the coelom is formed. The arthropods evolved from early protostomes. They were once thought to have arisen directly from annelid ancestors, but molecular evidence now indicates that both arthropods and annelids are separately derived from an earlier protostomate ancestor.

At present, a controversy exists as to whether the arthropods arose from a single ancestor and were thus **monophyletic**, or were derived from multiple ancestors and can be considered **polyphyletic**. Those favoring the polyphyletic hypothesis consider that the major groups of arthropods, the Trilobita, Crustacea, Chelicerata, and Uniramia, should be considered phyla in their own right. Many paleontologists, however, argue that even with their different body plans, these groups have so many similarities in both external and internal anatomy as to constitute strong evidence favoring a monophyletic origin, and therefore retention of the Arthropoda as a single phylum. The controversy instills uncertainty in any classification of arthropods. Here we follow a traditional and simple scheme that treats the Arthropoda as a phylum and permits treatment of each major group more or less separately with the tentative rank of superclasses.

SUPERCLASS TRILOBITA

The trilobite exoskeleton (Fig. 10.1) includes a dorsal shield and a thin ventral membrane. Both the membrane and other trilobite appendages are rarely preserved. The word *trilobite* means "three lobed," and refers to the division of the dorsal shield into three longitudinal segments. These are the **axial lobe**, and two lateral or **pleural lobes**. An **axial furrow** on either side of the axial lobe marks the boundaries of the three lobes. Transversely, the dorsal shield is differentiated into an anterior **cephalon**, followed by the **thorax**, and, when developed, the **pygidium**. Like other arthropods, trilobites grew by **molting**. As a result, any single individual may have contributed parts or all of several exoskeletons to the fossil record.

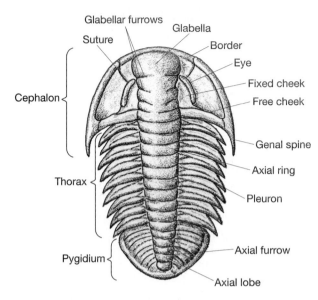

Glabellar furrows
Glabella
Suture
Border
Eye
Fixed cheek
Free cheek
Cephalon
Genal spine
Axial ring
Thorax
Pleuron
Axial furrow
Pygidium
Axial lobe

Figure 10.1   The Middle Cambrian trilobite *Bathyuriscus*, showing terminology for the dorsal exoskeleton. Specimen is about 4 cm in length.

THE CEPHALON

The morphologic features most useful in trilobite identification and phylogenetic studies are seen on the cephalon. It is usually semicircular in outline, and typically supports a raised axial region called the **glabella** (see Fig. 10.1). In many trilobites, the glabella may reveal its original segmentation by the presence of partial or complete **glabellar furrows**. The most posterior of these furrows (the **occipital furrow**) defines the final segment of the glabella, called the **occipital ring**. In some taxa, the occipital ring may bear a spine. On either side of the glabella are the **cheeks**. In certain groups of trilobites, each cheek bears a **facial suture** along which molting is facilitated. Facial sutures define two regions of the cheeks, namely the **fixed cheek** and the **free cheek**. The former is attached to the glabella, whereas the latter falls free during molting.

Facial sutures extend along the cephalon in a number of different ways. In the so-called **opisthoparian** trilobites, the facial suture extends from a point at the anterior margin of the cephalon, toward the inner side of the eyes, and then continues backward to intersect the posterior margin at some point between the lateral margin of the glabella and the **genal angle** or the **genal spine** (see Fig. 10.7). In the **proparian** condition, the posterior part of the facial suture diverges to the lateral margin (see Fig. 10.21). If the posterior limb of the facial suture intersects the genal angle, as in Figure 10.25, the condition is termed **gonatoparian**. Among some forms, the facial suture may extend along the margin of the cephalon. The **metaparian** condition consists of a fused suture that both begins and ends at the posterior margin of the cephalon.

The **eyes** are an often prominent feature of the trilobite cephalon. Most trilobites have eyes. Typically, these lie on the margin of the free cheeks and abut against raised portions of the fixed cheeks. The eyes may be simple or compound. In the simple eye, there is a single lens consisting of a small transparent calcite node on the exoskeleton. Beneath the node are sensory and pigmented cells. Within the compound eyes of trilobites are a large number of separate cylindrical visual bodies (**ommatidia**), each consisting of an outer lens and an underlying sensory region. The lenses are composed of crystalline calcite with the optical c-axis of the crystal oriented perpendicular to the lens surface, thus reducing the problem of double refraction in the calcite. Eyes consisting of great numbers of closely adjacent lenses are termed **holochroal**. The holochroal eye has a smooth appearance because the lenses are covered by a continuous thin chitinous cornea. Eyes composed of relatively thick individual lenses that are each covered by a separate cornea and bounded by an opaque chitinous membrane are termed **schizochroal** (Fig. 10.2). The lenses in schizochroal eyes are larger and fewer in number than in holochroal eyes. Because each lens bears its own cornea, schizochroal eyes are more appropriately considered aggregate rather than compound eyes. Of these two types, the holochroal eye is the most ancient. It is present in trilobites ranging in age from Cambrian to Permian. Schizochroal eyes are confined to the Phacopida (Ordovician to Devonian).

The location and form of trilobite eyes can provide significant evidence relating to the animal's visual perception of its surroundings and of its habits. Because each ommatidium is separated from its neighboring ommatidium by pigmented areas, each is presumed to have received its own impulses. Each ommatidium is also directed toward a different zone of the visual field. Schizochroal eyes were apparently able to provide overlapping visual fields and the capablility for stereoscopic vision that encompassed 360° of the trilobite's surroundings. It is interesting that the lens is composed of calcite oriented so that the principal optic axis is normal to the surface of the lens. In looking through calcite along any other axis, double refraction occurs, but with the optic axis normal to the surface of the eye, trilobites had no problem with double vision.

Trilobites with large eyes may have inhabited darker, deeper marine areas. Schizochroal eyes may have been particularly effective for nocturnal trilobites or those that lived in dimly illuminated habitats. Some trilobites had degenerate eyes or had become secondarily blind. One can infer that these animals lived in deep-water environments with low light intensities, or that they were burrowers. Some trilobites that appear not to have eyes at all may have had them on the rarely preserved underside of the cephalon. In this regard, eyes on the underside of the cephalon would seem a useful adaptation for trilobites that swam above the surface of the seafloor while scanning the area below for food. Eyes on stalks permitted trilobites to crawl through bottom ooze with only their periscopic eyes exposed, or to lie partially covered by a camouflage of sediment while on the watch for food or predators.

Figure 10.2   *Phacops*, a Devonian trilobite from the Silica Shale of Ohio, is known for its often well preserved schizochroal eyes. Specimen is 3.5 cm in length.

Trilobite eyes are the most ancient complex visual systems known, and they provide some of the best evidence for eye evolution. The eyes of trilobites hold a special fascination, as suggested by these lines from T.A. Conrad.

> The race of man shall perish, but the eyes
> of trilobites eternal be in stone,
> And seem to stare about in mild surprise
> At changes greater than they have yet known.

Although features of the ventral side of the cephalon are less commonly seen, one is sometimes fortunate in finding exceptional preservation that reveals the nature of the trilobite underside. The dorsal shield is often seen to wrap around the anterior margins of the cephalon to form a **doublure** (Fig. 10.3). Two or three small plates may be attached to the doublure on its anterio-ventral margin. One of these serves as a movable flap in front of the mouth. It is called the **hypostome**. The second, the **rostrum**, is attached to the doublure just in front of the hypostome. The hypostome varies in shape; it may have a small inflated area at its center, and it commonly bears a pair of small nodes or **maculae**. When trilobites are found with the hypostome intact, the feature is exceptionally useful in identification. Often species can be identified only from the hypostome when the remainder of the exoskeleton is not found. If the hypostome has little or no correspondence to the shape of the glabella, it is termed an **impendent** hypostome. If it corresponds to the shape of the front of the glabella and is attached to the doublure, the hypostome is called **conterminant**, a condition considered primitive. As a third condition, termed **natant**, the hypostome may be merely positioned on the ventral membranous surface so that there is a gap between it and the doublure.

The delicate appendages of trilobites are only rarely preserved. Careful dissections and the use of X-radiography, however, have provided knowledge of the appendages of about 20 species. These observations indicate a high degree of constancy in the form of appendages. Typically, there are five pairs of jointed appendages attached to the underside of the cephalon (Fig. 10.3). The most anterior pair are the antennae, each composed of an unbranched series of short segments. Each of the four remaining appendages begins with a large segment called the **coxopodite**, from which extend two branches. For this reason, these appendages are termed **biramous**. The more leglike branch, termed the **endopodite**, is constructed of a linear series of six segments. The second branch extends from the coxopodite on the dorsal side of the

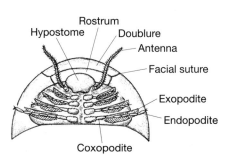

Rostrum
Hypostome
Doublure
Antenna
Facial suture
Exopodite
Endopodite
Coxopodite

Figure 10.3   Major features of the ventral surface of a trilobite cephalon.

endopodite. It is termed the **exopodite**, and is recognized by its many comblike threads. It appears likely that the exopodite was used in swimming, sweeping food toward the mouth, and possibly respiration as well.

THE THORAX

The thorax of a trilobite (see Fig. 10.1) is composed of a number of articulating segments, each including an **axial ring** and a lateral element on either side of the axial ring called the **pleuron** (plural, **pleura**). Typically, each pleuron is marked by an oblique groove termed the **pleural furrow** that strengthens the bladelike pleuron against bending. Thoracic segments (for all but one genus), range from 2 to 61, but more commonly the number does not exceed 10. The lateral extremities of pleura may taper to a point to give the entire thoracic margin a sawtooth outline. In other trilobites, the lateral extremities may be bluntly rounded. Thoracic segments are hinged to one another to provide flexibility to the exoskeleton and allow some species to protectively enroll. On their ventral sides, pleurons bear a pair of biramous appendages closely resembling each of the four pairs of biramous appendages beneath the cephalon.

THE PYGIDIUM

The "tail" or **pygidium** (plural, **pygidia**)of a trilobite is constructed along the same general plan as the thorax, except that the segments are fused to form a solid shield. Nevertheless, the positions of former segmental margins are often recognizable as furrows. In some forms, however, the axial lobe may merge imperceptibly with the surface of the pygidium, as in *Bumastus* (see Fig. 10.13). Pygidia may vary in size from very small to as large as the cephalon. Pygidium shape is also variable among genera, and may include semielliptical, quadrate, and attenuated V-shapes. The border of the pygidium can be smooth or bear spines. In general (except for the agnostid trilobites), Cambrian trilobites have small pygidia, whereas the pygidia are larger in post-Cambrian groups. Rarely, pygidia can be larger than the cephalon. Beneath the pygidium are biramous appendages that correspond in number and position to the fused segments.

The segmental furrows indicate that pygidia developed by fusion of post-thoracic segments present in early trilobites that lacked pygidia. The development of pygidia may have proceeded gradually, with the small pygidia characteristic of many Cambrian trilobites representing an early stage of pygidial evolution.

LARVAL STAGES

The developmental history of an organism is called its **ontogeny**. The ontogeny of trilobites, or at least certain aspects of their ontogeny, may be revealed by the fossil remains of larval molt stages, provided they can be found. When one is fortunate enough to find a complete sequence of molt stages or **instars**, these may be used to define taxonomic groups and show their evolutionary proximity and relationship to other groups, Larval molts can assist in distinguishing *monophyletic* taxa, which usually have similar larval stages, from *polyphyletic* taxa, which are likely to include dissimilar larval stages.

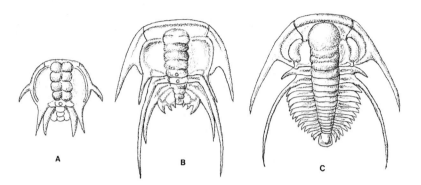

Figure 10.4   Ontogenetic stages of *Paradoxides*: protaspid (A); meraspid (B); and late meraspid (C) ontogenic stages of *Paradoxides*. (After H. Whittington. In R. C. Moore, (Ed.), *Treatise on Invertebrate Paleontology*, Pt. O, Arthropoda 1. Boulder, CO, and Lawrence, KS: Geological Society of America and University of Kansas Press, 1959.)

Trilobites passed through three larval stages during their postembryonic history, and each stage involved a succession of instars. During the earliest or **protaspis stage** (Fig. 10.4), the animal emerged from its egg as a tiny (0.5 to 1.0 mm) limbless larva covered by a uniform plate representing the beginning of the cephalon (i.e., the protocephalon). The protaspis stage ended and the next, or **meraspis**, stage began with an instar that possessed the first transverse suture separating the protocephalon from the protopygidium. In successive molts, thoracic segments appeared and the larva increased in size. When the adult number of thoracic segments is attained, the organism has reached the **holaspis** stage. Further molts are considered those of the adult organism and result only in an increase in size.

CATEGORIES OF TRILOBITES

Currently, the taxonomic status of various groups of trilobites has not been fully resolved. In the classification followed here, trilobites are divided into eight groups, each one approximating the taxonomic rank of either an order or a class. The Agnostida, Redlichiida, Corynexochida, Ptychopariida, Phacopida, Proetida, Lichida, and Odontopleurida comprise these eight major groups.

Agnostida (Cambrian to Ordovician)   Agnostids were very small (length about 8 mm) trilobites having similarly shaped cephalia and pygidia and only two or three thoracic segments (Fig. 10.5). Except in one family (the Pagetiidae), eyes were not present on the dorsal surface. Included within the Agnostida are two suborders, the Eodiscina (represented by the Middle Cambrian form *Eodiscus*) and the Agnostina (represented by *Agnostus* of the Early Cambrian). The Eodiscina possessed three thoracic segments and had transverse furrows on the pygidial axial lobe. Because the remains of agnostids occur in sedimentary rocks deposited in a variety of marine environments, it has been suggested that these tiny arthropods were pseudoplanktonic and crawled or swam about within masses of floating vegetation.

Figure 10.5   Agnostida.
(A) *Eodiscus*; (B) *Agnostus*. Length
0.7 mm.

Redlichiida (Early to Middle Cambrian)   These are relatively large trilobites noted for their distinctly semicircular cephalons and prominent genal spines, and the primitive kind of hypostome called *conterminant*. Spines often terminate the thoracic segments. Only a tiny pygidium exists. The eyes of redlichiids are usually large and crescent shaped, and sutures are of the opisthoparian type. The order includes such biostratigraphically useful fossils as *Redlichia* and *Paradoxides* (Figs. 10.6 and 10.7). In the Suborder Olenellina of the Redlichiida, facial sutures are of the fused or metaparian type. *Olenellus* (Fig. 10.8) is an example familiar to many geologists because of its importance in recognizing Lower Cambrian strata.

Corynexochida (Cambrian to Middle Devonian)   In general, species in this small group resemble olenellids except for their larger pygidia and fewer thoracic segments. Sutures are of the opisthoparian type. The glabella usually has parallel sides, but in a few species it expands anteriorly. In these trilobites, the hypostome is fused to the rostrum so as to form a single rostral hypostomal plate. Examples from the Middle Cambrian include *Zacanthoides* (Fig. 10.9) and *Bathyuriscus* (see Fig. 10.1).

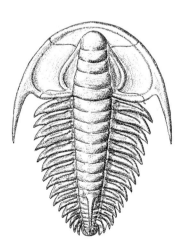

Figure 10.6   *Redlichia*. Maximum length about 10 cm. (This and all subsequent figures of trilobites are redrawn from H. J. Harrington et al. In R. C. Moore (Ed.), *Treatise on Invertebrate Paleontology*, Pt. O, Arthropoda 1. Boulder, CO, and Lawrence, KS: Geological Society of America and University of Kansas Press, 1959.)

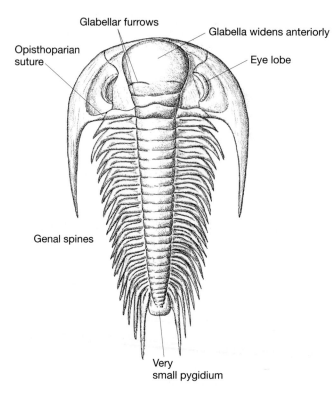

Glabellar furrows

Glabella widens anteriorly

Opisthoparian suture

Eye lobe

Genal spines

Very small pygidium

Figure 10.7   *Paradoxides.* Maximum length about 0.5 meter.

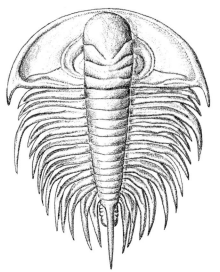

Figure 10.8   *Olenellus.* Maximum length about 40 cm.

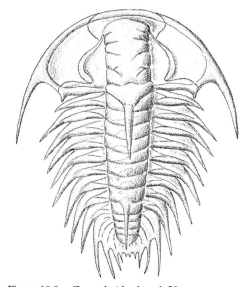

Figure 10.9   *Zacanthoides*: length 20 cm.

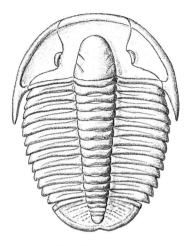

Figure 10.10   *Elrathia*: length 2.3 cm.

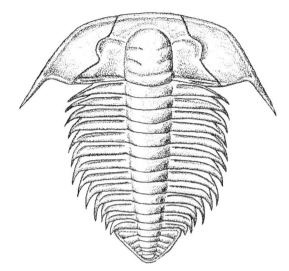

Figure 10.11   *Olenus*: length 1.4 cm.

Ptychopariida (Early Cambrian to Devonian) The Ptychopariida are the largest order of trilobites, and difficult to characterize as a group because of their considerable heterogeneity. Most have opisthoparian facial sutures and prominent, backwardly directed glabellar furrows. They are thought to be closely related to the Redlichiida, and may also be the ancestors of many nonagnostid post-Cambrian trilobites. The following five subdivisions (approximate suborders) of ptychopariids are recognized:

1. Ptychopariina (Early Cambrian to Devonian) Distinctive features of members of this group include a glabella that tapers anteriorly and that ends just short of the anterior margin of the cephalon; a prominent thorax; and straight glabellar furrows. Examples are *Elrathia* (Middle Cambrian, Fig. 10.10), *Olenus* (Late Cambrian, Fig. 10.11), and *Triarthus* (Ordovician, Fig. 10.12).

2. Illaenina (Ordovician to Devonian) These are trilobites with opisthoparian sutures, indistinct or lacking glabellar furrows, broad doublures, and pygidia about the same size or larger than the cephalons. *Bumastus* (Fig. 10.13), a Late Silurian guide fossil, displays the smoothly inflated appearance of many members of this group.

3. Asaphina (Late Cambrian to Late Ordovician) Typical members of the Asaphina have large, smooth exoskeletons with pygidia and cephalia about equal in size (a condition termed **isopygous**). Facial sutures are opisthoparian, and the thorax has six to nine segments. The hypostome is elongate and prominent. Such streamlined trilobites as *Asaphus* (Early to Middle Ordovician, Fig. 10.14), *Isotelus* (Middle to Late Ordovician, Fig. 10.15), and *Homotelus* (Late Ordovician, Fig. 10.16) exemplify this group.

4. Trinucleina (Early Ordovician to Middle Silurian) Most trilobites in this group have a distinctive appearance. The cephalon is large and semicircular. There are

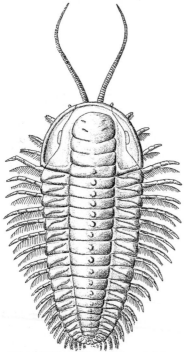

Figure 10.12   *Triarthus*: length 2.7 cm.

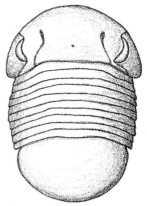

Figure 10.13   *Bumastus*: length 5.7 cm.

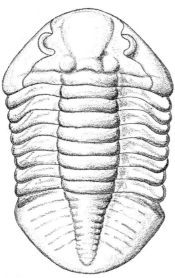

Figure 10.14   *Asaphus*: length up to 8 cm.

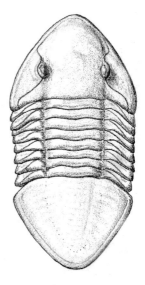

Figure 10.15   *Isotelus*: length varies from about 3 to 15 cm.

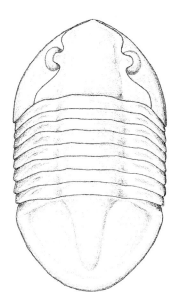

Figure 10.16   *Homotelus*:
maximum length about 15 cm.

long genal spines, and many forms have a cephalic border that is extensively pitted so as to resemble a lacy frill. Sutures are of the opisthoparian type. The thorax
is relatively short and contains five to seven segments. The pygidium is smaller
than the cephalon. *Cryptolithus* (Fig. 10.17) and *Trinucleus* (Fig. 10.18), both of
the Early to Middle Ordovician, are representative genera.

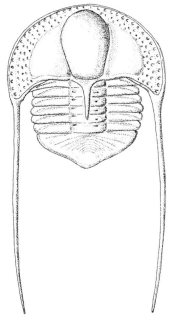

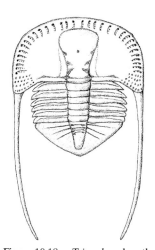

Figure 10.18   *Trinucleus*: length
1.7 cm.

Figure 10.17   *Cryptolithus*: length
2.7 cm.

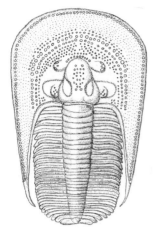

Figure 10.19   *Harpes*:
maximum length 4 cm.

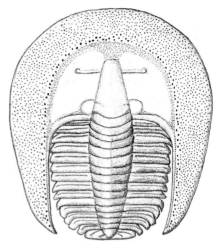

Figure 10.20   *Eoharpes*: length 1.8 cm.

5. Harpina (Early Ordovician to Middle Devonian) The Harpina include species having broad, often pitted cephalic borders. Long genal spines are often present on the cephalon, and these may extend the full length of the animal. The glabella stands high and narrows anteriorly. The eyes, which have only two to three lenses, are small. There are 12 or more thoracic segments. *Harpes* (Middle Devonian, Fig. 10.19) and *Eoharpes* (Fig. 10.20) are examples.

Phacopida (Ordovician to Late Devonian)   Phacopids have proparian or gonatoparian facial sutures and glabellae that tend to broaden anteriorly. They constitute a large and important order that includes the suborders Phacopina, Cheirurina, and Calymenina.

1. Phacopina (Ordovician to Late Devonian) Typically, phacopinids display proparian sutures, schizochroal eyes, and a thorax that includes 11 segments. Many Phacopina have proven valuable in biostratigraphic correlation, including Silurian to Devonian species of *Phacops* (see Fig. 10.2) and *Dalmanites* (Fig. 10.21). The former had well-developed enrollment ability and relatively small pygidia, whereas enrollment was less well developed in *Dalmanites* and the pygidium approximated the cephalon in size.

2. Cheirurina (Early Ordovician to Middle Devonian) Considerable morphologic variability exists among members of this suborder. Facial sutures, for example, may vary from the usual proparian condition to less common gonatoparian or even opisthoparian configurations. The glabella is prominent, expanded anteriorly, and has as many as four glabellar furrows. When present, eyes are holochroal and small. There are 8 to 19 thoracic segments. An overall spiny appearance resulting from spines on the pygidium and sawtooth margins of the thorax is characteristic of this group. Examples are *Cheirurus* (Late Ordovician to Middle Devonian, Fig. 10.22), *Deiphon* (Silurian, Fig. 10.23), and *Cerarus* (Middle to Late Ordovician, Fig. 10.24).

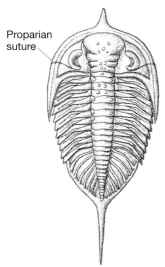

Proparian
suture

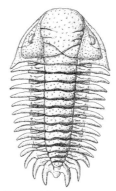

Figure 10.22
Cheirurus: length 5.5 cm.

Figure 10.21   *Dalmanites*: length
up to 10 cm.

3. Calymenina (Early Ordovician to Middle Devonian) In most members of this
   suborder, the glabella tapers toward the front and is deeply sculpted by four or
   five pairs of prominent glabellar lobes that diminish in size toward the anterior.
   Eyes are small and of the holochroal type. There are from 11 to 13 thoracic segments. Among the better-known calymeninids are *Calymene* (Early Silurian to
   Middle Devonian, Fig. 10.25), and *Trimerus* (Middle Silurian to Middle
   Devonian, Fig. 10.26). Compared to *Calymene*, *Trimerus* has a broader axial lobe
   and lacks the strongly lobate character of the glabella.

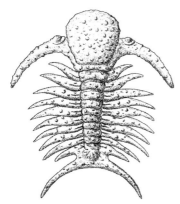

Figure 10.23   *Deiphon*: length 2.4
cm.

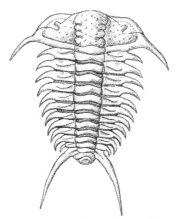

Figure 10.24   *Cerarus*: length 2.2
cm.

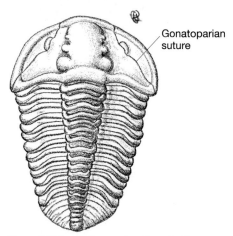

Gonatoparian
suture

Figure 10.25   *Calymene*: length up to 10 cm.

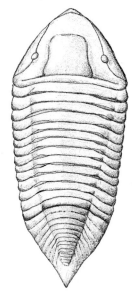

Figure 10.26   *Trimerus*: length up
to 20 cm.

Proetida (Late Cambrian?, Ordovician to Permian)   Trilobites in this long-ranging order typically have large glabellae, sturdy genal spines, and thickening of the margin of the cephalon. Opisthoparian sutures, large holochroal eyes, and eight to ten thoracic segments are characteristic. The pygidium is nearly the same size as the cephalon. *Proetus* (Ordovician to Devonian), *Bathyurus* (Middle Devonian), and *Phillipsia* (Early to Middle Mississippian) are examples (Fig. 10.27, 10.28, and 10.29).

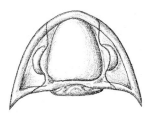

Figure 10.27   *Proetus*: cephalon
and pygidium. Complete specimens
measure up to 3.0 cm in length.

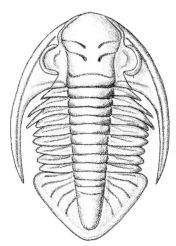

Figure 10.28   *Bathyurus*: length
4.0 cm.

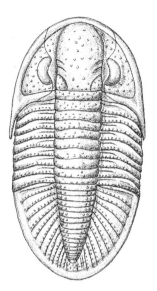

Figure 10.29   *Phillipsia*: length 2.8 cm.

Lichida (Upper Cambrian to Upper Devonian)   The Lichida are medium to large, distinctive trilobites. The large glabella extends to the forward edge of the cephalon, facial sutures are opisthoparian, and the pygidium is usually as large or larger than the cephalon. Spinosity is developed in many genera of lichids, and the exoskeleton is often covered with small tubercles. Conterminant hypostomes are characteristic of this group. *Lichas* (Late Ordovician to Middle Silurian) and the highly spinose *Terataspis* are examples (Figs. 10.30 and 10.31).

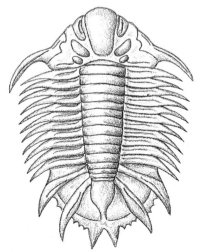

Figure 10.30   *Lichas*: length 1.5 cm.

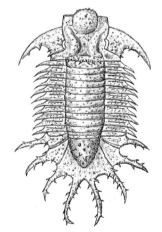

Figure 10.31   *Terataspis*: length up to 45 cm.

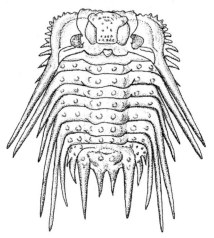

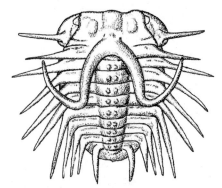

Figure 10.33   *Dicranurus*: length 4.0 cm.

Figure 10.32   *Odontopleura*: length up to 2 cm.

Odontopleurida (Late Cambrian?, Early Ordovician to Late Devonian)   The Odontopleurida (Figs. 10.32 and 10.33) are probably the most spinose order of trilobites. Most have long genal spines, spines at the lateral margins of thoracic segments, and spines along the border of the cephalon. Small spines, tubercles, and granules often cover the entire exoskeleton. Facial sutures are of the opisthoparian type. There are eight to ten thoracic segments. The pygidium is usually small and triangular in outline. Examples are the Silurian guide fossil *Odontopleura*, and the bizarre Devonian form *Dicranurus*.

TRILOBITES THROUGH TIME

Although trilobites are not the first animals to develop mineralized exoskeletons, they are at least prominent among the Early Cambrian invertebrates that achieved this condition. The Early Cambrian has been divided into five stages, designated the Nemakit-Daldynian (the oldest), Tommotian, Atdabanian, Botomian, and Toyonian (the youngest). The oldest trilobites are found in rocks of the Early Cambrian Atdabanian stage in northern Europe, Morocco, and Siberia. The host rocks for these early trilobites are about 530 million to 525 million years old. Slightly older Early Cambrian "small shelly fossils" occur in rocks of the Tommotian that began about 534 million years ago. The Tommotian fossils (see Fig. 3.12) include sponge spicules, primitive gastropod and monoplacophoran-type mollusks, mollusklike hyoliths, brachiopods, the first echinoderms, archaeocyathids, and the toothlike phosphatic conodonts that are the focus of an ongoing debate about their chordate affinities. Body fossils of trilobites are not known from the Tommotian, but tracks present along bedding planes of some strata suggest they may have been present. Perhaps trilobites lacked a mineralized skeleton at this time, and hence were not preserved.

Following their appearance in the Atdabanian, trilobites expanded rapidly (Fig. 10.34). The exoskeletons of many of these early trilobites are characterized by numerous thoracic segments, and pygidia that were either very small or completely absent. These early trilobites include *Olenellus* (see Fig. 10.8), a genus widely used in correlation of Early Cambrian rocks of the northeastern United States, Newfoundland, Scotland, and Greenland. Somewhat later in the Early Cambrian, the Agnostida made their appearance, and these were soon joined by members of the Corynexochida, Redlichiida, and Ptychopariida. During Middle Cambrian time, such genera as *Paradoxides*, *Ptychoparia*, and *Olenoides* became abundant. They were followed during the Late Cambrian by species of the "scoop-tailed" trilobite *Dikelocephalus* and *Olenus*. As the Cambrian drew to a close, however, trilobites underwent a widespread decline in both numbers and diversity. It was clearly a time of crisis, and only relatively few survived into the next geologic period. Paleontologists are uncertain about the cause of the extinctions, but contributing factors may have been the rise of predatory cephalopods and the global restriction of epeiric seas near the end of the Cambrian.

The survivors of the Late Cambrian episode of extinctions provided the stock for a second expansion of trilobites that began in the early stages of the Ordovician. Members of the Illaenina, Trinucleina, Cheirurina, and Lichida became particularly numerous. Two new orders, the Proetida and Lichida, made their appearance. The basic designs developed in these Ordovician groups persisted throughout most of the remainder of the Paleozoic.

With the exception of the Redlichiida, Corynexochida, and Agnostida that became extinct by the end of the Ordovician, Silurian and Devonian trilobites represent a continuation of Ordovician lineages. Calymenids were particularly abundant during the Silurian, whereas the most ubiquitous Devonian trilobites were the Phacopida. For a time, they were quite successful, but by the Middle to Late Devonian, trilobites experienced another episode of hard times. Only a few groups were able to carry on into the Carboniferous. The final blow for the trilobites came during the Middle Permian when all of the remaining families disappeared. Eustatic decline in sea level and consequent loss of epeiric sea, shelf, and reef habitats may have hastened their demise.

As frequently stated, the very best fossils for biostratigraphic correlation have short geologic ranges and wide geographic distribution. The former criterion was amply met by early Paleozoic trilobites, but the latter was not always achieved. Most trilobites were benthic creatures that lived in shelf areas. Thus, they tended to be provincial (confined to a particular region). For this reason, trilobite biozones are often coordinated with biozones of more widely distributed organisms such as graptolites.

Although the provinciality of trilobites may have diminished their usefulness in far-ranging correlations, it enhanced their value in paleogeographic studies. For example, the provinciality of trilobite faunas in Early Paleozoic rocks of western Ireland, Scotland, part of eastern North America, Greenland, and Newfoundland indicates that these areas were once joined. An ancient oceanic tract called Iapetus separated these faunas from the distinctly different faunas of what is now England, Wales, southeast Ireland, and New Brunswick. Iapetus, however, was subsequently to close, and as it narrowed, the once widely separated regions approached one another. Eventually shelf area met shelf area. The trilobite faunas intermingled, lost their distinctive traits, and produced a generally cosmopolitan assemblage.

As one surveys the general history of trilobites, certain morphological trends become apparent. These seem to occur in several independently evolving lineages. The first is the previously noted reduction in the number of thoracic segments. This may represent a trend away from multisegmented ancestors of both trilobites and annelids. A second evolutionary trend was the development of distinct pygidia as a result of fusion of posterior segments and gradual weakening of segmental borders. The process also appears to have involved a loss of pleural pygidal segments, so that the axial lobe possessed more segments than did the pleural lobes. Cephalization, or the fusion of furrows on the glabella, constitutes a third trend among trilobites. Glabellar furrows tend to become progressively weaker in trilobites of successively younger geologic ages until they ultimately disappear.

TRILOBITE HABITS AND HABITATS

A measure of uncertainy accompanies attempts to infer the living habits of animals having no living representatives, as is the case with trilobites. Nevertheless, on the basis of associated organisms, one can at least state that trilobites were entirely marine, and that they moved about in search of food on, in, or just above the ocean floor. Most lived in the nearshore shelf environment, and some were successful in colonizing reefs within that relatively shallow realm. Fewer species lived in deep water. Morphological features of particular groups provide the basis for a few additional inferences about trilobite habits. Thin-shelled forms with large pygidia may have been able to swim for short distances above the ocean floor. Those with heavy exoskeletons and small pygidia probably favored crawling over swimming. Those with scoop-tailed pygidia and crescentic cephalons were likely to have been burrowers. Spines may have had more than one function. In some cases they appear to have been used to moor the animal to the seafloor to provide stability against strong currents. In the exceptionally spinose forms, it seems likely that the spines helped to spread body weight over a greater area to improve flotation. In a similar way, laterally directed spines may have acted somewhat like pontoons in preventing the animal from sinking into seafloor oozes.

Blind or nearly sightless trilobites like the trinucleids were probably infaunal organisms, meaning that they searched for food within the soft upper layers of sediment on the seafloor. The living horseshoe crab *Limulus* seeks its food in this way. Discovery of a few specimens preserved in their original living positions indicates that some species of *Bumastus*, *Illaenus*, and *Asaphus* burrowed "tail first" into sediment with their cephalons poised horizontally above the substrate to detect and consume food.

Trilobite crawlers and burrowers have sometimes left trace fossils that provide clues to their habits. Ribbonlike markings with oblique lateral impressions are commonly interpreted as the trails of crawling trilobites. They are termed *Cruziana* (Fig. 10.35). Shallow pits called by the trace fossil name *Rusophycus* mark the former resting places of trilobites like *Calymene*, some of which have been found preserved in their own excavations.

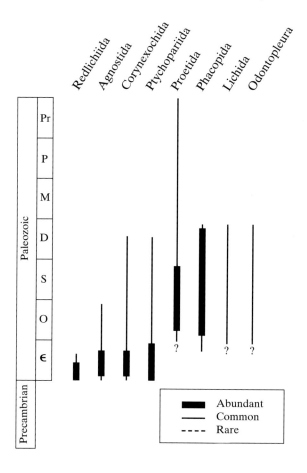

Figure 10.34   Geologic ranges of major groups of Trilobita.

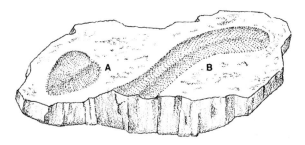

Figure 10.35   Trilobite trace fossils. (A) *Rusophycus* (a trilobite resting trace) and.(B) *Cruziana* (a tribolite trail).

SUPERCLASS CRUSTACEA, CLASS OSTRACODA

The small, bivalved crustaceans known as ostracodes may exceed even the trilobites in their value for biostratigraphic studies. Not only are they exceptionally abundant and diverse, but their small size also provides an additional advantage. Most ostracodes do not exceed 3 or 4 mm in length, and are therefore recovered fully intact from well cuttings and cores that contain only fragments of large fossils. As a result, ostracodes are extensively used in correlation and paleoecologic studies of subsurface strata.

Ostracodes first appear in Cambrian rocks, and they have survived to the present day. They have successfully invaded every aquatic habitat, including freshwater lakes and streams, transitional environments of brackish water, and nearly all depths of the ocean. Some living species of the ostracode *Mesocypris* can even crawl out of ponds to seek its food in damp plant debris. Specimens obtained from ponds are easy to maintain in small aquarium tanks like those used for tropical fish. Freshwater ostracodes are sometimes called "seed shrimp." Although planktonic species are known, most have been benthic animals. They crawl, swim, or burrow, and they seem to prefer relatively quiet waters having bottom sediment rich in organic matter. Ostracodes consume a variety of nutrients, including algae, smaller animals and larvae, and the tissue of dead marine animals of all kinds.

THE LIVING ANIMAL

The body of an ostracode (Fig. 10.36), along with its appendages, is enclosed in a bivalved exoskeleton. It is suspended from the dorsal region of the shell with lateral attachments to either valve for additional support. During growth, the bivalved exoskeleton or **carapace** is shed many times, resulting in a sequential series of **instars** (molted carapaces). The carapace partially hides seven pairs of segmented appendages that mark the ostracodes as *bona fide* members of the arthropod clan. The fragile appendages of ostracodes do not preserve well, but a remarkable discovery of ostracodes with phosphatic preservation has greatly improved our understanding of ostracode appendage morphology. The phosphatic fossils occur in Cambrian rocks of Sweden. At the time of deposition, ostracodes at that fossil locality slipped into an oxygen-deficient zone near the seafloor where they were covered by phosphatic bacteria. The bacteria formed a thin coating around the ostracodes that precisely duplicated the tiny components of the appendages. The head region comprises much of the body of an ostracod and bears antennae and antennules. The trunk is greatly reduced, and all traces of trunk segmentation have disappeared.

An ostracode chews its food with the aid of a pair of biramous **mandibles** alongside the mouth, and biramous **maxillae** just behind the mouth. In most ostracodes there are no gills. Respiration is accomplished by diffusion of gases through their thin outer body wall. Such a system appears to work very well in animals as small as ostracodes.

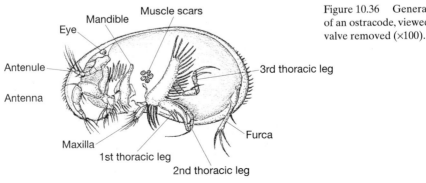

Figure 10.36   General morphology of an ostracode, viewed with left valve removed (×100).

The sexes in ostracodes are separate, and males and females exhibit sexual dimorphism. Males have paired sperm ducts and clasp the female during copulation. Ostracode spermatozoa are enormous relative to the size of the adult male. An adult *Pontocypris monstrosa*, for example, is only 1 mm long, yet produces sperm six times that long. Eggs produced by the females are usually shed directly into the water, or attached in small clusters to aquatic plants. In some species, the eggs are retained in the female's brood pouch. The hatched larvae already have tiny bivalved carapaces resembling those of their parents.

THE OSTRACODE CARAPACE

Both the identification and the classification of ostracodes are based on the size, shape, and ornamentation of the carapace, the manner in which the valves overlap, characteristics of the hingement, and the general structure and surface texture of the valves. The carapace has a generally hydrodynamic, bean-shape, or ovate form. The two valves (as in the Bivalvia) are designated *right* and *left*. They are held together dorsally by a ligament and transversely by adductor muscles. These muscles leave a cluster of scars that can be seen on the interiors of well-preserved valves. The number and arrangement of muscle scars are important criteria for classification of ostracodes.

Valves are secreted by two lateral sheets of soft tissue that partially enclose the body. The valves are of unequal size, so that one slightly overlaps the other. Each consists of an **inner lamella** and an **outer lamella** (Fig. 10.37). The chitinous base of the outer lamella is usually heavily calcified, except along the dorsal margin where an uncalcified zone exists that serves as the ligament. The inner lamella is chitinous except toward the periphery where it is calcified. A **line of concrescence** or fusion is formed where the inner and outer lamellae are joined. The space enclosed by the two lamellae is designated the **vestibulum**, and the area between the line of concrescence and the edge of the valve is simply called the **marginal zone**. The way in which the valves are hinged is important in the classification of ostracodes. The hingement is termed **adont** if teeth and sockets are lacking, but articulation occurs where a single groove along the margin of the larger valve meets a corresponding ridge on the smaller valve. In **merodont** hingement, teeth and sockets exist at either end of the hinge, and between these is an elongate groove and corresponding ridge. When that ridge and groove median element is more strongly crenulated or lobed, the hingement is termed **entomodont**. Finally, in **amphidont** hingement, the median portion contains another tooth and socket. There are, however, many variations of these more or less basic hingement structures.

The bean-shape of many ostracode carapaces makes recognition of the anterior and posterior ends difficult. The hinge, of course, identifies the dorsal margin. Viewing the ostracode from the side with the hinge uppermost, the more rounded extremity is usually anterior, and the more pointed end posterior. Muscle scars are located slightly anterior to the center of the valves. A further aid to orientation are the eye spots, which are in the anterodorsal position. As a general rule, spines and winglike extensions of the carapace are directed toward the posterior. Finally, in most Paleozoic ostracodes, the posterior end of the carapace is wider.

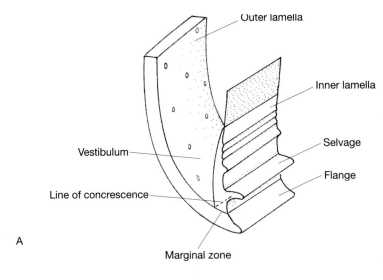

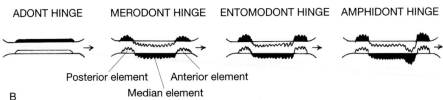

Figure 10.37   (A) Diagram of the wall structure of a podocopid ostracode. (B) Schematic drawings of dorsal views of some ostracode hinge types. Black areas represent grooves or sockets. There are many variations on these basic patterns. (Wall structure modified from R. H. Benson et al. In R. C. Moore (Ed.), *Treatise on Invertebrate Paleontology*, Pt. Q, Arthropoda 3. Boulder, CO, and Lawrence, KS: Geological Society of America and University of Kansas Press, 1961. Hinge structures after F. van Morkhoven. 1962. *Post-Paleozoic Ostracoda*, Vol. 1. Amsterdam: Elsevier.)

MAJOR GROUPS OF OSTRACODES

Although biologists emphasize appendage morphology as a basis for classifying extant ostracodes, paleontologists of necessity concentrate on features of the carapace. In a workable but not universally accepted scheme, both extant and extinct ostracodes are placed in one of the following five orders: Archaeocopida, Leperditicopida, Palaeocopida, Podocopida, and Myodocopida.

Archaeocopida (Early Cambrian to Ordovician)   Archaeocopida are primitive ostracodes that are considered ancestral to the Leperditicopida. They are known to have existed in Early Cambrian time. These ancient ostracodes have weakly calcified, flexible carapaces. A few species have carapaces composed of calcium phosphate. The dorsal margin is straight and lacks hinge elements. The outer surface of the carapace may be smooth, wrinkled, punctate, or ribbed. It bears an eye tubercle. *Bradoria* (Fig. 10.38A) is an example of a Cambrian archaeocopid.

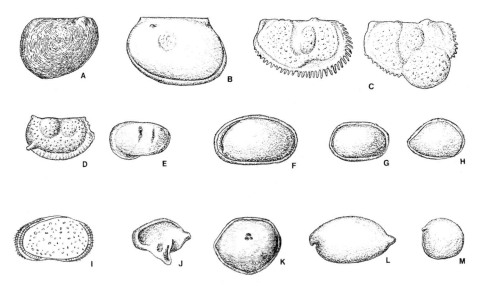

Figure 10.38   Representatives of the major orders of Ostracoda. (A) *Bradoria* (Archaeocopida), right valve, ×8. (B) *Leperditia* (Leperditicopida), left valve, ×1.5. (C) *Beyrichia* (Palaeocopida), right valve of male (right) and right valve of female (with brood pouch), ×1.8. (D) *Hollinella* (Palaeocopida), right valve, ×20. (E) *Kloedenella* (Palaeocopida), right valve, ×20. (F) *Healdia* (Podocopida), right valve, ×60. (G) *Cytherella* (Podocopida), left valve, ×30. (H) *Bairdia* (podocopida), right valve, ×45. (I) *Cyprideis* (Podocopida), left valve, ×30. (J) *Cytheropteron* (Podocopida), left valve, ×18. (K) *Polycope* (Myocopida), left valve, ×30. (L) *Cypridina* (Myocopida), left valve, ×20. (M) *Entomoconchus* (Mycocopida), left valve, ×0.8.

Leperditicopida (Ordovician to Devonian)   The leperditicopids are exceptionally large ostracodes, ranging in length between 5 and 30 mm. The carapace is smooth, thick, and exhibits an astonishing array of about 200 muscle scars. The dorsal margin is long and straight, whereas the ventral border is convex. Leperditicopids are a step ahead of the archaeocopids in their development of adont hingement. *Leperditia* (Fig. 10.38B), which was abundant and widespread during the Silurian and Devonian periods, serves as a good example of this group.

Palaeocopida (Early Ordovician to Early Triassic)   The carapaces of palaeocopid ostracodes usually have long, straight hinge lines, as well as distinctive sulci and lobes. **Sulci** are troughs or grooves indenting the outer surface of the valves, whereas **lobes** are variously shaped, rounded protuberances. Females are recognized by their prominent brood pouches. Among the more widely distributed genera (Fig. 10.38 C,D,E) are *Beyrichia* (Silurian to Devonian), *Hollinella* (Devonian to Permian), and *Kloedenella* (Silurian and Devonian).

Podocopida (Early Ordovician to Holocene)   Most of the Mesozoic and Cenozoic ostracodes are members of the Podocopida (Fig. 10.38F–J). The valves of podocopids are well calcified and unequal in size. As a result, the slightly larger valve overlaps the smaller valve. The dorsal margin is straight or arcuate, and the central margin concave or straight. Hingement may be adont, merodont, or amphidont, with the latter two types more prevalent. The number of muscle scars rarely exceeds ten. This large order of ostracodes includes many genera whose names are very familiar to micropaleontologists. Among these are *Healdia* (Devonian to Permian), *Cytherella* (Jurassic to Holocene), *Cyprideis* (Miocene to Holocene), *Bairdia* (Silurian to Holocene), and *Cytheropteron* (Jurassic to Holocene).

Myodocopida (Ordovician to Holocene)   The myodocopids are a rather heterogeneous, mostly planktonic group. In most, the carapace is thin, and the interiors of the valves display many elongate muscle scars. There is usually a notch developed in the carapace through which the second antennae protrude. The valves, which are nearly equal in size, have no overlap. Hingement is of the adont type. Examples (Fig. 10.38K–M) are *Polycope*, which occurs in Jurassic to Holocene sediments (except for questionable Devonian occurrences); *Cypridina*, a Late Cretaceous to Holocene genus with a prominent anterior projection called a **rostrum**, *Entomoconchus*, a Mississippian genus characterized by an opening at the posterior margin termed a **siphonal gap**.

OSTRACODES THROUGH TIME

Of the five orders of ostracodes described above, only the Archaeocopida and Leperditicopida are found exclusively in Paleozoic rocks (Fig. 10.39). The Palaeocopida, which appeared in the Ordovician, include a few species that survived until Triassic time, but most members of this order are restricted to the Paleozoic. Members of the Myocopida are found in rocks of Ordovician to Holocene age. The Podocopida are the dominant ostracodes of the Mesozoic and Cenozoic.

The oldest ostracodes are the archaeocopids, which appear along with trilobites at the beginning of the Cambrian Period. They were largely supplanted in late Cambrian time by the Leperditicopida, which continued with moderate success until the end of the Devonian. The heyday for all Paleozoic ostracodes appears to have been in the Ordovician, as indicated by a dramatic increase in ostracode abundance and diversity. At that time, the Palaeocopida and Podocopida underwent major expansions.

The first freshwater ostacodes (Cypridacea of the Order Podocopida) evolved during the Devonian. Mesozoic and Cenozoic podocopids and myodocopids provided richly diverse faunas that occupied brackish water, lacustrine (lake), and marine environments.

OSTRACODE HABITS AND HABITATS

In addition to their use in delimiting biostratigraphic zones, ostracodes are valuable indicators of ancient environmental conditions, including water depth, salinity, bottom sediment, and temperature. Their value as environmental indicators has been enhanced by comparison of fossil forms to species living today.

With regard to freshwater ostracodes, most living and fossil species are characterized by smooth or slightly punctate carapaces having adont hingement and lacking eye spots. Live species can be observed swimming a few centimeters above the substrate, digging through organic-rich bottom debris, or feeding in clusters on bits of carrion. Brackish-water ostracodes tend to have similar habits. Ostracodes inhabiting marine waters of normal

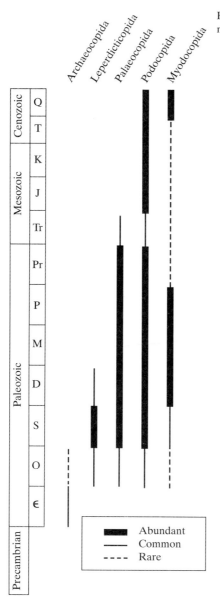

Figure 10.39   Geologic ranges of major groups of Ostracoda.

salinity are generally more robust than their freshwater or brackish-water contemporaries. As a group, they are also far more diverse. Those living in turbulent nearshore areas have sturdy carapaces that are often coarsely sculptured with heavy spines and reticulated valve patterns. Benthic ostracodes inhabiting quieter, deeper areas are smaller and have smoother carapaces. Although able to swim, they spend most of their time crawling on, or burrowing into, silt and mud on the seafloor. Areas of coarse sand or oxygen-poor muds are generally avoided by ostracodes. In addition to the marine bottom dwellers, many species of ostracodes are lifelong swimmers. These truly pelagic forms tend to thrive in areas where nutrients are brought to the ocean surface by upwelling currents.

SUPERCLASS CRUSTACEA, CLASS MALACOSTRACA

Lobsters, shrimps, crabs, and crayfish are all members of the Class Malacostraca. As one might judge from the frequency with which they appear on the dinner table, malacostracans are abundant aquatic arthropods. Although their fossil record extends from Late Devonian to the present, they are not common as fossils. The two principal subdivisions of the Malacostraca are the Phyllocarida (Cambrian to Holocene) and the more advanced Eumalacostraca (Devonian to Holocene). In the Phyllocarida, the carapace is either sharply folded or bivalved with a dorsal hinge. *Echinocaris* (Fig. 10.40A) is a representative from the Devonian, whereas *Nebalia* (Fig. 10.40B) is widely distributed in today's ocean.

Eumalacostracans have abdomens constructed of six segments, all of which bear appendages. There are several taxonomic divisions of this group of crustaceans, but only one, the Decapoda, has a significant fossil record. That record includes rare body fossils of lobsters, crabs, and shrimps. The decapods have three pairs of thoracic appendages modified for feeding. The remaining five pairs of thoracic appendages function as legs. They are the basis for the name Decapoda, meaning ten legs. The first (and sometimes the second) pair of these thoracic legs are usually enlarged and modified to form the pincers or **chelae** so prominently developed in crabs and lobsters. Behind the thoracic legs are smaller abdominal appendages that function in walking and swimming, and behind these are the final segments and the telson that forms a paddle for swimming. Lobsters use the telson (terminal segment) to swim rapidly backward. In this way, a pursuing predator is confronted by the chelae with their wound-inflicting potential. The chelae, however, are not only used in defense but also in feeding and courtship.

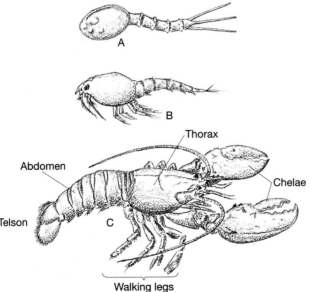

Figure 10.40   Three crustaceans of the Class Malacostraca. (A) *Echinocaris*, ×0.4; (B) *Nebalia*, ×5; and *Homarus*, ×0.3.

Among the decapods, the lobster *Homarus americanus* is well known because of its commercial importance along the New England coast of the United States. The body of this benthic animal is composed of 21 segments, the first 14 of which are united into a large cephalothorax—a combined head and thorax. *Homarus* is dark green when alive, but when boiled for a meal the calcium carbonate impregnated carapace turns red. The earliest fossil record for lobsters is Jurassic. Splendidly preserved fossils of lobsters, as well as shrimps and crayfish, are recovered from the famous Solenhofen Limestone of Bavaria (also spelled "Solnhofen").

Because lobsters are rather secretive animals that inhabit subtidal waters, we rarely see them on excursions to the beach. Crabs, however, are frequently observed skittering over sand and rock along coastlines. The distinctive pill-box shape of crabs is the result of flexing of the abdomen so that it lies beneath the broad cephalothorax, and does not trail behind as in lobsters. Like lobsters, however, crabs have prominent chelae. The fossil record for crabs extends back to the Triassic.

There are approximately 4500 species of crabs living within a broad range of habitats. Many are benthic and reside at great depths in the ocean. Others prefer wave-washed shorelines. There are also marine planktonic crabs, and crabs that live on land. There is even a tree-climbing crab that feeds on coconuts. Several species live in freshwater lakes and streams. Many of us are familiar with such edible species as the East Coast blue crab (the "soft-shell crab" if molting has just occurred), and the West Coast Dungeness crab.

Shrimps (prawns) are the third group of familiar decapods. Most shrimps have cylindrical to somewhat compressed abdomens and thin, flexible exoskeletons. The first three pairs of legs are chelate, but the chelae are not prominent. Some shrimps are pelagic, living primarily in the upper 1000 meters of the ocean. Among these are species with luminescent organs located internally or on the body surface. Most shrimps, however, are not truly pelagic, but live on the sea bottom where they can use their legs for crawling or intermittent swimming. Species of *Penaeus* are the most important commercially, and they are gathered by the millions in nets that are towed across the seafloor. Shrimps are rarely found as fossils. Their meager fossil record, however, does extend back to the Permian. Shrimplike crustaceans probably existed much earlier.

SUPERCLASS CRUSTACEA, CLASS BRANCHIOPODA

Branchiopods are more familiarly known as "fairy shrimps" and "water fleas." They are mostly small crustaceans having shrimplike bodies and appendages that are flattened and leaflike in shape. Although the body of most branchiopods is enclosed within a carapace, the fairy shrimps lack this covering. They are recognized by their multisegmented, elongate trunk, as seen in *Branchinecta* (Fig. 10.41A).

The only group of branchiopods having a noteworthy fossil record are the conchostracans, or clam shrimps. *Cyzicus* (Fig. 10.41B) is a conchostracan found in certain Mesozoic rocks. Conchostracans take their name from the resemblance of the carapace to that of a small clam. The carapace encloses the entire body, but it is not actually composed of two separate valves. Rather, it consists of two flaps that are thin and flexible, but continuous across the dorsal margin. Also, the carapace is not shed during molting. The animal molts only the exoskeletal elements of the body. Each molting event, however, is recorded by a concentric ridge on the carapace, reflecting a thickening of the margin at each time of molting.

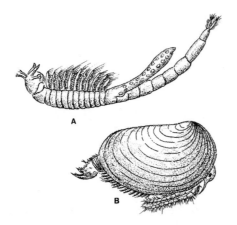

Figure 10.41   Two members of the Branchiopoda. (A) The fairy shrimp *Branchinecta* in normal position, swimming on its back (×2.5). (B) The conchostracan *Cyzicus*, a branchiopod. (Modified from R. C. Moore, C. G. Lalicker & A. G. Fischer, *Invertebrate Fossils*. New York: McGraw-Hill, 1952.)

Branchiopods are rather unusual among crustaceans in that there are relatively few marine species. Fossil branchipods are found in sediments deposited in brackish or freshwater bodies. Brine shrimp do very well in saline lakes. The remains of fossil conchostracans are sometimes found in Pennsylvanian, Triassic, and Cretaceous strata.

SUPERCLASS CRUSTACEA, CLASS CIRRIPEDIA

Cirripeds are barnacles. They are a persistent group that has survived from Cambrian to the present (*Priscansermarinus* is a presumed fossil barnacle from the Burgess Shale). Cirripeds take their name from the curly tufts of appendages that can be extended from the open end of their exoskeletons. Barnacles (cirripeds) have always been ocean dwellers. They also have the distinction of being the only crustaceans that are exclusively sessile. One usually finds them firmly attached to rocks, shells, coral, driftwood, and other objects such as the hulls of ships. Although adult barnacles do not appear to resemble other crustaceans, a closer examination (as described by Louis Agassiz) reveals that "a barnacle is nothing more than a little shrimplike animal standing on its head in a limestone house and kicking food from its mouth."

According to the manner in which barnacles attach themselves to other objects, one can distinguish three groups of living barnacles. Pedunculate forms have a stalk surmounted by an almond-shaped crown of calcareous plates. Because of their prominent stalk or peduncle, they are sometimes called "gooseneck barnacles." *Lepas* (Fig. 10.42A) is a typical gooseneck barnacle. Pedunculate barnacles are usually found attached to floating objects, and are thus pseudoplanktonic. The pedunculate barnacles range in age from Silurian to Holocene, with questionable occurrences in the Cambrian. *Praelepas*, from the Pennsylvanian of Russia is representative. Although rare in Paleozoic rocks, pedunculate barnacles become more abundant during the Mesozoic where they are represented by species of *Archaeolepas* (Jurassic) and *Stramentum* (Cretaceous).

The second major group of barnacles are the volcano-shaped forms exemplified by *Balanus* (Fig. 10.42B). Balaniform barnacles lack a fleshy stalk. They attach the broad base of their skeletons directly to a host surface. This may be a firm object on the

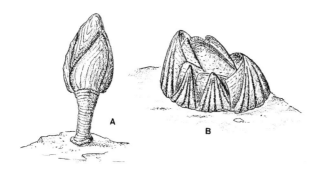

Figure 10.42 (A) The pedunculate or gooseneck barnacle *Lepas* (×0.5). (B) The volcano-shaped barnacle *Balanus* (×0.5).

seafloor, floating debris, or even the skin of a whale. The oldest balaniform barnacles are Jurassic (questionably Triassic), but they are not common until Cretaceous time.

The third major group of barnacles are the acrothoracians, represented today by about 30 species. Usually only a few millimeters in length, these are the smallest cirripeds. Acrothoracians bore into coral or mollusk shells and live in the burrows.

SUPERCLASS CHELICERATA

Horseshoe crabs, scorpions, spiders, ticks, and mites are examples of living members of the Chelicerata. The extinct eurypterids are probably the most widely known fossil chelicerates. However, the earliest known chelicerate is *Sanctacaris*, from the Middle Cambrian Stephen Formation of the Burgess Shale in British Columbia. This predatory creature had a wide head shield, six pairs of biramous head appendages, a trunk composed of 11 segments, and a paddlelike telson to assist in steering and swimming. The number and structure of head appendages and body divisions in *Sanctacaris* occur somewhat modified in such later aquatic chelicerates as the eurypterids and horseshoe crabs.

Chelicerates are set apart from other arthropods by their lack of antennae. Most members of this group possess a pair of jointed chelae as their first appendages. These structures are found in front of the mouth, and are followed successively by a pair of **pedipalps** that are used in food gathering, and four pairs of walking legs. The body of a chelicerate is divided into a forward portion or **prosoma** (formed by coalescence of six cephalothoracic segments bearing appendages), and an abdominal **opisthosoma** (composed of as many as a dozen segments without appendages). The Chelicerata can be divided into two taxonomic classes, namely the Merostomata (which has the most paleontological importance), and the Arachnida (which although diverse has provided few fossils).

MEROSTOMATA

Included within the Merostomata are the horseshoe crabs and eurypterids. Horseshoe crabs are not crabs at all, but are in fact more closely related to spiders. Two representative genera are *Limulus*, which lives today, and *Mesolimulus*, fossils of which are found in the Jurassic Solenhofen Limestone. Like *Mesolimulus, Limulus* has a strikingly large prosoma that bears compound eyes and dorsal, light-sensitive organs called **ocelli**. The opisthosoma includes nine segments, and is followed by a sturdy, pointed telson. The animal lives in shallow coastal waters of the Atlantic and Gulf of Mexico. *Limulus* prefers muddy and sandy substrates into which it burrows by pushing through sediment with telson and appendages. The shape of the prosoma facilitates burrowing and provides protection for the ventrally located legs. Horseshoe crabs are able to walk along the seafloor, swim by flapping their abdominal plates, and push themselves along with the aid of their telson.

Eurypterids (Fig. 10.43) are merostomates that may have the most malevolent appearance of all Paleozoic invertebrates. Not only did some reach spectacular size, but many bore vicious-looking claws. Eurypterids, or water scorpions as they are sometimes called, were already present during the Ordovician, but apparently in only modest numbers. Their numbers increased during the Silurian when they probably presented a threat to contemporaneous invertebrates occupying the same environment. Their success, however, was not sustained, for the group declined during the Devonian. In Permian time the eurypterids became extinct.

The largest eurypterid known from North America is *Pterygotus buffaloensis*. With its nearly 2 meters of length, it is the largest arthropod of all time. Most eurypterids, however, were less than a meter in length. They apparently evolved in the ocean, but subsequently invaded brackish, freshwater, and hypersaline environments. Their protected book gills may even have permitted brief incursion on land. (Book gills consist of parallel laminae that provide for gas exchange between the blood circulating in the laminae and the surrounding water).

Compared to horseshoe crabs, the prosoma of an eurypterid is small and narrow. The opisthosoma is constructed of 12 articulating segments that terminate in a postanal telson. Visual organs include two large compound lateral eyes and a pair of median simple eyes. Six pairs of appendages surrounded the mouth. The most anterior of these were chelate, whereas the last pair resembled a powerful paddle that could have been

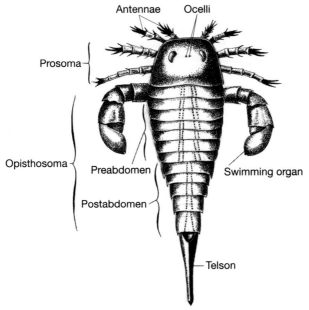

Figure 10.43 Dorsal view of *Eurypterus*, a Silurian eurypterid. Average length 30 cm. (From H. L. Levin, *The Earth Through Time*, 5th ed. Philadelpia: Saunders College Publishing, 1996.)

used both for swimming and digging. Indeed, the winglike aspect of these legs is the source of the name eurypterid. *Eurys* is from the Greek for "broad," and *pteron* means "wing." This resemblance to wings was noticed over a century ago by Scottish quarrymen who came upon eurypterids in the lower Old Red Sandstone. According to an account by Charles Lyell, the workers called these fossils "seraphim," although any resemblance to angels would seem to require an extraordinary stretch of imagination.

ARACHNIDA

Few invertebrates are regarded with as much alarm as spiders and scorpions (Fig. 10.44) of the Class Arachnida. Perhaps this is because some of these arthropods live in dark places and trap their prey in webs. Their sinister reputation is enhanced by an ability to inject poison into their victims, to make a meal of blood, or to spread disease, yet many spiders are predators that perform a service to humans by dining on insects that are a greater nuisance than the spiders themselves. Like eurypterids, the two prominent divisions of the arachnid body are the prosoma and opisthosoma. There are also six pairs of appendages emanating from the prosoma. These include a pair of wound-inflicting **chelicerae**, followed by a pair of **pedipalps** (serving both a grasping and sensory function), and four pairs of walking legs. (In contrast, insects have only three pairs of walking legs.) The chelicerae of spiders may be modified for the injection of poison. Spiders do not swallow solid food. Rather, the animal inflicts a wound with its chelicerae and then injects digestive juices into the wound. It then sucks up the liquefied and predigested tissues of its victim. Other characteristics of living arachnids include simple eyes (rather than compound), no antennae (only sensory bristles), and no true jaws. Respiration is accomplished by book lungs (lungs that like book gills are composed of multiple laminae, suggestive of book pages) or by **tracheae** (air conducting tubes).

Arachnids have a sparse but interesting fossil record that begins in the Late Silurian. Excellent collections of fossil arachnids have been obtained from Devonian, Carboniferous, and Miocene terrestrial sediments. Tertiary amber deposits have provided splendidly preserved fossils. Scorpions are known from Silurian and Devonian rocks, as indicated by *Palaeophonus* (Fig. 10.44) from the Late Silurian.

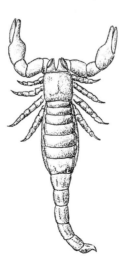

Figure 10.44   The Silurian scorpion *Palaeophonus*: length 3 cm.

SUPERCLASS MYRIAPODA

Myriapods are many-legged arthropods with long, slender bodies, a single pair of antennae, one pair of mandibles, tracheal respiration, and unbranched appendages. The two principal groups are the centipedes and millipedes. Nearly 3000 species of living centipedes are known, all of which feed on slugs, worms, snails, and other arthropods.

Each of the trunk segments except the first and last in myriapods has one or two pairs of walking legs. These are moved in regular order, each slightly out of phase with the preceding pair so that waves of movement seem to traverse the body. Millipedes ("thousand-leggers") can have even more legs than centipedes. About 700 legs appears to be the maximum number ever counted. In the adult millipede, there are two pairs of legs to each of the long series of cuticular rings that extend down the body. Millipedes are burrowers, and the majority are herbivores that feed on decaying plant debris. As such, they contribute significantly to soil formation.

Because they are terrestrial creatures with delicate exoskeletons, myriapods are rare as fossils. The oldest millipede fossils are obtained from Silurian rocks of Scotland. The probability of finding myriapods improves somewhat in terrestrial rocks of Pennsylvanian age. The best-preserved fossil centipedes are those recovered from Oligocene amber.

SUPERCLASS HEXAPODA, CLASS INSECTA

About 7000 species of fossil insects have been described. Although this is an impressive number, it probably represents only a small proportion of the insects that have inhabited Earth since their appearance in the Devonian. Animals that live mostly on land and that have relatively delicate exoskeletons often require exceptional conditions for preservation. Today, insects are the most abundant and widespread of all land animals, and they are the only arthropods to have achieved flight.

Although the diversity of insects may appear bewildering, members of the class adhere rather well to a fundamental morphology. The body is divided into three major components: head, thorax, and abdomen. The head is composed of six fused segments bearing uniramous appendages modified into antennae and mouth parts. The latter are mostly adapted for biting and chewing (as in cockroaches and grasshoppers), or for impaling and sucking (as in mosquitoes). The thorax consists of three segments, each having a pair of legs (hence the name "Hexapoda"). Usually, two pairs of wings are attached to the second and third thoracic segments.

Except for a few insect taxa that are either primitively or secondarily without wings, wings are a common feature of insects. Insect wings are composed of two thin sheets of chitin that enclose a network of hollow veins that serve as blood vessels or as tracheae. Knowledge of the pattern of wing venation is essential for classification and identification of insects. In some insects such as beetles, the front wings are modified to form wing covers that protect the rear pair when the animal is not in flight. In the Diptera (mosquitoes and flies), the rear wings are reduced to knobs called **halteres**. These beat with the same frequency as do the wings, and they serve rather like gyroscopes in helping to maintain stability during flight.

The third major morphological division of the insects, the abdomen, is notable because of its lack of appendages. In typical adult insects, the abdomen is composed of

11 segments and a terminal element around the anus that is analogous to the telson of crustaceans.

One may initially classify insects into those that are primitively wingless, and the wing-bearing (or secondarily wingless) groups. The former are placed in the Subclass Apterygota, and are represented today by the silverfish and springtails. Included in the Apterygota is the oldest known insect, *Rhyniella*, from the Middle Devonian Rhynie Chert of Scotland. The winged group is placed in the Subclass Pterygota, and has a fossil record that extends back into the Early Pennsylvanian. Among the pterygots are palaeopterans such as dragonflies and damselflies. The giant dragonfly *Palaeodictyoptera*, which inhabited Pennsylvanian coal swamps, had a 50-cm wingspan, and is frequently depicted in textbooks and museum dioramas. *Mischoptera* (Fig. 10.45) is another form that has been recovered from Pennsylvanian rocks. Like other palaeopterans, *Mischoptera* carried its wings straight out from the body. Another important group of insects, the orthopteroids, were able to fold their wings back over their bodies. Included here are grasshoppers, crickets, earwigs, cockroaches, mantids, and termites. The enormous swamp-dwelling cockroaches of the Pennsylvanian, some of which attained lengths of over 10 cm, are certainly impressive. Although some Pennsylvanian cockroaches were formidable in size, the extraordinary diversity of the group is more important. More than 800 species are known from the Pennsylvanian and Permian rocks.

Other important insect groups are the hemipteroids (lice, cicadas, aphids, and mealybugs), which left a fossil record that begins in the Permian. Lacewings and mantispids (like the so-called praying mantis) are neuropteroids, whereas beetles are

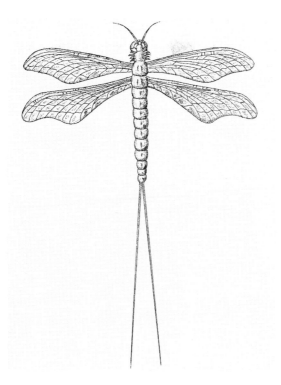

Figure 10.45   *Mischoptera*, a fossil pterygot insect of Pennsylvanian age. (After F. M. Carpenter, Studies on Carboniferous insects from Commentary, France, Pt. II. *Journal of Paleontology* 25(3):336–355, 1951.)

coleopteroids. Mecopteroid insects include fleas, butterflies, and moths, whereas ants, bees, and wasps are hymenopteroids. The neuropteroids and mecopteroids appear to have arisen near the end of the Paleozoic. Excellent fossils of some of the smaller species are sometimes found in amber.

Many gaps exist in the insect fossil record, and this has resulted in areas of uncertainty with regard to insect evolution. Nevertheless, it is apparent that the history of insects was punctuated by four important events. The first of these was the evolution of wings, an event that provided for the passage from the Apterygota to the Pterygota. Wings first appeared during the Late Devonian. Initially, short wings may have had a thermoregulatory function, and were subsequently modified for flight. Tall trees, present by Devonian time, may have provided some of the selective stimulus for flight. The second evolutionary milestone was the development of wing covers. Insect wings are fragile structures, and unless protected they are subject to damage when the animals are required to pursue prey or escape predators by crawling into abrasive crevices and rough litter. Protective wing covers thus had an important survival function.

The next evolutionary event was the attainment of complete insect metamorphosis. By "complete" is meant a metamorphosis in which the larva that emerges from the egg is distinctly different from the adult in both structure and habit. The most familiar example, of course, is the caterpillar and butterfly. This type of metamorphosis removes the immature larval form from direct competition with adults. In addition, it may assist the insect in surviving stressful seasonal conditions. Insects having this type of metamorphosis appear for the first time in the Permian. The fourth important event in the history of insects occurred during the Cretaceous and Cenozoic. It was the coevolution of insects and flowering plants, whereby insects were able to benefit from the highly nutritious nectar obtained from flowers, while at the same time the plants benefited from insect dispersal of their pollen.

MINOR ARTHROPOD GROUPS

ONYCHOPHORA

Onychophorans, sometimes called by such names as velvet worms or walking worms, superficially resemble worms with legs. In fact, the name of the principal genus, *Peripatus*, means "the walker." The term Onychophora has Greek roots and translates as "claw-bearer" in reference to the curved claws on the feet of these interesting animals. Onychophorans range in age from the Cambrian to the Holocene. Extant species live only in the Southern Hemisphere and the West Indies, prompting the hypothesis that they are the descendants of onychophorans that were residents of Pangaea before that supercontinent broke up. Living onychophorans are terrestrial animals that inhabit moist areas beneath leaf litter and decaying logs. Most species are predacious. They feed on small snails, insects, and worms. Onycophorans are able to discharge an adhesive liquid that entangles their prey. Once the prey is captured, salivary secretions are passed into the body of the victim and the partially digested tissues are then sucked into the mouth.

As one considers the morphology of onychophorans, it becomes apparent that they possess an interesting mixture of annelid and arthropod characteristics, as well as features that are uniquely their own. This mix of morphological traits suggests these

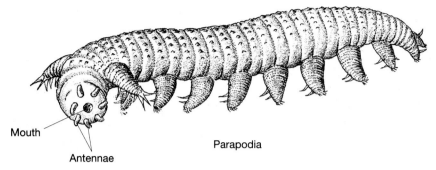

Mouth

Antennae

Parapodia

Figure 10.46   *Aysheaia* a Burgess Shale onychophoran (×1.8).

invertebrates are close to the ancestral stock that was ancestral to both annelids and arthropods. Annelidlike traits include the wormlike form, undifferentiated segments, soft covering, simple eyes, and unstriated muscles. However, the animal is arthropod-like in having sequentially arranged appendages, an anterior pair of appendages modified for handling food, a pair of large annulated antennae, and trachaea resembling those in insects.

The uniquely onychophoran characteristics of *Peripatus* include a distinctive arrangement of apertures for the tracheae, the velvetlike smoothness of its skin, and paired nerve cords that lack true ganglia. The stubby legs (**parapodia**) of *Peripatus* are not jointed. They are inflated with blood to provide for rhythmic walking.

Fossil onychophorans are rare. *Aysheaia* (Fig. 10.46) and *Hallucigenia* are onychophorans from the Middle Cambrian Burgess Shale, and *Luolishania* is a probable onychophoran from the Early Cambrian Chengjiang fossil site of China. These animals were marine, indicating that today's terrestrial onychophorans may have evolved from forms that made the transition to land initially by colonizing of damp soil and litter along coastal tracts.

TARDIGRADA

The tardigrada are tiny animals (less than a milimeter long) that can sometimes be found in droplets or films of water on plants or in pore spaces of soil. Commonly called "water bears," they are great fun to watch through a microscope as they lumber along over plant debris on their four pairs of stubby, clawed legs (Fig. 10.47). Tardigrades elicit particular interest because of their ability to survive for years without food or water, or even under frigid conditions. During dry spells they shrink, and their rate of

Figure 10.47   Tardigrades crawling on a plant filament. The length of each animal is about 0.7 mm. (After E. Marcus. In H. G. Bronn (Ed.), *Klassen und Ordnungen des Tierreichs*, Vol. 5, Pt. 4. Frankfurt: Akademische Verlagsgesellschaft.

metabolism severely decreases. At such times they have entered a state of suspended animation or **anabiosis**. When a moist environment returns, they require only a few hours to restore themselves to their former healthy state.

The body of a tardigrade is rather indefinitely divided into a head and four trunk segments, each with a pair of short legs. The body is covered with a delicate cuticle that is periodically molted. The mouth is anterior and bears two sharp stylets used to puncture the cell walls of mosses and algae in order to suck their nutritive fluids.

Until recently, tardigrades were almost unknown as fossils, except for a single discovery in rocks of Cretaceous age. In 1994, however, fossils of these fascinating little animals were found in Cambrian strata of Siberia. Thus, tardigrades can be traced back to that time of rapid diversification of life sometimes called the "Cambrian explosion."

PYCNOGONIDA

The pycnogonids (Fig. 10.48A) or "sea spiders" are a class of unusual marine arthropods that appear to be related to the chelicerates. Rarely seen as fossils, these animals have a primitive body plan that may have originated in the Cambrian, but the only unquestioned remains of sea spiders are from the Devonian of Germany. Today, however, pycnogonids are fairly common, with over a hundred species described from Antarctic waters alone. Most are inhabitants of deep, cold bottom environments of the ocean, although some seek their food in shallow intertidal areas. Their small (1 to 10 cm long) bodies bear an anterior proboscis (used to suck the tissues of their prey) and four to six pairs of exceptionally long legs. A few forms recovered from abyssal depths are truly giants, and have legs that extend as much as 74 cm from the body. Pycnogonids are thought to be descendants of an early group of marine arachnids.

PENTASTOMIDA

The Pentastomida (Fig. 10.48B) are a group of animals having affinities to arthropods but rarely seen as fossils. They have a wormlike body that is 2 to 12 cm in length. The body bears five short protuberances at the anterior end. All are parasitic and feed on blood and tissue fluids in the lungs and air passages of vertebrates. Reptiles living in tropical realms are commonly hosts for pentastomids.

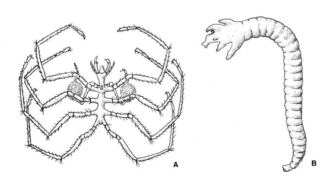

Figure 10.48 (A) A living pycnogonid or "sea spider"; length of body about 4 mm. The shaded spherical bodies are egg masses that are carried by the male. (B) *Cephalobaena*, a living pentastomid (length 6 cm).

A   B

ARTHROPOD COLONIZATION OF THE LAND

The fossil record indicates that arthropods originated in the ocean. As we have seen, many remained for all time within the marine realm, but others made the transition to life on land. The earliest completely terrestrial arthropods are found in Silurian and Devonian rocks. Their appearance can be correlated with the advent of early land plants, which provided food and habitats. It seems probable that small aquatic arthropods initially colonized piles of wet seaweed, algal flotsam, and possibly land-derived organic debris from early plants. Then, by means of a succession of adaptations, arthropods in the transitional environments attained ever greater independence from their original aquatic habitat. These early colonizers of the land faced many problems, including those involving respiration, reproduction, the senses, and protection against both desiccation and wetting.

The marine ancestors of terrrestrial arthropods depended upon gills for respiration. Certain terrestrial crustaceans were able to continue the use of gills for obtaining oxygen by keeping the gills constantly moist. Chelicerates were able to breath by means of **book lungs** (such as those mentioned in the discussion of Arachnida). In living scorpions, these consist of a pocket like structures, one wall of which is folded into leaflike lamellae that are held apart by bars that permit air to circulate. Diffusion of gases occurs within the blood in the lamellae and the air in the interlamellar spaces. Another solution to the problem of respiration on land involved the evolution of **branchial lungs**, in which lungs are developed from gill chambers. Branchial lungs characterize those living land crabs that do not need to return to water in order to wet their gills periodically. A more common method of respiration seen in insects and arachnids employs a system of tubes called **tracheae**. Tracheal tubes carry on gas exchange at their terminations where they are not enclosed by a cuticle and are in contact with tissues. Blood circulation is not needed in this respiratory mechanism because the tracheal tubes circulate oxygen to all parts of the body.

The last method of respiration employed in terrestrial crustaceans such as wood lice or pill bugs utilizes a system of branching tubular structures called **pseudotracheae**. Pseudotracheae rather resemble tracheae but are not ancestral to the tracheae of other arthropod groups. They are, however, an interesting example of convergent evolution.

The colonization of land also required changes in the methods of arthropod reproduction. Aquatic arthropods need only release their eggs and sperm into the surrounding water for external fertilization. However, fertilization in terrestrial arthropods is usually internal, either directly by transfer of a packet containing sperm from the male to the female, or by copulation. Associated with internal fertilization was the appearance of insect eggs, with their resistance to destruction and capacity for gas exchange with the surrounding air.

Coming ashore also necessitated changes in the way arthropods perceived their surroundings. One can generalize that visual and auditory perceptions were more highly developed in terrestrial arthropods than in their marine precursors. A particularly distinctive sensory structure in many terrestrial arthropods are fine hairs of great length called **trichobothria**. The trichobothria arise from an expanded socket or base that contains a sensory cell stimulated by even the smallest vibration or air current.

Another problem to confront the first land-living arthropods was desiccation, or drying up. This was actually less of a problem for arthropods than one might think because of the protection afforded by their chitinous exoskeleton. In many land-dwelling arthropods today, the cuticle of the exoskeleton has an added waxy **epicuticle** that resists drying. Most arthropods lacking this covering live in damp and humid settings. In contrast to the role of the epicuticle in preventing desiccation, it may also have served as a water repellent. Without such a repellent, tiny arthropods would readily adhere to water films because of molecular adhesion, and they would not have the strength to free themselves.

In their conquest of the land, arthropods have solved many lesser problems relating to locomotion, feeding, temperature control, and larval development. With each solution, many new pathways for evolution were opened, each in turn producing an extraordinary diversity of animals. Indeed, the arthropods are the most diverse group of organisms to have ever inhabited our planet.

REVIEW QUESTIONS

1. Provide the geologic range for each of the following arthropod groups:
   a. Phylum Arthropoda
   b. Superclass Trilobita
   c. Superclass Hexapoda
   d. Class Ostracoda

2. What are the distinguishing characteristics of the Phylum Arthropoda?

3. How do protostomes differ from deuterostomes in their embryologic development? In which group are the Arthropoda placed? Which group includes the Echinodermata?

4. What are the three longitudinal divisions of a trilobite carapace? What are the three major transverse divisions?

5. Distinguish between the following features of trilobites:
   a. free cheek and fixed cheek
   b. opisthoparian and proparian facial sutures
   c. holochroal and schizochroal eyes
   d. endopodite and exopodite

6. What major group of Trilobita are characterized by:
   a. very small size, two or three thoracic segments, and pygidia similar to cephalia in size and shape.
   b. many thoracic segments that terminate in spines, and tiny pygidia.
   c. border of cephalon pitted, long genal spines, and large semicircular cephalon.
   d. schizochroal eyes, proparian sutures, 13 thoracic segments.
   e. exceptionally spinose, long genal spines, spines on distal ends of thoracic segments and along the border of the cephalon.

7. How does adont dentition in ostracodes differ from amphidont dentition?

8. Which orders of ostracodes are confined to the Paleozoic?

9. What sexual characteristics and method of growth may result in misidentification of ostracodes?
10. Match the terms in the column on the right to the taxonomic group listed at the left.

_____ Malacostraca	A. barnacles
_____ Merostomata	B. scorpions, spiders
_____ Branchiopoda	C. lobsters, crabs, shrimp
_____ Cirripedia	D. insects
_____ Arachnida	E. millipedes
_____ Onycophora	F. clam shrimp (conchostracans)
_____ Myriapoda	G. eurypterids
_____ Hexapoda	H. velvet worms

11. What adaptations for life on land were acquired by formerly aquatic insects in their colonization of land areas?
12. Nearly all arthropods are capable of locomotion. Name one group composed of sessile arthropods.

SUPPLEMENTAL READINGS AND REFERENCES

Barnes, R. S. K., Calow, P. & Olive, P. G. W. 1993. *The Invertebrates: A New Synthesis*. London: Blackwell Scientific.

Bignot, G. 1985. *Elements of Micropaleontology*. London: Graham and Trotman.

Brasier, M. D. 1980. *Microfossils*. London: George Allen and Unwin.

Briggs, D. E. G. & Fortey, R. A. 1989. The early radiation and relationship of major arthropod groups. *Science* 246: 241–243.

Clark, K. V. 1973. *The Biology of the Arthropoda*. New York: Elsevier.

Culver, D. G. (Ed.). 1990. *Arthropod Paleobiology*. Paleontologic Society Short Course in Paleontology, No. 3, convened by D.G. Mikulic. Knoxville: The Paleontological Society, Dept. of Geological Sciences, University of Tennessee.

Manton, S. M. 1977. *The Arthropoda: Habits, Functional Morphology, and Evolution*. Oxford: Clarendon Press.

Moore, R. C. (Ed.). 1959. *Treatise on Invertebrate Paleontology, Pts. O, P, Q, and R; Arthropoda*. Boulder, CO, and Lawrence, KS: Geological Society of America and University of Kansas Press.

Ruppert, E. E. & Barnes, R. D. 1994. *Invertebrate Zoology*, 6th ed. Philadelphia: Saunders.

Shear, W. A. 1990. Silurian–Devonian Terrestrial Arthropods. In D. G. Culver (Ed.), *Arthropod Paleobiology*. Paleontological Society Short Course in Paleontology, No. 3. Knoxville: The Paleontological Society, Dept. of Geological Sciences, University of Tennessee.

Wilmer, P. 1990. *Invertebrate Relationships: Patterns in Animal Evolution*. Cambridge: Cambridge University Press.

CHAPTER 11

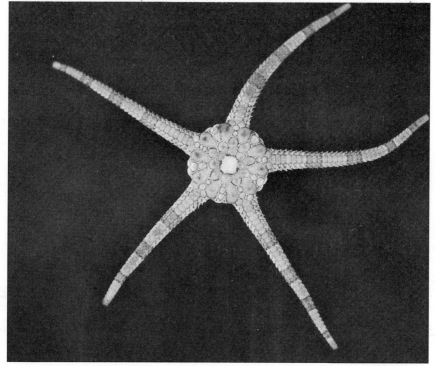

An ophiuroid (brittle star or serpent star). Over 2000 species of these echinoderms have been described.

Echinodermata

The sea urchin's mouth is from beginning to end a continuous structure; but in surface view is not continuous, but looks like a lantern with the horn-panes left out all around.

Aristotle (384–322 B.C.E.) *Historia Animalium*

Included within the Phylum Echinodermata are invertebrates having such nicknames as sea stars (starfish), serpent stars, sea urchins, sea cucumbers, sea lilies, and several other groups, all of which are quite different from the animals of any other invertebrate phylum. Indeed, if starfish were not so familiar to us, they might seem creatures from another world. Their distinctive body plan and evidence of their relationship to our own phylum render echinoderms particularly interesting to biologists and paleontologists. Ancient people were also intrigued by these animals, believing that certain parts could cure illnesses or improve sexuality, or simply be eaten as a delicacy. Even today, echinoderm eggs are found on the menus of some Italian restaurants under the name *rizzo di mer* ("sea rice"), and sea urchin gonads are served in many Japanese sushi bars.

Echinoderms are mostly benthic animals. Their calcitic skeleton favors preservation, except in those groups whose skeletons are readily disaggregated upon death. There are three particularly distinctive traits of echinoderms. The first is an array of canals, bladders, and tubelike processes that serve such functions as locomotion, respiration, and prey capture. This is the **water vascular system**. Second, most adult echinoderms possess pentameral symmetry (although with superimposed fundamental bilateral symmetry). Third, they have a skeleton composed of calcareous plates that are either articulated for flexibility or constructed rigidly, and a skeleton covered by soft tissue of the epidermis. Warty or spiny projections extend from the skeleton and provide the basis for the name Echinodermata, which means "spiny skinned."

GENERAL CHARACTERISTICS OF ECHINODERMS

The sea star (starfish) serves as a very suitable animal for describing the water vascular system (Fig. 11.1) that characterizes members of the echinoderm phylum. The water vascular system is an arrangement of fluid filled canals and appendages that are derived from the coelom and are lined with ciliated epithelium. Fluid within the system has a composition similar to seawater, but also contains coelomic cells, small amounts of protein, and potassium ions. The canal system has a connection to the exterior through tiny openings in a plate called the **madreporite**, a porous, buttonlike structure located near the juncture of two arms on the animal's aboral surface. From the madreporite, a canal called the **stone canal** descends toward the oral side of the animal. Calcium carbonate is present in the walls of the stone canal, which accounts for its name. Near the oral surface of the sea star, the stone canal joins the circular **ring canal**. Ciliated radial canals extend from the inner side of the ring canal into each of the sea star's arms, and each of these bears numerous small side branches that connect to paired **tube feet**. A bladderlike muscular **ampulla** extends from the top of each tube

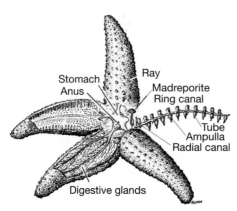

Figure 11.1   Partially dissected starfish or sea star showing elements of the water vascular system.

Stomach
Anus
Ray
Madreporite
Ring canal
Tube
Ampulla
Radial canal
Digestive glands

foot. Contraction of circular muscles in the ampulla forces water into the tube foot, causing it to extend. At the same time, a valve prevents water from flowing back into the radial canal. The sucker at the end of the tube foot is shaped to adhere to objects over which the sea star is moving or which it is attempting to capture. As a further aid in capturing prey, the ends of the tube feet secrete a sticky adhesive substance. To shorten the tube foot, longitudinal muscles contract to force water back into the ampulla. The alternate contraction and extension of hundreds of tube feet act in concert to propel the animal about. In soft sediment, suckers are of little use, and the tube feet simply act as tiny walking legs. Indeed, some sea stars lack suckers altogether.

The mouth is located at the center on the underside of a sea star. That side is therefore referred to as the *oral* surface. Extending radially from the mouth are furrows called *ambulacral grooves*, and each of these contains two or four rows of tube feet. The margins of the ambulacral grooves may be protected by spines that close over the groove when protection is required. Ambulacral areas are also readily seen as meridional bands extending from the oral to the aboral sides of the globular shells (called *tests*) of sea urchins. In fact, if one were to bend the arms of a sea star upward, then fasten their tips together at the top and fill the intervening space with a mosaic of calcareous plates, one would have produced the basic skeletal anatomy of a sea urchin.

Although sea stars consume a variety of marine animals as food, many, such as *Asterias*, feed on bivalves. They mount their prey and use their tube feet to pry the valves apart. The bivalve resists the effort to pry it open by contracting its adductor muscles. These muscles, however, eventually fatigue and the valves open. The sea star then everts the lower part of its stomach into the open shell to envelop and digest the soft tissues of its prey.

Another interesting characteristic of sea stars as well as other echinoderms is their ability to regenerate lost parts. Many sea stars inhabit turbulent shoreline areas, and it is not unusual for them to lose an arm or two if they have loosened their hold on the substrate and are bashed about by waves. If at least one arm and a part of the central disk remain intact, however, the missing parts may be regenerated to form a fully constituted animal.

The digestive system of sea stars extends from the mouth to an esophagus, which in turn leads into a stomach. Above the stomach, five branches extend outward to paired digestive glands located in each arm. A short, tubular intestine brings waste from the stomach to the anus where it is expelled. Undigested materials, including small shells and particles of sediment, are often ejected through the mouth. In most echinoderms, sexes are separate. Fertilization is external, and this results in the development of a swimming dipleurulid larvae. Dipleurulid larvae are cylindrical with a ventral anterior mouth and posterior anus.

CLASSIFICATION

The Echinodermata can be divided into five major groups that can be recognized as subphyla. The subphyla include 20 groups ranking as taxonomic classes.

Subphylum Asterozoa
   Class Asteroidea
   Class Ophiuroidea

Subphylum Echinozoa
   Class Echinoidea
   Class Holothuroidea
   Class Edrioasteroidea
   Class Ophiocistioidea
   Class Helicoplacoidea
   Class Cyclocystoidea
   Class Edrioblastoidea

Subphylum Crinozoa
   Class Crinoidea

Subphylum Blastozoa
   Class Blastoidea
   Class Rhombifera
   Class Diploporita
   Class Eocrinoidea
   Class Parablastoidea
   Class Paracrinoidea

Subphylum Homalozoa
   Class Stylophora
   Class Homoiostelea
   Class Homostelea
   Class Ctenocystoidea

SUBPHYLUM ASTEROZOA

CLASS ASTEROIDEA (EARLY ORDOVICIAN TO HOLOCENE)

The Class Asteroidea contains the free-moving sea stars. The body is composed of **rays** or **arms** that radiate from a **central disk**. The **mouth** is at the center of the underside of the disk, and the anus is on the aboral side near the juncture of two arms. As compared to ophiuroids, the arms of asteroids are thick, and the central disk is not as distinctly set apart from the arms. Most asteroids have five arms, but the number may vary even to as many as 40. Within the arms are paired, elongate masses of glandular digestive cells, tuftlike clusters of gonads, and components of both the water vascular and nervous systems.

From the mouth of the sea star, the **ambulacral groove** extends radially toward the termination of the arm. At the tip of the arm, the water vascular system ends with a small tentacle having a simple eye at its base.

The asteroid skeleton lies beneath the thin, ciliated epidermis. It consists of calcareous plates bound together by flexible connective tissue. Variously shaped spines protrude through the epidermis. Among these are movable, jawlike **pedicellaria**, which function in the removal of debris, in defense, or in the capture of food. The ambulacral grooves (Fig. 11.2) are bordered on either side by ambulacral plates that end with a single terminal plate at the outer extremity of each arm.

Asteroids have changed relatively little since they first came upon the scene during the Ordovician. Although they have a lengthy geologic range, they are not common fossils except at a few localities where populations were smothered by rapid deposition of mud and silt. Their lack of preservation at other sites is probably related to the relative ease by which plates are disarticulated after death. Fossil sea stars such as *Hudsonia* are found in Ordovician rocks quarried for building stone in New York State, and both sea stars and serpent stars occur well-preserved in the famous

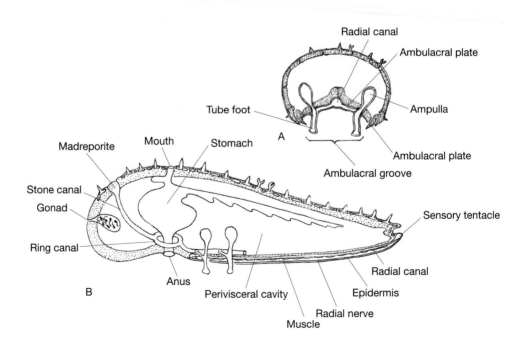

Figure 11.2 (A) Diagrammatic cross section of the arm of a starfish showing plates of the ambulacrum. (B) Simplified diagrammatic section through the central disk and one arm of a starfish.

Bundenbach Slate (Devonian) of Germany. Today, *Asterias* is probably the most famous asteroid, whereas *Acanthaster* is the most infamous because of its voracious attacks on reef corals in the Pacific.

Sea stars of uncertain taxonomic position are the somasteroids. In fact, they are sometimes designated a separate subclass of the Asterozoa. Somasteroids have a pentagonal shape, but they lack definite arms. Ambulacra are confined to the disk area. These echinoderms have several features that suggest they may be ancestral to both asteroids and ophiuroids. They occur rarely in rocks that range from Ordovician to Devonian in age.

Asteroids occur abundantly in tropical waters but are also numerous in temperate and even polar seas. They are found at depths from tide level to the abyss. Ocean water of normal salinity and sufficient food appear to be the only critical requirements of this class of Echinodermata.

CLASS OPHIUROIDEA (EARLY ORDOVICIAN TO HOLOCENE)

Ophiuroids (Fig. 11.3) take their name from the Greek root *ophi*, meaning snake, an obvious reference to their snakelike arms. In addition to being called "serpent stars," they are called "brittle stars" because of the ease with which arms can be broken off (particularly in dried specimens). Unlike asteroids, the arms of ophiuroids do not have ambulacral grooves. Each serpentine arm emerges abruptly from the rounded or pentagonal central disk. The arms are supported centrally by a single series of cylindrical ambulacral skeletal elements that have been formed by fusion of originally paired ossicles. Within the arms are muscle strands, radial nerves, and parts of the coelom. Dorsal, ventral, and lateral plates support movable spines, and the entire arm is enveloped in a thin epidermis. Tube feet along each arm have no ampullae. As in the asteroids, they provide for locomotion and have a sensory and feeding function as well. The sticky tube feet pass food particles along to the star-shaped mouth. On leaving the mouth, food is passed into a gut and saclike stomach in the central disk. As there is no anus, waste is expelled through the mouth. The madreporite lies adjacent to the mouth.

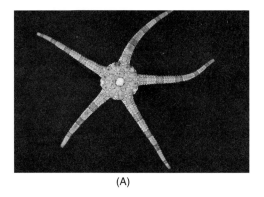
(A)

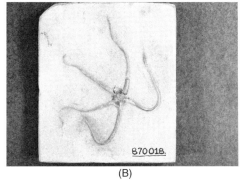
(B)

Figure 11.3   (A) Dried specimen of a serpent star or brittle star. (B) Fossil serpent star from the Triassic of England.

Like asteroids, ophiuroids are not common fossils, although they achieved global distribution in the Paleozoic. They survived the terminal Paleozoic mass extinction with little change and increased in number during the Mesozoic and Cenozoic. Living species are mostly benthic detritus or suspension feeders, with fewer numbers of carnivores.

SUPHYLUM ECHINOZOA

Within the Subphylum Echinozoa are globular, discoidal, and cylindrical echinoderms whose skeletal elements may be either tightly arranged in wall-like patterns, or consist merely of calcareous sclerites embedded in a leathery covering. The Echinozoa include sea urchins and sea cucumbers. Members feature a water vascular systems, tube feet that protrude through small holes in ambulacral areas, and a complete digestive tract with mouth and anus at opposite ends. Arms supported by calcareous ossicles are not present in this group.

CLASS ECHINOIDEA (LATE ORDOVICIAN TO HOLOCENE)

Although sea urchins and sand dollars may seem initially quite unlike their relatives the sea stars, they are nevertheless basically similar. A living sea urchin resembles an animated pin cushion because of its thick cover of sharp, movable spines. The spines provide protection and also assist the podia in feeding and locomotion. The skeleton of an echinoid is, by convention, referred to as a **test** (Fig. 11.4). It is constructed of closely fitted calcareous plates that form a globular or flattened container for the animal's soft parts. For study, the spines can be removed in order to see the pattern of plates that

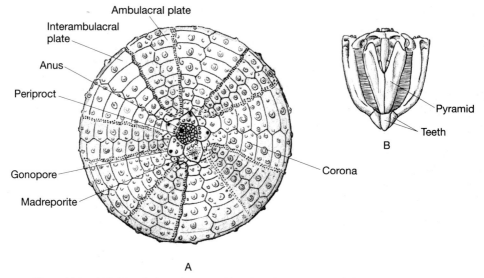

Figure 11.4   (A) Aboral view of a sea urchin with spines removed. (B) Aristotle's lantern. (Redrawn from R. T. Jackson, Phylogeny of the Echini, with revision of Paleozoic species. *Boston Society of Natural History*, Memoir 7.)

make up the test. When this is done, one can observe that the main part of the test, the **corona**, is surmounted by an **apical disk** containing a double ring of ten plates that enclose a tough, plate-studded skin, the **periproct**. The anus is centrally located in the periproct, and is surrounded by a ring of plates each pierced by an opening called the **gonopore**. Sex cells produced in the gonads pass through a duct and are extruded at the gonopore. Appropriate to this function, this inner ring is named the **genital ring**. One plate in the genital ring is riddled with pores, permitting its recognition as the **madreporite**. The smaller plates of the outer circle of the periproct comprise the **ocular ring**. Each bears a pore that serves as an opening to the water vascular system.

In a sea urchin like the living *Echinus*, the mouth is at the opposite pole from the corona. It is surrounded by a structure similar to the periproct, but it is called the **peristome**. In sea urchins, the mouth contains a remarkably contrived chewing apparatus composed of five sharp, magnesium-strengthened teeth radially arranged in a complex structure called **Aristotle's lantern** (see Fig. 11.4B). As indicated in the quote at the beginning of this chapter, Aristotle first noted the similarity in outward appearance of this structure to a Greek lantern. In addition to its five teeth, Aristotle's lantern contains prominent elements called **pyramids**, which are joined in pairs by a deep notch between their upper ends. The pyramids are held at the top by pairs of calcareous arches, and the entire complex of calcareous elements is worked by powerful muscles attached to a series of pedestals and arches called the **perignathic girdle**. There are some 35 to 40 skeletal elements and 60 muscles in Aristotle's lantern. It is a powerful tool for boring and preparing food for digestion.

The large corona of the sea urchin is composed of ten meridional rows of plates. Five of these bear fewer spines and are bordered by rows of tube feet. These are the **ambulacral areas**. When the sea urchin is stripped of its spines, the ambulacral areas are easily recognized by the pairs of pores for the tube feet that are aligned along the margins of the ambulacra (see Fig.11.4A). The meridional areas between the ambulacra are termed **interambulacra**. In *Echinus*, there are only two rows of plates in each interambulacrum, but the number of rows in other genera may range from 1 to 20. Regardless of their number, one cannot help but be impressed by the precision of the mosaic they form. The interambulacral plates may be simple, or they may bear tubercles. Tubercles are small, conical protuberances surmounted by a knob. The concave base of a spine fits neatly around this knob in a ball and socket arrangement so that the spine can be moved in many directions by muscles attached to its base. The spines are often quite sharp and may be equipped with poison glands. Pedicellaria are also present on sea urchins, and these too may bear poison glands.

From the madreporite located in the genital ring of the periproct, a stone canal passes downward to the ring canal that lies near the top of Aristotle's lantern. A radial canal extends from the ring canal along the underside of each ambularcral area. Ampullae and tube feet are connected to the radial canal by short lateral canals, and tube feet extend to the exterior through pores in the ambulacral plates. The pores in most, but not all, echinoids occur in pairs. This is because two canals penetrate the ambulacral plate from each ampulla, and these merge on the outside of the test as a single pseudopodium.

Although the sea urchin *Echinus* is a marvel of symmetry, other echinoids are less symmetrical. In the most symmetrical forms, the mouth lies at the lower pole and the anus at the opposite pole of a spherical or globular test (see Figs. 11.4 to 11.7). Such echinoids with opposed oral and anal openings are spoken of as **regular** or **endocyclic**. In other echinoids, the anal opening has migrated from the top of the test to a posterior position (see Figs. 11.9 to 11.11) The change is often accompanied by a corresponding shift of the mouth to a more anterior location. Such echinoids are termed **irregular** or **exocyclic**. One can use this distinction in a simple descriptive way to describe the more important groups of echinoids, as is done below. In this scheme, used in 1987 by Porter Kier, seven major groups having the approximate taxonomic grade of orders are discussed. One of these divisions includes only post-Paleozoic non-cidaroids (see below).

Regular Echinoids   As noted above, a polelike opposition of mouth and anus is the distinguishing mark of the regular echinoids (Fig. 11.5). In plan view they are circular. Although the oral surface may be somewhat flattened, the dorsal surface is strongly convex. Pentameral symmetry is nearly perfect. Regular echinoids range from the Ordovician to the Holocene, and include most of today's common sea urchins. Within this group are the orders Echinocystitoida, Palaechinoida, Cidaroida, as well as an informal group of Triassic to Holocene regular non-cidaroid echinoids.

Echinocystitoids (Ordovician to Permian)   The ambulacra of these regular echinoids contain 2 to 20 columns of plates. One to 14 rows of plates may fill the interambulacral area. The plates of the interambulacral areas are often lacking in precise uniformity of shape. They overlap, rather like roofing tiles, and are joined in a manner that provides flexibility to the test. *Aulechinus* (Fig. 11.6), from the Ordovician of England, is an example.

Palaechinoids (Silurian to Holocene)   These echinoids have rigid tests constructed of sturdy plates. There are two or more columns of plates in the ambulacral

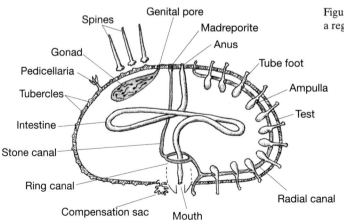

Figure 11.5   Major components of a regular echinoid.

Figure 11.6   *Aulechinus*, a representative Echinocystoid (×3). (Redrawn from R. C. Moore (Ed.), *Treatise on Invertebrate Paleontology*, Pt. U, Echinodermata. Boulder, CO, and Lawrence, KS: Geological Society of America and University of Kansas Press.)

areas. Interambulacra may have one or move columns of plates, often exhibiting remarkable uniformity of shape. There are five genital plates. The genus *Melonechinus* (Fig. 11.7) is about the size of a melon and may occur locally in large numbers in Mississippian limestones.

Cidaroids (Devonian to Holocene)   Cidaroids are thought to be the ancestral stock from which post-Paleozoic echinoids evolved. Their ambulacra contain two columns of plates. In Paleozoic forms, the interambulacra have one or more columns of plates, but post-Paleozoic cidaroids have only two. The plates are simple and may or may not overlap. Each of the interambulacral plates bears an exceptionally large central tubercle that serves to support robust spines, as seen in the Cretaceous genus *Dorocidaris* (Fig. 11.8).

Triassic to Holocene non-Cidaroids   Non-cidaroid regular echinoids comprise the most abundant echinoids in today's oceans. The ambulacra in this group are composed of two columns of usually compound plates. There are two columns of plates in the interambulacra as well, and these characteristically bear more than one tubercle.

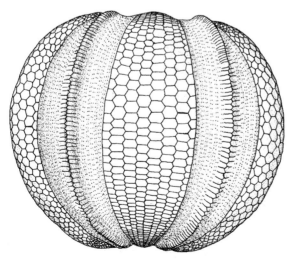

Figure 11.7   *Melonechinus*, average diameter 7.8 cm. (Redrawn from Keyes, R.C., *Paleontology of Missouri*, Bull Missouri Geological Survey, IV, 1894.

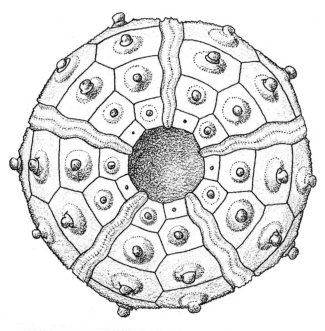

Figure 11.8   *Dorocidaris*, average diameter 3.0 cm. (Redrawn from Clarke, W.B. & M. W. Twitchell, *Mesozoic and Cenozoic Echinodermata of the U.S.* U.S. Geol. Survey Mongraph 54, 1915.)

The plates vary in thickness, and in most species they are not imbricated. Tubercles are smaller and more numerous than in cidaroids.

Irregular Echinoides   The irregular echinoids include the heart urchins and sand dollars. In these common echinoids, the anus has migrated toward the rear, sometimes even reaching the oral side, and the mouth may have moved toward the front. The test is bilaterally symmetrical and often sometimes elongated and flattened on the underside. Ambulacral areas contain specialized podia for gathering particles of food, and are commonly formed into petaloid patterns confined to the upper surface. The posterior interambulacrum is distinct from the other four, and there are two columns of nonimbricating plates in the ambulacra and interambulacra. Most irregular echinoids are adapted for burrowing in sandy substrates; consequently, they have smaller and more numerous spines than do regular echinoids. The spines are used in locomotion and to keep debris off the surface of the test.

Irregular echinoids evolved from regular echinoids by acquiring adaptations for living and feeding on organic matter in soft, loose sediment. Adaptations for an infaunal mode of life were already evident by Early Jurassic time. Thereafter, irregular echinoids became abundant and exceptionally diverse. The principal groups to be described below are the spatangoids (heart urchins), cassiduloids, and clypeasteroids (sand dollars).

Spatangoids (Cretaceous to Holocene)   Spatangoids are generally oval in shape, and the entire center of the oral surface, including the mouth and peristome, has migrated to an anterior location. There is no Aristotle's lantern. In addition, the anus

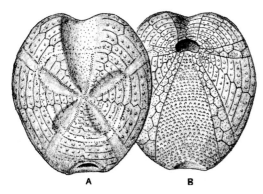

Figure 11.9   *Hemiaster,* ×1.
(A) Dorsal view showing depressed
petaloid ambulacral areas and mouth
at the truncated posterior margin.
(B) Ventral view showing large
opening for the mouth. (Redrawn
from Clarke, W.B. & M. W. Twitchell,
*Mesozoic and Cenozoic
Echinodermata of the U.S.* U.S.
Geol. Survey Mongraph 54, 1915.)

and periproct have moved to the posterior end and occupy what can now be designated the posterior interambulacrum. The forward ambulacrum is depressed and bears
smaller pores. Like most irregular echinoids, the spatangoids are bulk sediment swal-
lowers. *Hemiaster* (Fig. 11.9) is a common Cretaceous to Holocene spatangoid with
depressed petals.

Cassiduloids (Jurassic to Holocene)   Cassiduloids have round to oval tests
with often poorly developed pentaloid ambulacra. As juveniles, cassiduloids
develop an Artistotle's lantern, but this structure is lost by the time they attain
maturity. There are paired ambulacral pores, and a groove commonly connects the
two pores of each pair. The pores accommodate tube feet which on the aboral surface, function in respiration. *Cassidulus* (Fig. 11.10) serves as an example of this
group.

Clypeasteroids (Cretaceous to Holocene)   Because they are round and flat
like a large coin, clypeasteroids have taken on the nickname "sand dollars." They are
burrowers that sieve food particles from silty sediment with the aid of minute spines.
The flat test bears five aboral petaloid ambulacra and contains an Aristotle's lantern.
Some species develop marginal indentations (**notches**) as well as enclosed perforations
called **lunules**. Lunules provide a short-cut for moving food from the aboral surface
to the mouth. Along with the notches, they also increase the marginal zones used in
gathering food. Perhaps also, these features help to prevent overturning by sudden

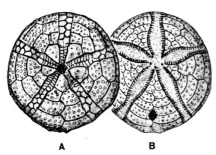

Figure 11.10   Dorsal (A) and
ventral (B) views of *Cassidulus*
(×0.5). (Redrawn from Clarke, W.B.
& M. W. Twitchell, *Mesozoic and
Cenozoic Echinodermata of the
U.S.* U.S. Geol. Survey Mongraph
54, 1915.)

currents. In clypeasteroids, the test is strengthened by a system of internal supports so that the weight of sediment or force of an impact can be transmitted not only to adjacent plates but also through the supports to the other side. Such supports were essential to sand dollars, for a flat test is weaker than a globular test of similar diameter. *Clypeaster* (Fig. 11.11) is not as flat as many clypeasteroids. This globally distributed Cretaceous to Holocene echinoid has a pentagonal shape with rounded margins and a depressed oral surface. *Dendraster*, a nonlunate clypeasteroid, formed the bulk composition of entire limestone formations during the Pleistocene. *Dendraster exocentricus* is common today along the western coast of the United States where it feeds on suspended food particles by positioning itself with half of its body projecting vertically above the substrate in order to intercept current-transported small organisms and algal fragments.

Echinoids Through Time   The total geologic range for echinoids is Late Ordovician to Holocene. Following their initial appearance they diversified slowly during the Silurian and Devonian, reaching their highest levels for the Paleozoic era during the Early Carboniferous. During the Carboniferous, palaechinoids, echinocystitoids, and some cidaroids were locally abundant. For the remainder of the Paleozoic and into the Triassic, echinoids esperienced a marked decline in species diversity. However, during the Late Triassic, a major adaptive radiation of echinoids had begun that continued throughout the remainder of the Mesozoic and into the Tertiary. By Jurassic time, the majority of orders of both regular and irregular echinoids were abundantly represented in the fossil record. The rich diversity of Mesozoic and Cenozoic echinoids is a reflection of their success in exploiting new sources of food and a greater range of marine environments than was achieved by their Paleozoic precursors. In addition to their abundance, flat irregular echinoids have had a greater potential for preservation than have surface-dwelling regular echinoids. Clypeasteroids, for example, radiated into soft substrates where burial was

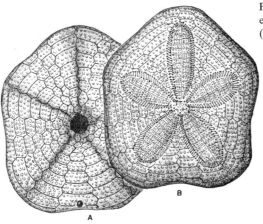

Figure 11.11   The exocyclic echinoid *Clypeaster*. (A) Oral and (B) aboral views (×0.5).

enhanced and tests were less likely to be transported and broken after death. Also, their disklike shape limits damage from rolling.

Echinoid Habits and Habitats   Paleozoic echinoids were mostly inhabitants of quiet nearshore environments, whereas post-Paleozoic groups expanded into a wide range of oceanic environments. Today, echinoids are found from the shoreline to the abyss in all latitudes. They are, however, more numerous in shallow, warmer seas. Many species tend to be gregarious, perhaps to facilitate fertilization during spawning. Echinoids live freely on the seafloor, burrow into soft sediment, or occupy crevices in rocks, reefs, and pilings. Many species are able to enlarge the cavities they live in by using their spines as digging tools.

With regard to food, echinoids tend to be dietary generalists. Although algae are an important food for many, others feed on higher plants as well as animal matter. In general, regular echinoids use their dental structures to rasp or scoop up such food as drifting plant particles, small invertebrates, or drifting bits of various kinds of organic detritus. Many have the chemosensory receptors that permit them to detect food at a distance and move toward it. Podia are adapted to "taste" potential food before passing it on to the mouth. In contrast to the regular echinoids, irregular groups swallow sediment in bulk and extract tiny particles of organic matter. In a way, we are indebted to echinoids for the role they play in helping to keep the ocean clean.

Predators on echinoids include fish, certain gastropods, sea stars, and sea otters. As noted earlier, even some humans take pleasure in an occasional meal of echinoid roe or gonads (*uni sushi*).

CLASS HOLOTHUROIDEA

The holothuroids or sea cucumbers (Fig. 11.12) are elongate echinoderms with a mouth at one end of a cylindrical body and an anus at the other. Most crawl about with one median and two latero-ventral ambulacra in contact with the seafloor. Suckered tube feet along these ambulacra provide for locomotion and attachment. The remaining two ambulacra are latero-dorsal. They contain small, pointed tube feet that function primarily in respiration. At the oral end of the sea cucumber, a ring of retractable buccal tentacles (actually modified podia) frame the mouth and assist in food gathering. In most holothuroids, the body wall is thick and flabby and is supported by radial muscles.

Holothuroids lack large covering plates such as those in echinoids. The major skeletal element is an internal calcareous ring of plates, usually ten, that surrounds the pharynx. This structure serves as a support for the ring canal, for muscles, and for the nerve ring. The only other mineralized hard parts are microscopic calcareous elements called **holothurian sclerites** (Fig. 11.12B). These are embedded in the sea cucumber's skin. Sclerites are thought to be the embryological remnants of an earlier and more extensive skeleton. They take the form of tiny spines, wheels, anchors, antlers, and pedestals. Several fossil species of holothuroids can be identified from the characteris-

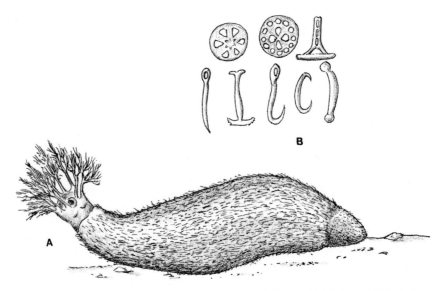

Figure 11.12   (A) External view of a sea cucumber or holothuroid (×0.4). (B) Holothurian sclerites (×30).

tics of their sclerites alone. They are occasionally discovered in Carboniferous, Jurassic, and Cenozoic rocks that have been prepared for study of microfossils.

Holothuroids feed on both plankton and organic matter, both living and dead. The majority are benthic crawlers, but there are a few pelagic genera that can swim in a pulsating manner rather like jellyfish and some that burrow through soft sediment on the seafloor. Most burrowing varieties feed by extracting nutrients from ingested sediment. In this way, they consume and rework large amounts of sediment, and they effectively destroy primary sedimentary structures and textures. The process whereby sediment is disturbed and reworked by organisms is called **bioturbation**. Because of their effectiveness in causing bioturbation, holothuroids have been called "earthworms of the sea." Holothuroids range from polar to tropical waters and occur at all ocean depths. As with sea stars, they have extraordinary powers of regeneration. For example, they occasionally eviscerate their digestive tract, parts of the respiratory system, and their gonads in order to provide a meal for a threatening predator, and they subsequently regenerate these organs. In addition, when irritated or attacked, some species turn their posterior sides toward intruders, contract their body walls, and by rupturing the cloaca, shoot a mesh of sticky tubules out of the anus. The detached tubules entangle predators, leaving them helpless. The sea cucumber then crawls away and regenerates a new set of tubules.

Except for their mineralized sclerites, fossil remains of holothuroids are rare. Their fossil record extends back to the Early Cambrian. A likely holothuroid has been found among the fauna of the Chengjiang fossil site in Yunan, China, and an apparent pelagic holothuroid named *Eldonia* occurs in the Middle Cambrian Burgess Shale. Its shape, which resembles that of a jellyfish, is the principal reason for the belief that it was a pelagic creature. Within the disklike body, fossils reveal a coiled gut, grapelike clusters of tentacles adjacent to the mouth, and gonads beneath the anterior end of the gut. Aside from these extraordinary fossils, occasional holothuroid fossils have been recovered from Devonian, Carboniferous, and Jurassic strata.

CLASS EDRIOASTEROIDEA

Edrioasteroids are an ancient group of echinoderms that made their appearance during the Early Cambrian. The group beccame extinct by the end of the Late Carboniferous. *Walcottidiscus* is an edrioasteroid found in the Burgess Shale. They are not common as fossils and are therefore usually not suitable for biostratigraphic studies.

Edrioasteroids constructed globular, discoidal, or cylindrical tests. Many forms developed concave lower surfaces, which were apparently attached to hard substrates or the shells of larger invertebrates. Typically, edrioasteroids have five curved or straight ambulacral areas. The plates of the ambulacral areas bear small openings for the tube feet, as well as cover plates that provided protection for these openings when the tube feet were withdrawn. A pyramidal circlet of plates occurs around the anus, which is located between two ambulacra. The mouth is centrally located on the oral surface, and it also is protected by several plates. A third aperture, interpreted as an opening for the water vascular system, can sometimes be discerned between the mouth and anus. Edrioasteroids like *Edrioaster* (Fig. 11.13) were benthic creatures that preferred shallow, nearshore environments.

CLASS OPHIOCISTIOIDEA

Ophiocistioids are a small but interesting fossil group known only from about six genera. Their geologic range is Ordovician to Mississippian. Ophiocistioids rather resemble brittle stars that have lost their arms. In lieu of arms, they have exceptionally long ventral tube feet. The body is pentaradiate and has a depressed domal shape. Most genera appear to have been epifaunal, mobile animals. *Volchovia* (Fig. 11.14) is a representative genus.

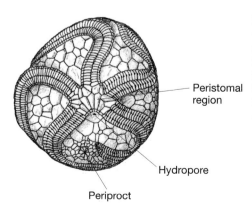

Peristomal region

Hydropore

Periproct

Figure 11.13   *Edrioaster* (×2.5) with cover plates removed from three of the ambulacra to reveal floor plates and pores. The pores probably mark the positions of tube feet. (Redrawn from F. A. Bather, *Treatise on Zoology*, Pt. III London: Adam and Charles Black, 1900.)

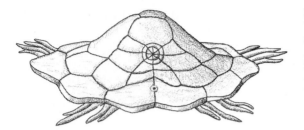

Figure 11.14   The ophiocistoid
Volchovia. (After J. W. Durham et
al. In R. C. Moore (Ed.), *Treatise on
Invertebrate Paleontology*, Pt. U.
Boulder, CO, and Lawrence, KS:
Geolocial Society of America and
University of Kansas Press, 1966.)

CLASS HELICOPLACOIDEA

The Helicoplacoidea are fusiform or spindle-shaped echinoderms whose remains
occur only in the Lower Cambrian of western North America. In helicoplacoids, the
interambulacral plates and ambulacral areas spiral upward around the test, giving
these fossils a distinctive spindlelike appearance. The mouth is uppermost on the
rounded end, and the anus is at the opposite end. Very likely, the animal lived vertically in sandy substrates. Helicoplacoid plate arrangement allowed for considerable
flexibility. The animal could expand and contract its body from the inside and thereby
adjust its position in the enclosing sediment. *Helicoplacus* (Fig. 11.15), from the White
Mountains of California, is representative of this class of echinoderms.

CLASSES CYCLOCYSTOIDEA AND EDRIOBLASTOIDEA

Cyclocystoids are small, discoidal echinoderms that probably attached themselves to
the substrate by suction on their flat aboral surface. There are two rings of plates
around the peripheral margin of the test, a submarginal ring composed of large plates
and a flexible marginal ring composed of tiny imbricated plates.. The mouth and peristome are centrally located, whereas the periproct is off-center. The class appeared in
the Middle Cambrian and peristed into the Devonian.

The Edrioblastoidea contains only a single genus, *Astrocystites*, from the
Ordovician of North America.

Figure 11.15   The spiralled,
spindle-shaped Early Cambrian
echinoderm *Helicoplacus*. Height
2.6 cm. (Redrawn from J. W. Durham
& K. E. Caster, *Science* 140:820–822,
1963.)

SUBPHYLUM CRINOZOA

The Crinozoa are echinoderms characterized by globular or cup-shaped calyces composed of several circlets of plates, and arms that extend from the calyx for food gathering. Although the majority of Crinozoa have been sessile in habit and possessed a stem and holdfast, pelagic forms such as the "feather stars" are also common. Crinozoans have a lengthy range, extending from the Cambrian to the Holocene.

CLASS CRINOIDEA

The Living Animal   One has only to glance at a crinoid to see that is has three major components. It is attached by a stalk or **stem**, the stem is surmounted by the main part of the body, or **calyx**, and extending from the calyx are the **arms** (Fig. 11.16).

The crinoid stem consists of a columnar series of disk-shaped plates called **columnals**. These may be circular, elliptical, stellate, or pentagonal. Each has a variously shaped central performation or **lumen** (plural, **lumina**) from which radiate ridges

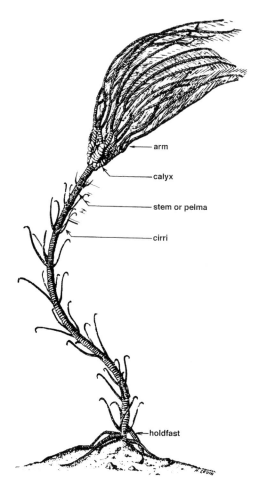

Figure 11.16   General external morphology of a stalked crinoid.

arm

calyx

stem or pelma

cirri

holdfast

(**crenellae**), which interlock to provide greater strength to the stem. A column of soft tissue extends upward through the lumen, between columnals, and around the exterior of the stem. The stem may also bear slender, movable branches called **cirri**. In stemless forms like the feather stars, long, slender cirri branch from the bottom of the calyx. These branchlets are used on occasion to hold the crinoid onto the substrate.

At its lower end, the stem might be modified in various ways. The most common terminal portion consists of rootlike branches. In other forms, the terminus of the stem bears an anchorlike structure. Some crinoid stems simply tapered to a point and were coiled around other objects for anchorage and support. With regard to length, fossil species are known with stems over 20 meters long, although most living and extinct crinoid stems rarely exceed a meter in length.

Because it contains most of the vital organs, the calyx is a crinoid's most important component. The calyx is composed of a **cup** attached to the stem aborally, and a capping portion, the **tegmen**, on which the mouth and anus are located. The mouth is placed in or near the center of the tegmen, whereas the anus is in a slightly off-center position. Five ambulacral areas extend from the mouth into the arms. Tube feet are present, but as there are no ampullae, these are extruded by contraction of the radial canals. The cup usually has the general shape of the bowl of a wine glass, and is sturdily constructed of polygonal plates. Near the top of the cup there is a circle of five plates called the **radials**. The radials are aligned with the ambulacra, and each serves as a base for the proximal plates of the arms. Beneath the radials are five **basal plates**. These alternate in position with the radials. Crinoids having only radial and basal plates in the cup are termed **monocyclic**. The name refers to their having only one row of plates below the radials. Other crinoids have a circlet of three to five **infrabasal plates** below the basals. Crinoids having two rows of plates below the basals are called **dicyclic** (Fig. 11.17).

In addition to the radials, basals, and infrabasals, some crinoids exhibit interradial plates between the radials, and in many groups the lowest plates of the arms, the **brachials**, are incorporated into the calyx. Brachial plates that have become part of the calyx do not meet laterally, and therefore additional **interbrachial plates** are formed to fill the space between brachials.

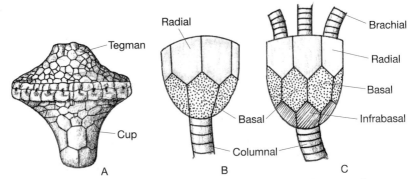

Figure 11.17   Crinoid calyx nomenclature. (A) Crinoid calyx showing cup and tegmen. (A) Diagrammatic lateral view of a monocyclic crinoid cup. (B) Dicyclic crinoid cup.

In many crinoids, there is a large extra plate developed in the radial cycle. This plate, called the **anal plate**, can be recognized in part because it does not lead into an arm. Some crinoids have two or more plates in the same position as the anal plate. The anal plate is used in orienting the calyx, for it defines the posterior inter-radius. Above and to the right of the anal plate there is a **radianal plate**.

The tegmen, which covers the cup of the calyx, may have the form of a plate-stud-ded membrane, or a rigid structure that extends over the mouth and adjacent food grooves. In its simplest form, the tegmen is composed of only five plates. More com-monly, however, it will contain additional accessory plates. It may also bear a tall **anal tube** through which waste can be discharged at a more hygenic distance from the mouth.

The third major component of crinoids are the arms. Arms not only function in food gathering but they also provide propusion for the free-living feather stars. As mentioned above, the arms extend from the cup and are composed of plates called brachials. In most living crinoids, arms branch shortly after leaving the calyx, forming ten branches, and then branch still further to provide an impressive food-gathering array. A row of small, jointed branchlets called **pinnules** extend from the sides of the arms. Ambulacral grooves extend along the oral surface of each arm as well as along the pinnules. These grooves have tiny cover plates that can be lifted to admit particles of food. Food particles that are caught by cilia on the arms and pinnules are trans-ported along the ambulacral grooves and ultimately passed to the mouth.

The arms of crinoids can be unbranched (**atomous**), equally branched (**isotomous**), or unequally branched (**heterotomous**). If the arms or branch contains closely spaced branchlets or pinnules like the lateral vanes of a feather, the structure is called **pinnulate**. Arms can be further described as uniserial or biserial. **Uniserial arms** consist of a single column of brachial plates, whereas **biserial arms** are constructed of a double column of alternating brachials (Fig. 11.18).

The branching of arms occurs at special **axillary plates** that have sloping upper surfaces to accommodate the initial plates of the branches. Arm plates that extend immediately above the radials of the calyx are termed **primibrachials**. Plates in the sec-ond segment are appropriately called **secundibrachials**, and those successively above the secundibrachials are the **tertibrachials** and **quadribrachials**. Minor branches of arms called **ramules** occur in some crinoids. Ramules are less regular in occurrence than are pinnules, and they may even bear pinnules themselves.

MAJOR CATEGORIES OF CRINOIDS

Characteristics such as the kind and arrangement of plates in the calyx, construction and manner of branching of the arms, and nature of the tegmen are used to divide the Class Crinoidea into four subclasses. These are the Inadunata, Camerata, Flexibilia, and Articulata. The first three of these subclasses were primarily Paleozoic, whereas the articulates survive today as the descendants of the inadunates.

Inadunata (Ordovician to Triassic)   The Latin word *inadunatus* means "not united," and it refers to the inadunate characteristic of having arms free above the

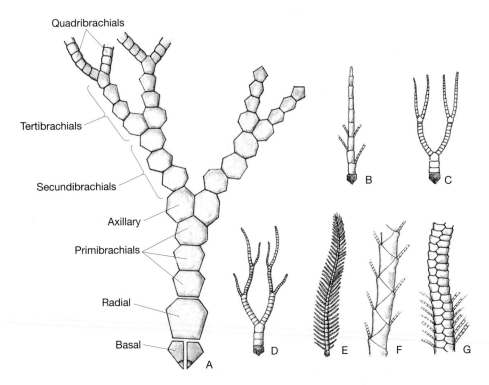

Figure 11.18   Some of the structures seen in the arms of crinoids, all diagrammatic. (A) Basic arm nomenclature; (B) an atomous arm; (C) isotomous; (D) heterotomous; (E) pinnulate; (F) zigzag arrangement of arm plates; (G) biserial arrangement of arm plates.

radials. The cup, which may be either monocyclic or dicyclic, is composed of firmly connected plates. A tegmen covers the mouth, and anal plates are present. There are no interradial plates. Inadunates are the oldest of the crinoid subclasses. *Cyathocrinites* (Fig. 11.19), with species from the Silurian to the Carboniferous, serves as an example of the inadunate group. All of these species possessed bowl-shaped cups, stout tegmens, three to five infrabasals, and isotomous branching of the arms. The stem varies in length, and can be circular, eliptical, or pentagonal in transverse section.

Camerata (Ordovician to Permian)   In the camerates, plates of the calyx are rigidly joined to one another. Fixed brachials and interbrachials are incorporated into the upper part of the cup, which is bowl-shaped. The tegmen provides a firm roof over the mouth and ambulacral grooves. Many species have an anal tube and lateral spines. Arms are uniserial or biserial and always bear pinnules. *Platycrinites* (Fig. 11.20) is a good example of a camerate crinoid from the late Paleozoic.

Flexibilia (Ordovician to Permian)   This subclass takes its name from the flexible manner in which plates of the calyx are held together. The Flexibilia have dicyclic calyces characterized by three plates in the infrabasal circlet or having an infrabasal circlet fused into a single plate. The lower brachials are incorporated into the calyx, and there is an anal tube. The uniserial arms of Flexibilia lack pinnules and often coil inward.

Figure 11.19 *Cyathocrinites*, an inadunate crinoid with isotomous branching. First anal plate is stippled, above which lies the anal sac.

In transverse section the stem is usually round or elliptical and lacks small, jointed branchlets called *cirri*. Flexibilia are believed to have evolved from dicyclic inadunates during the Ordovician. The Devonian to Mississippian genus *Taxocrinus* (Fig. 11.21) is a representative of this group.

Articulata (Triassic to Holocene) The Subclass Articulata includes nearly all the Mesozoic to Holocene crinoids, including both today's stalked sea lilies (Order Isocrinida) and the stemless pelagic feather stars (Order Comatulida). The calyx is dicyclic, or it may be pseudomonocyclic in that the infrabasal plates are sometimes atrophied or fused. In general, the calyx in these crinoids is small, and the tegmen is leathery and only loosely plated. There are five infrabasals, five basals, and five radials, although some relatively rare species exhibit variation in the number of plates in each circlet. The arms are long and always uniserial. They are also pinnulate, and thus distinct from the arms of the nonpinnulate Flexibilia.

Triassic articulates resemble inadunates, from which they may have evolved. They also share certain characteristics with Flexibilia, prompting speculation that they

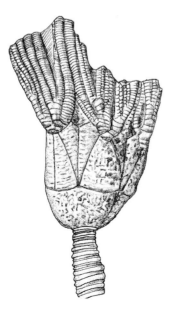

Figure 11.20 The camerate crinoid *Platycrinites* (×0.8). (Redrawn from R. C. Moore & C. Teichert (Eds.), *Treatise on Invertebrate Paleontology*, Pt. T, Echinodermata 2, Vol. 1. Boulder, CO, and Lawrence, KS: Geological Society of America and University of Kansas Press, 1978.)

Figure 11.21 *Taxocrinus*. Specimen is 7 cm in height.

may have had a polyphyletic origin. *Pentacrinus* (Fig. 11.22) is a widely dispersed Mesozoic genus found commonly in western North America. *Saccoma* (Fig. 11.23) is an unusual-looking unstalked Jurassic and Cretaceous genus. Excellently preserved specimens have been extracted from the Solenhofen Limestone (Bavaria).

CRINOIDS THROUGH TIME

The fossil record for crinoids begins in the Cambrian, but many of the earliest forms are so primitive and enigmatic that they are difficult to place in the subclasses described above. By the Ordovician, however, the class was well established, and throughout the remainder of the Paleozoic period, crinoids were consistently present in shallow-water carbonate environments. The climax of their Paleozoic evolution occurred in Early to Middle Mississippian. Indeed, because of the abundance of

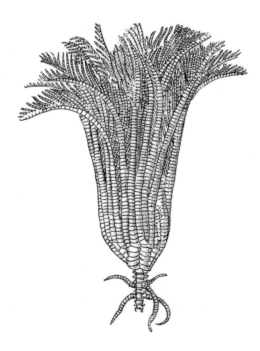

Figure 11.22 *Pentacrinus*, a genus characterized by many cirri on its stalk or column, often stellate columnals, and multiple branching of the arms.

Figure 11.23   The small, unstalked, free-swimming crinoid *Saccoma*, from the Jurassic Solenhofen Limestone. Diameter is about 4 cm.

crinoid remains in certain crinoidal limestones, the Mississippian has been dubbed the "age of crinoids." Inadunates and camerates formed extensive "crinoid gardens" (Fig. 11.24) on the seafloor during the first two epochs of the Mississippian, but by the Late Mississippian their numbers began to dwindle. The decline continued into the Late Permian when extinction finally overtook the Camerata and Flexibilia. A few inadunates managed to survive into the Triassic, and these were the probable ancestors of the articulates. Articulates appear in the Early Triassic and have been represented in marine invertebrate faunas since that time. Early Mesozoic crinoids were mainly stem-bearing forms like *Pentacrinites*, but stemless forms had become abundant by Jurassic time. These continue to be abundant today.

CRINOID HABITS AND HABITATS

Throughout most of their evolutionary history, crinoids have had worldwide distribution. These exclusively marine animals have occupied seafloor habitats as shallow as a meter or so, and depths of the ocean in excess of 4000 meters. Although today they tolerate water as cold as that in the Ross Sea of Antarctica, they are more abundant and diverse in tropical to subtropical realms. Like many of the extant echinoids, living crinoids tend to be gregarious, and they are frequent denizens of reefs. Clear water of normal salinity is their preference. Many Paleozoic crinoids, however, provide exceptions to generalizations based on living forms.

It has been observed that the number of living species of crinoids decreases with depth in the ocean as well as toward higher latitudes. Another interesting observation is that the species with many branches to their arms are most prevalent in warmer seas. In contrast, cold-water dwellers generally lack luxuriant branching. In general, attached forms with long stalks occur at water depths well below wave base where damage by turbulence is less likely. Heavy robust crinoids with short, stout stalks occur in shallow, more turbulent zones.

The principal predators on crinoids are worms, other echinoderms, and carnivorous gastropods. Some Paleozoic marine snails survived on crinoid excrement, and are found firmly attached over the anal opening of the calyx. Polychaete worms adhere to plates adjacent to food grooves, where they intercept food particles being swept by cilia toward the crinoid's mouth.

Figure 11.24   Reconstruction of a
Mississippian crinoid garden.
(Courtesy of the United States
National Natural History Museum.)

As indicated by the stomach contents of specimens dredged from the ocean
floor, crinoids consume all sorts of organic particles, as well as diatoms, dinoflagellates,
radiolarians, foraminifers, and bacteria.

SUBPHYLUM BLASTOZOA

The Subphylum Blastozoa embraces five classes of mostly stemmed echinoderms.
They take their name from the Greek word *blastus*, meaning bud. In fact, many of the
calyces of echinoderms in the Class Blastoidea do have a superficial resemblance to
flower buds. Blastozoans are noted for their pentameral symmetry and compact architecture. The term **theca** (rather than calyx) is used for the plate-covered enclosure containing the internal organs. Like crinoids, most blastozoans were stemmed suspension
feeders. Blastozoans occur in rocks of all Paleozoic systems, but are not known in
younger rocks.

CLASS BLASTOIDEA (ORDOVICIAN TO PERMIAN)

Blastoids are characterized by compact thecae that often exhibit an attractive pentamerous symmetry. Five ambulacral areas are symmetrically spaced around the sides of the theca. The mouth and anus are located at the summit of the theca, with the latter somewhat to one side in an interambulacral location. Blastoid stems are relatively thin, and a few recumbent blastoids were stemless. Unlike crinoids, blastoids had no arms. Instead, the food grooves of the ambulacra were flanked by rows of tiny armlets called **brachioles**.

The theca of a blastoid is composed of only 13 or 14 plates. These are arranged in three circlets (Fig. 11.25). At the base of the calyx, encircling the topmost (proximal) columnal of the stem, are three **basal plates**. Two of these are similar in shape, and the third is smaller and referred to as the **azygous basal**. Above the basals in the second circlet are five deeply notched **radials**, each forked so as to provide space for an ambulacrum. The topmost circlet of plates is set between the ambulacra and the mouth. These are the smaller **deltoid plates**. At the upper extremity of each deltoid is an opening called the **spiracle**, which serves as an exhalent pore for the water vascular system. One of these openings encompasses the anus, and is therefore somewhat larger. Thus, as one looks down upon the summit of the theca, one can observe six openings: five spiracles (one of which contains the anus) and a central mouth.

Each of the five ambulacra is occupied by an elongate, spear-shaped plate bearing many transverse grooves and a single longitudinal groove on the upper surface. These are the **lancet plates**. The grooves may be roofed over by tiny covering plates, but such plates are rarely seen except on well-preserved specimens. Pairs of small side plates occupy the spaces along the margin of each ambulacrum, and each pair of these side plates bears a socket for the attachment of a brachiole. Along each brachiole there is a narrow food groove that lies beneath delicate cover plates. Food particles captured by the brachiole are propelled by cilia down the brachiole food groove to the ambulacrum where they move across one of the transverse grooves of the lancet plate to the central longitudinal groove, and thence to the mouth.

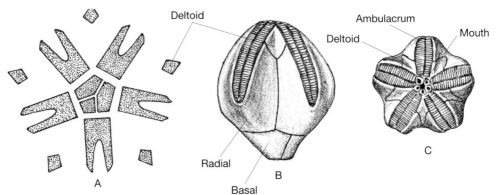

Figure 11.25   (A) Typical plate arrangement in the calyx of a blastoid. (B) Side and (C) oral views of the Mississippian blastoid *Pentremites* (×2.3).

A distinctive feature of blastoids was their system of longitudinal calcitic folds called **hydrospires**, which hang inward toward the center of the calyx beneath or adjacent to the margins of each ambulacrum. With ciliary action as the driving mechanism, oxygenated water was drawn into the hydrospires through small pores aligned along the ambulacral margins. The water then flowed upward through the hydrospire folds to the spiracles where it was discharged. The folds of the hydrospires provided ample surface area for the exchange of gases.

There are a few poorly preserved and somewhat questionable remains of blastoids in rocks of Ordovician age, but forms with fully developed blastoid traits did not appear until the Silurian. From that time until their demise in the Late Permian, blastoids achieved a global distribution. Silurian genera developed distinctly conical thecae with short ambulacra. Examples are *Codaster* and *Troostocrinus* (see Fig. 11.26). By Devonian, however, blastoids with bud-shaped or barrel-shaped thecae had become common. The heyday of blastoid evolution occurred during the Mississippian period when species of *Pentremites* (see Fig. 11.25) became exceptionally abundant and widespread. *Cryptoblastus* (Fig. 11.26) typifies the barrel-shaped blastoids found commonly in Mississippian crinoidal limestones. The Permian blastoid *Calycoblastus* (Fig. 11.27) is noted for its symmetry, long ambulacra, and large thecae.

The habits of this extinct class of echinoderms must be inferred from morphology, associated fossils, and the nature of enclosing sedimentary rock. It is a reasonable inference that their habits generally resembled those of crinoids, and that they were exclusively marine animals requiring waters of normal salinity. The blastoids commonly found in limestones lived in warm, carbonate-precipitating seas in association with crinoids, corals, fenestellid bryozoans, and brachiopods. The observation that great

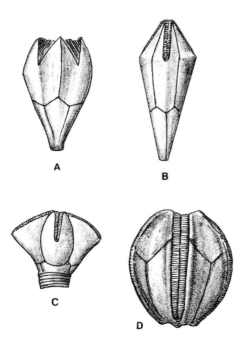

Figure 11.26   Paleozoic blastoids.
(A) *Codaster;* (B) *Troostocrinus;*
(C) *Orophocrinus;* (D)
Cryptoblastus. (All ×0.5.)

A

B

C

D

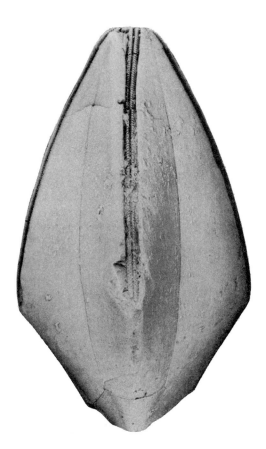

Figure 11.27   *Calycoblastus*, an exceptionally large blastoid from the Permian of Timor. The theca of this blastoid is 5.0 cm in length. (From D. B. Macurda, *Journal of Paleontology* 46(1):97 1972.)

numbers of blastoids are sometimes areally concentrated in patches suggests that, like many other echinoderms, they were gregarious.

CLASS RHOMBIFERA

Until recently, certain stalked echinoderms with distinctive pore patterns were considered members of the Class Cystoidea. That taxonomic designation has been dismantled and its members placed in either the Class Rhombifera or the Class Diploporita. Both of these groups are characterized by variously placed tubes or grooves that traverse thecal plates. In some members, they were open to the outside, whereas in others they opened only as pores located along plate margins.

The pores in Rhombifera are concentrated into rhombic patterns that are called **pore rhombs**. On close examination, these pore rhombs are seen to consist of thin grooves or tubes arranged in adjoining pairs of plates. Each of the plates comprising the pair contained half of the rhomb. In some forms, the ends of the grooves bend outward and exit as pores on the outer surface of the theca. Often the pores are developed in aligned rows that outline polygonal areas. Evidently, these complex systems of tubes and pores were components of the water vascular system and also had a role in respiration. In addition to their distinctive pore rhombs, the Rhombifera have four or five circlets of plates in the theca. The theca has brachioles, but these are delicate and rarely preserved. Two to five ambulacra are present. The rhombiferan stem is usually larger just below its point of attachment to the theca. The lumina in the columnals were large and may have contained strands of muscle for moving the stem and, consequently, the theca as well.

The Rhombifera began their evolution in the Early Ordovician. They achieved global distribution before becoming extinct late in the Devonian. Perhaps they lost in the competition with the blastoids. The Early Paleozoic genera *Caryocrinites* (Fig. 11.28) and *Echinosphaerites* (Fig. 11.29) are representatives.

CLASS DIPLOPORITA

Whereas pore rhombs were the most distinctive trait of rhombiferans, the most diagnostic feature of the Diploporita are paired pores contained within elliptical depressions. In some species, these so-called **diplopores** are distributed over the entire theca,

Figure 11.28   *Caryocrinites* from the Silurian Lockport Formation of New York. Scale divisions are in millimeters. (Courtesy of J. Sprinkle, from J. Sprinkle, *Journal of Paleontology* 49(6):1062.)

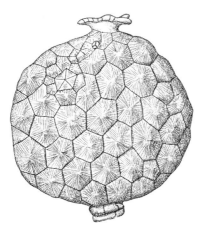

Figure 11.29   *Echinosphaerites*, a rhombiferan of Silurian age (×0.75). (After H. H. Beaver et al. In R. C. Moore (Ed.), *Treatise on Invertebrate Paleontology*. Boulder, CO, and Lawrence, KS: Geological Society of America and University of Kansas Press, 1967.)

but in others they are sparingly present in one or more localized areas. The theca is globular to elongate in diploporites, and may or may not have a stem. Small, brachiole-like armlets are attached around the mouth or along the three to five food grooves that extend downward from the mouth. The Diploporita range from Early Ordovician to Early Devonian. Two Ordovician examples are *Sphaeronites* and *Glyptosphaerites* (Fig. 11.30).

CLASS EOCRINOIDEA

Eocrinoids are not well represented in the fossil record. Nevertheless, they are an interesting class of echinoderms because they may include the ancestral stock for the Blastoidea, Rhombifera, and Diploporita. They first appear in the Early Cambrian and were present until the Early Silurian. Eocrinoids were the first echinoderms to have evolved brachioles and stems composed of columnals. Some species, however, were stemless and developed instead a multiplated holdfast.

The eocrinoid theca is globular to flattened in shape. Thecal plates are arranged in regular patterns in some forms, but rather randomly in others. As seen in *Gogia* (Fig. 11.31), some had long brachioles that arose from two to five ambulacra.

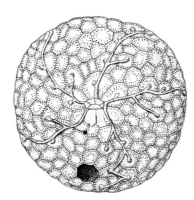

Figure 11.30   *Glyptosphaerites*. (Redrawn from H. H. Beaver et al. In R. C. Moore (Ed.), *Treatise on Invertebrate Paleontology*. Boulder, CO, and Lawrence, KS: Geological Society of America and University of Kansas Press, 1967.)

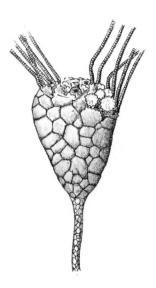

Figure 11.31   The Cambrian eocrinoid *Gogia* (×1).

CLASS PARABLASTOIDEA

The Parablastoidea represent a small class of Blastozoa and they contain only three genera. They are thought to be an aberrant side branch in the early evolution of blastozoans. Superficially, they resemble blastoids and have well-developed pentamerous symmetry. However, they differ from true blastoids in the number of calyx plates they possess and in the structure of their ambulacra. Parablastoids are confined to the Early and Middle Ordovician. *Blastoidocrinus* (Fig. 11.32) is an example.

CLASS PARACRINOIDEA

This extinct, small class of echinoderms is known only from a few Ordovician and Silurian fossils. Paracrinoids have globular to lense-shaped calyces composed of plates that are not arranged in circlets. There is no distinct differentiation of cup and tegmen, and there are three to five free or recumbent, uniserial, pinnulate arms. Typical forms also have a system of subepithecal pores. The mouth opens at the top of the calyx, and

Figure 11.32   *Blastoidocrinus* (×2).

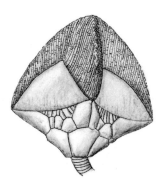

Figure 11.33   *Comarocystites* (×1).
A distinctive feature of this
Ordovician paracrinoid is the
hollowed central area of the plates.
(Redrawn from H. H. Beaver et al.
In R. C. Moore (Ed.), *Treatise on
Invertebrate Paleontology*, Pt. S,
Echinodermata 1. Boulder, CO, and
Lawrence, KS: Geological Society of
America and University of Kansas
Press, 1967.)

the anus is in an adjacent eccentric position. One of the better-known paracrinoids is the Ordovician genus *Comarocystites* (Fig. 11.33).

SUBPHYLUM HOMALOZOA

The Homalozoa are a small but interesting subphylum of echinoderms that lived in shallow marine seas of the Cambrian to the Devonian. Many had a stalklike process extending from the theca so that the animal took on the appearance of an armorplated tadpole (Fig. 11.34). In some homalozoans, the thecae bear pores that may represent the position of tube feet. There are also stemless species, and species having a solitary arm that bears an ambulacrum.

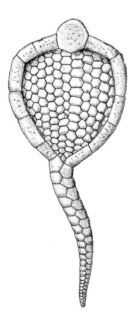

Figure 11.34   *Trochocystites*, a
homalozoan or carpoid. The theca is
1.5 cm in length.

Homalozoans are of particular interest because at least some species may not be echinoderms at all, but rather chordates. Proponents of this hypothesis interpret the stalklike process of these peculiar organisms as a tail, and they favor placing them in a separate subphylum of their own called the Calcichordata. The more conservative opinion is that they are true echinoderms that are related to chordates. The Subphylum Homalozoa contains four classes: the Stylophora, Homoiostelea, Homostelea, and Clenolystoidea. *Trochocystites* in Fig. 11.34 is a member of the Homostelea. Because of their rarity, the classes are not individually described here.

ECHINODERMS THROUGH TIME

Except for a possible edrioasteroid form named *Arkarau* from the Ediacaran fauna, fossils of echinoderms older than Cambrian are lacking. It is likely that the phylum originated late in the Proterozoic from soft-bodied bilaterally symmetrical metazoans that had progressed in their evolution to the level at which a true body cavity or coelom existed. The embryology of living echinoderms indicates that their coelom develops from pouches that grow outward from the sides of the embryonic gut (as is the case in primitive chordates) and not as a split in a mass of mesodermal cells derived from the gut wall (as in annelids and mollusks).

The echinoderm fossil record began in the Early Cambrian, at which time they were already highly diversified. Their initial radiation produced helicoplacoids, edrioasteroids, and homalozoans. By Middle Ordovician time (Fig. 11.35), a second more diverse evolutionary radiation was underway that produced a variety of Crinozoa, Blastozoa, and Asterozoa (both asteroids and ophiuroids). Holothuroids, Rhombifera, and echinoids made their entrance during Middle Ordovician time. During the Late Ordovician, however, echinoderm diversity diminished as Parablastoidea experienced apparent extinction and Rhombifera, Eocrinoidea, and Paracrinoidea declined.

During the Silurian and Devonian, many echinoderm families declined, although crinoids continued to be locally abundant. It was during the Mississippian, however, that crinoids, as well as blastoids and echinoids, became exceptionally prolific and diverse. In some parts of formations like the Mississippian Burlington Limestone, every cubic meter of rock contains the remains of over 10,000 crinoids. The compact, symmetrical thecae of the blastoid *Pentremites* are the delight of amateur fossil collectors because of their abundance and good preservation in some Late Mississippian strata.

The good times for stalked echinoderms ended near the Permian–Triassic boundary. Blastoids (which had been declining since the Mississippian), as well as many orders of crinoids, disappeared forever. A few species of inadunate crinoids managed to survive the crisis. These became the ancestors to today's crinoids. Similarly, only a few species of asteroids, ophiuroids, echinoids, and holothuroids were able to cross into the Mesozoic. These provided the ancestral stock for echinoderm expansion during the Jurassic, Cretaceous, and Cenozoic. Echinoids were the dominant echinoderms during the Jurassic and Cretaceous, but they were overtaken during the Cenozoic by asteroids and ophiuroids. The Asterozoa are now the most diverse echinoderm group in today's ocean.

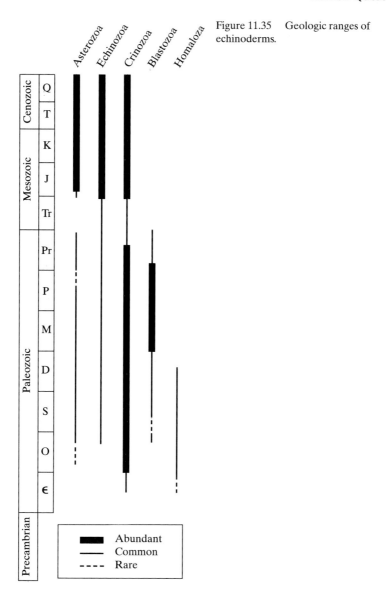

Figure 11.35   Geologic ranges of echinoderms.

REVIEW QUESTIONS

1. What morphological features are shared by Blastozoa and Crinozoa?
2. When were the Crinoidea most successful in terms of abundance and diversity? When were the irregular echinoids most successful?
3. Name the major morphological features that differentiate the following:
   a. Asteroidea and Ophiuroidea
   b. Regular and irregular echinoids
   c. dicyclic and monocyclic crinoids

4. What is the function of Aristotle's lantern?

5. Which group of echinoids were primarily burrowers? In what ways were they adapted for this habit?

6. What is the relation between test shape and the potential for preservation in echinoids?

7. What conditions would enhance the preservation of crinoids?

8. Of all the major groups of echinoderms, which have been most effective in causing bioturbation?

9. Which groups of crinoids incorporate arm plates into their calyces?

10. Describe the plates of a typical blastoid calyx.

11. Distinguish among atomous, isotomous, and heterotomous crinoid arms.

SUPPLEMENTAL READINGS AND REFERENCES

Binyon, J. 1972. *The Physiology of Echinoderms.* Oxford: Pergamon Press.

Clark, A. M. 1962. *Starfishes and Their Relatives.* London: The British Museum.

Kier, P. M. 1987. The Class Echinoidea. In R. S. Boardman (Ed.), *Fossil Invertebrates,* Ch. 18:596–611. Palo Alto, CA: Blackwell Scientific.

Nichols, D. 1972. *Echinoderms,* 4th ed. London: Hutchinson University Library.

Seilacher, A. 1979. Constructional morphology of sand dollars. *Paleobiology* 5(3):191–221.

Smith, A. 1984. *Echinoid Paleobiology.* London: George Allen and Unwin.

Sprinkle, J. 1987. Phylum Echinodermata. In R. S. Boardman (Ed.), *Fossil Invertebrates,* Pt. I, Ch. 18:550–595. Palo Alto, CA: Blackwell Scientific.

Tasch, P. 1980. *Paleobiology of the Invertebrates,* 2nd ed. New York: John Wiley.

Waters, J. A. & Maples, C. G. 1997. Geobiology of echinoderms. *The Paleontological Society Papers* 3:205–224.

CHAPTER 12

Branches (stipes) of the graptolite *Diplograptus*. *Diplograptus* is a
common Middle and Upper Ordovician genus. (Base of specimen 6 cm)

Graptolites and Other Hemichordates

*The tubes of the tubarium in the Rhabdopleuridae and the tubes of the coenosteum
in the Cephalodiscoidea have in general a structure identical to the thecae of the
Graptolithina, and a structure of such type does not exist in any other animal either
Recent or fossil.*

R. Kozlowski, 1966

The Phylum Hemichordata is composed of two groups of marine animals that differ considerably in body form and life styles. Members of the Class Enteropneusta comprise the more numerous and familiar group. They are commonly called *acorn worms*. Enteropneustans are solitary creatures that live under stones and shells or burrow into soft nearshore sediment. Although many of us have not observed a living acorn worm, we may have seen their coiled castings on recently drained tidal flats. In contrast to the acorn worms, the more rarely seen members of the Class Pterobranchia are sessile, colonial inhabitants of relatively deep water.

As is suggested by their name, which means "half chordates," the Hemichordata share certain developmental and anatomical features with the Chordata. Among these is a pharynx that is perforated to the exterior by one or more slits (gill slits) or pores. This feature exists in the Enteropneusta and in the genus *Cephalodiscus* of the Pterobranchia. Another chordate feature is a hollow dorsal nerve cord. The cord exists in the collar region of acorn worms, but not in pterobranchs. In both Enteropneusta and Pterobranchia, the opening in the early embryo known as the *blastopore* develops into the anus, as in chordates. The Graptolithina are allied with the hemichordates because the unique structure of the tubes in which they dwell is virtually identical with that of the pterobranchs.

The hemichordates have not progressed sufficiently to be considered members of the Chordata. They remain invertebrates, for they lack the notochord that is essential for membership in the Chordata. A **notochord** is a slender rod that extends along the back of a chordate from head to tail. In higher chordates, it is supplanted by the vertebral column, but is present in the embryo. Although hemichordates lack this feature, they are nevertheless interesting as the invertebrate group closest to our own.

CLASS ENTEROPNEUSTA

Enteropneustans (Fig. 12.1) are relatively large, elongate amimals. The majority are between 10 to 40 cm in length. An exception is the giant species *Balanoglossus gigas*, which lives in the shallow waters along the coast of Brazil, and may exceed 1.5 meters in length. The body of an enteropneustan has three major components. At the anterior end, there is an elongate or ovoid **proboscis** that the animal uses to burrow in the sand or mud. This is followed by a thickened **collar**, which can be used as an anchor during burrowing. A wormlike **trunk** extends posteriorly from the collar. Behind the collar are two rows of pharyngeal clefts, and in some species there is a tubular dorsal nerve cord in the collar. Most Enteropneusta derived their nourishment by ingesting particles of organic matter within the sediment that they pass through their digestive tract. An alternate method of feeding involves collection of organic particles that stick to the mucous that envelops the proboscis and are then propelled to the mouth by cilia. The process is assisted by water currents that enter the mouth and flow out through the pharangeal slits.

Enteropneusta like *Balanoglossus* and *Saccoglossus* are sluggish anmimals that live in near shore sediments where they often construct **U**-shaped burrows having two openings to the surface. Other acorn worms inhabit masses of seaweed or live under rocks and other seafloor objects. Studies of *Saccoglossus horsti* indicate that larvae of

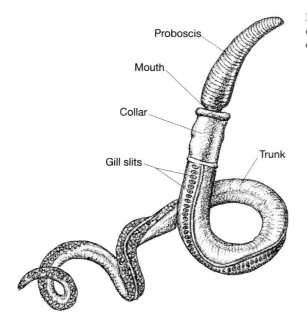

Proboscis

Mouth

Collar

Gill slits

Trunk

Figure 12.1   *Balanoglossus*, an
enteropneustan that is less
commonly called an "acorn worm."

this species possess a post-anal tail (considered a characteristic of chordates). It has
been suggested that this larval tail may be homologous to the stolon of the pterobranchs.

Body fossils of enteropneustans are not known. Their presence in ancient environments, however, may be indicated by castings and burrows found in nearshore
deposits.

CLASS PTEROBRANCHIA

Pterobranchs have a superficial resemblance to bryozoans in that they are colonial,
they can expand their colonies by budding, they have **U**-shaped digestive tracts, saclike
bodies, live in tubes, lack chitin in their exoskeletons, and are crowned by food-gathering tentacles. On closer examination, however, their nonbryozoan status is fully justi-
fied. Like enteropneustans, their bodies are composed of a **proboscis**, **collar**, and **trunk**
(Fig. 12.2). The proboscis, however, is shield-shaped and is used as a movable foot to
adjust the body within its housing. The collar is extended to form one or more pairs of
movable arms that function primarily in food gathering. *Rhabdopleura* has only one
pair of arms, but *Cephalodiscus* has five pairs. Each of the five pairs of arms contains
extensions of the coelum, suggesting a phylogenetic link to echinoderms. Among hemichordates, pterobranchs are more primitive than enteropneustans. *Rhabdopleura*, for
example, has no pharyngeal clefts, and there is only a single pair of such openings in
other members of the Pterobranchia. Also, the dorsal nerve cord is not hollow.

An individual pterobranch, one of many within the colony, is called a **zooid**.
Zooids are housed in zooidal tubes and are joined to other zooids in the colony by a

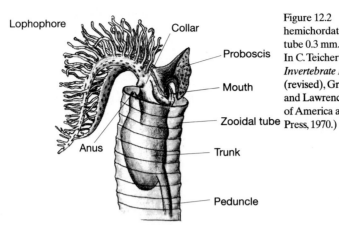

Figure 12.2   The pterobranch hemichordate *Rhabdopleura*; width of tube 0.3 mm. (After O. M. B. Bulman. In C. Teichert (Ed.), *Treatise on Invertebrate Paleontology*, Pt. V (revised), Graptolithina. Boulder, CO, and Lawrence, KS: Geological Society of America and University of Kansas Press, 1970.)

tubular **stolon**. The stolon is generated from the initial sexually produced zooid of the colony. From the stolon, later zooids form by asexual budding. Growth is accomplished as a special zooid known as the **terminal bud** proceeds to secrete a **creeping tube** (Fig. 12.3) that is extended along the substrate. The creeping tube lengthens itself by adding proteinaceous annular bands constructed of two half segments joined so as to form a zigzag pattern. New zooids sequentially bud from the creeping tube, each building its own zooidal tube. As the zooidal tube grows, it is sealed off from the creeping tube, and the stolon in the creeping tube turns black. It is now referred to as the **black stolon**. The zooidal tubes are secreted by glands on the zooid. Like the creeping tubes, they are constructed of annular rings; however, these are not in two parts. Rather, each ring is a single element marked by an oblique suture at the place where secretion of the ring began and ended.

These distinctive traits of pterobranchs favor the hypothesis that pterobranchs are closely allied to graptolites, and that both groups belong in the Hemichordata. Graptolites and pterobranchs both secrete and live within proteinaceous tubes. The general structure of these tubes is similar in the two groups. Both have a stolon system that links zooids to one another and that functions in the production of buds.

It is not easy to find fossil pterobranchs. Sporadic remains are known from Early Paleozoic, Jurassic, Cretaceous, and Eocene rocks. Because pterobranchs are rare,

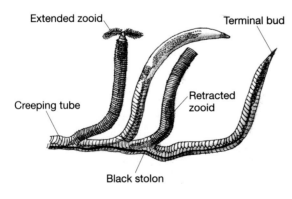

Figure 12.3   Part of a colony of *Rhabdopleura* (×25). (Redrawn from O. M. B. Bulman. In C. Teichert (Ed.), C. Dawiff, Classé des Pterobranches. N: P. Grasse, *Traité de Zoologie*, Vol. 12. 1948.)

paleontologists are more interested in them because of pterobranch affinities to grapto-lites than for their use in biostratigraphy.

CLASS GRAPTOLITHINA

One is not likely to be delighted by the appearance of graptolites. They occur as black, shrubby markings in rock, and they lack the esthetically pleasing form of many other fossil groups. Nevertheless, they are both intriguing and valuable fossils. Intriguing because as hemichordates they are allied to the chordates, and valuable because they are of great value in the biostratigraphic correlation of Ordovician and Silurian rocks where they are abundant, globally distributed, and easy to identify.

The usual host rocks for graptolites are gray to black shales. In such rocks, they are usually preserved as flattened carbonaceous films. The resemblance of these fossils to writing in pencil on rocks is the source of their name: in Greek, *graphos* means writ-ing. As for the dark color of the rock in which graptolites are commonly (but not always) found, this may reflect a low-oxygen environment that favored their preserva-tion by repulsing scavengers and retarding oxidation.

MORPHOLOGY

Graptolites were colonial animals. Sessile species tend to form bushy colonies, whereas the colonies of planktonc graptolites are usually composed of threadlike branches that radiate from a central point of origin. Graptolite nomenclature is relatively simple. The entire colony is the **rhabdosome**. Each branch of the rhabdosome is a **stipe**, and along each stipe are the individual housings or **thecae** (singular, **theca**) in which resided the living **zooids**. Growth of the colony begins with the development of a 1-to-3-mm-long conical tube called the **sicula** (Fig. 12.4). In planktonic forms, the sicula is directed downward like an inverted ice cream cone. Extending from its pointed end is a thin hollow thread called the **nema**. In some forms, the rhabdosome is suspended from the nema. If the nema is located within the wall of the colony, it is referred to as the **virgula**.

In sessile graptolites, the sicula is inverted, and its pointed base is expanded to form a holdfast. The sicula then represents the first-formed chamber of the graptolite colony. It is thought to have been formed by sexual reproduction. All subsequent mem-bers of the colony developed by budding, and were therefore asexually produced. Close examination of the sicula reveals it has two components. At the pointed end is the **prosicula** (see Fig. 12.4). It has a distinctive appearance that includes a spiral thread and longitudinal rodlike structures. The open or apertural end of the sicula is termed the **metasicula**. In general, the wall structure of the metasicula resembles that of the subsequently formed thecae.

The thecae of the rhabdosome are connected to one another by the threadlike **stolon**. As in pterobranchs, individual zooids developed at the ends of the branches of the stolon and secreted tube or cup-shaped thecae. The thecae formed are the basis for two major orders of graptolites, the Dendroidea and Graptoloidea. Colonies of the more primitive sessile dendroid graptolites have considerable diversity in the forms of thecae. The larger **autothecae** probably housed female zooids. Zooids that functioned

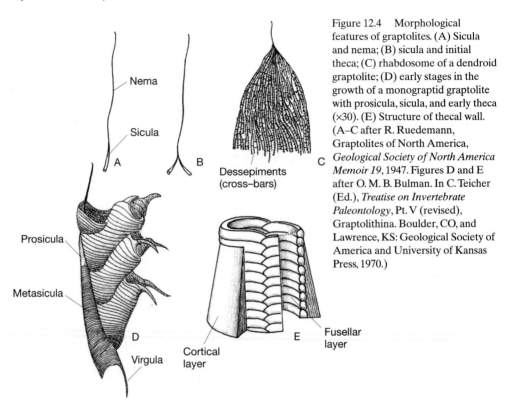

Figure 12.4   Morphological features of graptolites. (A) Sicula and nema; (B) sicula and initial theca; (C) rhabdosome of a dendroid graptolite; (D) early stages in the growth of a monograptid graptolite with prosicula, sicula, and early theca (×30). (E) Structure of thecal wall. (A–C after R. Ruedemann, Graptolites of North America, *Geological Society of North America Memoir 19*, 1947. Figures D and E after O. M. B. Bulman. In C. Teicher (Ed.), *Treatise on Invertebrate Paleontology*, Pt. V (revised), Graptolithina. Boulder, CO, and Lawrence, KS: Geological Society of America and University of Kansas Press, 1970.)

as males occupied the more slender **bithecae**. The **stolothecae**, a third type, were not thecae at all, but parts of autothecae or bithecae that enclosed the stolon. In life, the stolothecae may have contained immature zooids that subsequently developed into autothecal zooids.

In the planktonic graptoloid graptolites, only one type of theca is present: the autotheca. Cup-shaped and arranged in series along stipes, it has been suggested that these graptoloid autothecae housed hermaphroditic zooids.

STRUCTURE OF THE SKELETON

Because most graptolites are found as flattened carbonizations, one does well to discern thecal shape, or recognize a sicula or nema. To learn about the finer characterizations of the exoskeleton, it is necessary to extract uncompressed graptolites from limestone or chert by dissolving the rock in solutions that do not corrode the fossils themselves. Once free of their rock matrix, the less distorted remains can be mounted in a transparent medium for examination with the light microscope, or prepared for examination with the electron microscope.

Studying uncompressed graptolites prepared in the above manner permits one to ascertain the structure of the graptolite periderm. The **periderm** is the layer that forms the graptolite exoskeleton. Although once thought to be composed of chitin, recent chemical analyses have shown that the periderm is made of a scleroprotein similar to that found in the skeletons of pterobranchs. Transverse sections of the graptolite stipes reveal that the periderm is constructed of two layers (see Fig. 12.4E). The inner or **fusellar layer** consists of stacked half-rings or **fuselli** that interlock uniformly along two longitudinal zigzag lines. The pattern is virtually identical to that found in the zooidal tubes of pterobranchs. The outer or **cortical layer** is composed of several thin, transparent lamellae that surround the fusellar tissue. In dendroid graptolites, the cortical tissue is extended from the rhabdosome to form a holdfast. The more interesting feature of the cortical layer, however, is the montage or matwork of thin, flat, bandagelike elements that adorn its surface. Successive layers of these **bandages** give the cortical tissue its laminated appearance. With the high magnifications made possible by the electron microscope, one can see the bandages are composed of long fibers called **fibrils**. Both strength and flexibility are afforded the graptolite skeleton by the fibrilar bandages.

ORDERS OF GRAPTOLITHINA

The Class Graptolithina has been divided into as many as six orders. Only two of these, the Dendroidea and Graptoloidea, however, are abundant and useful in biostratigraphy. The less important orders are the Tuboidea, Stolonoidea, Camaroidea, and Crustoidea.

ORDER DENDROIDEA

The dendroids are at once the most primitive and the most complex of graptolites. They grow in small, shrubby, or fan-shaped colonies that range from 2 to 8 cm in length (Fig. 12.4C). Linnaeus had seen these fossils, and he believed they were merely inorganic encrustations, as indeed they often appear to be. Other early workers assumed they were the remains of small mossy plants. By the turn of the century, however, most paleontologists had recognized that dendroid graptolites were true colonial animals.

The rhabdosome of a dendroid graptolite colony is typically attached to the substrate by the apex of the sicula, which may be strengthened with cortical tissue to form a thin stalk and holdfast. Less commonly, the rhabdosome is suspended by the nema. The dendroid rhabdosome has many stipes. Hundreds of tiny thecae may be present on each stipe. Also, stipes may be connected laterally by branches called **dissepiments**. Autothecae, bithecae, and stolothecae are all prominent members of dendroid rhabdosomes. These tend to occur in regular sequence along the length of a stipe. The sequence consists of repeated triads of autothecae, bithecae, and stolothecae, with the autothecae always at the center and the bithecae alternately to the right and left (Fig. 12.5). The sequence begins with an initial stolotheca that buds from the side of the sicula and in turn buds to produce the larger autotheca and smaller bitheca. The stolotheca then buds again to sequentially produce autotheca-bitheca pairs along the length of the stipe. To form branches, the stolotheca divides and budding continues on each branch in the same way. The budding sequence was first recognized by the Swedish paleontologist Carl Wiman, and is known as Wiman's rule. A thin tube, the **stolon**, extends through the stipe and branches into each successive bud. Dendroid autothecae are characteristically inflated or upwardly expanded, and may have oddly sculpted apertural openings. They may also bear an apertural spine. Bithecae are

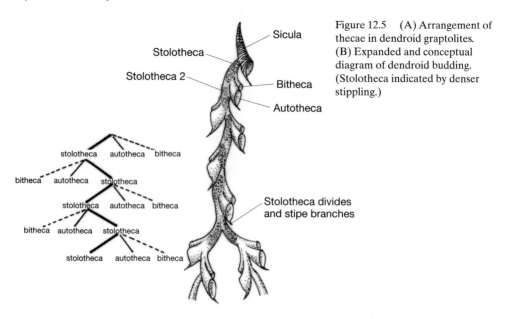

Figure 12.5   (A) Arrangement of thecae in dendroid graptolites. (B) Expanded and conceptual diagram of dendroid budding. (Stolotheca indicated by denser stippling.)

shorter and narrower than autothecae, and are therefore not as conspicuous. The dendroid stolon system is well developed, but is rarely seen except in specimens having excellent preservation.

Dendroid graptolites are an ancient group that appear in the Middle Cambrian. They were the ancestors of the graptoloids. Because of their relatively conservative evolution, they have not been as useful as the graptoloids in stratigraphic studies.

ORDER GRAPTOLOIDEA

Graptoloids are a primarily planktonic group having bilaterally symmetrical rhabdosomes and distinct thecal patterns. Typically, they lived attached to floating objects by their threadlike nema. Some, however, were free-floating. In general, graptoloids have fewer stipes as compared with dendroids, and only autothecae are present. These, however, may change progressively in size or shape along the stipe. No stolon system has ever been detected in graptoloids, and it is therefore assumed that the stolon was composed of soft tissue too delicate for preservation. Also, the graptoloid skeletal wall is thinner than in dendroids.

The first formed element of the graptoloid colony is the sicula, which is oriented with its aperture facing downward. At the conical upper end of the sicula is the nema. A notch, called the **primary notch**, provides the site for growth of the first theca. Subsequently, additional thecae are added progressively in one of five major patterns (Fig. 12.6). The **pendent** arrangement has downward directed stipes. In the **horizontal pattern**, the stipes extend laterally from the sicula. **Reclined** stipes grow obliquely upward on either side of the nema, whereas in the final **declined pattern**, the stipes grow upward. In the **scandent pattern**, stipes grow upward and in contact with the nema.

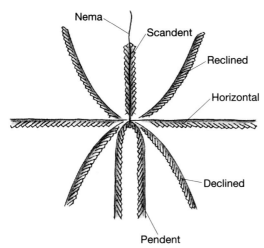

Figure 12.6   Diagram illustrating different patterns of stipe growth in graptoloids. (Adapted from O. M. B. Bulman. In C. Teichert (Ed.), *Treatise on Invertebrate Paleontology*, Pt. V (revised), Graptolithina. Boulder, CO, and Lawrence, KS: Geological Society of America and Univeristy of Kansas Press, 1970.)

The various families of graptoloids exhibit considerable variety in rhabdosome form. Some are composed of many stipes and some only a few. An important evolutionary trend among graptoloids is a reduction in the number of stipes with the passage of geologic time. Early Ordovician graptoloids, for example, have up to eight stipes, Middle and Late Ordovician genera often have two, and many Silurian species have only a single stipe. Differences also occur in the shapes of the autothecae. Some are simple, tubular, and closely spaced in a linear or *en echelon* series. Others are twisted so that apertural openings tilt upward, downward, or to the side. In some graptoloids, the thecae are in contact with one another, whereas others have uniform spacing between thecae. The apertures of thecae can be simple, or variously modified with spines, collars, hoods, or flanges. *Glossograptus*, an Early to Middle Ordovician graptoloid, is distinctive because of its long lateral and apertural spines (see Fig. 12.8A).

Clues to the phylogeny of graptoloids can be seen in the manner in which thecae are added during growth of the rhabdosome. Three of the more commonly seen patterns of thecal additions are called *dichograptid, diplograptid*, and *monograptid* (Fig. 12.7). In **dichograptid** development, the initial theca emerges from the side of the sicula and in turn produces two more thecae. One of these crosses over the sicula, and

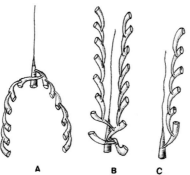

Figure 12.7   (A) Dichograptid, (B) diplograptid, and (C) monograptid budding.

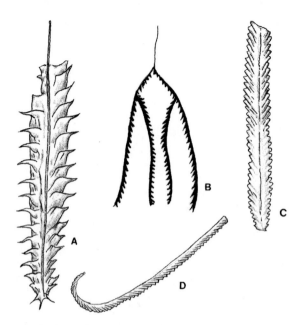

Figure 12.8 Representative graptoloids.(A) *Glossograptus* (Ordovician), ×5. (B) *Tetragraptus* (Ordovician), ×1. (C) *Diplograptus* (Ordovician), ×2. (D) *Monograptus* (Silurian), ×1.

each then initiates budding that produces two stipes that extend in opposite directions. These stipes then branch dichotomously. An example of dichograptid development is seen in *Tetragraptus* (Fig. 12.8B) from Early Ordovician strata. Some dichograptids have stipes that radiate from a central body surrounded by capsules containing siculae. The rhabdosomes of such graptoloids are termed **synrhabdosomes** (Fig. 12.9). The central body in synrhabdosomes may have been a flotational device, and the siculae contained in the capsules may have been in temporary storage before being released to initiate the growth of new rhabdosomes.

In **diplograptid** development, the initial theca begins to grow downward from the sicula, but then curves back upward rather like a Sherlock Holme's pipe. The next few

Figure 12.9 Synrhabdosome of *Orthograptus*. (After O. M. B. Bulman. In C. Teichert (Ed.), *Treatise on Invertebrate Paleontology*, Pt. V (revised), Graptolithina. Boulder, CO, and Lawrence, KS: Geological Society of America and University of Kansas Press, 1970.)

thecae are added on alternate sides, and then serially so that the stipes are extended upward parallel to, or in contact with, the nema. When a stipe actually includes the nema, it is then called the **virgula**. An example is *Diplograptus* (Fig. 12.8C), which ranges from Middle Ordovician to Early Silurian.

Monograptid development is the most advanced and the simplest. The intial theca springs from the sicula with an immediate, obliquely upward direction of growth (Fig. 12.7C). The single stipe is then formed as subsequent thecae are added in series, backed by the virgula. The stipe may extend in a straight line, a curve, or it may coil. *Monograptus* (Fig. 12.8D), which ranges from the Silurian to the Early Devonian, and *Rastrites* (Early Silurian) are examples of graptoloids with monograptid developmental patterns.

MINOR ORDERS OF GRAPTOLITES

Tuboidea   Like the dendroids, members of the Tuboidea have autothecae, bithecae, and stolothecae. They are sessile benthic graptolites that sometimes construct erect rhabdosomes, but they are more commonly encrusting. In tuboids with the encrusting habit, thecae arise from a basal discoidal layer that once adhered to the substrate. This order of graptolites appeared in Late Cambrian and survived until the Silurian, but with a poor fossil record. *Idiotubus* (Fig. 12.10A) from the Early Ordovician is an example.

Stolonoidea   Although known from Lower Cambrian and Ordovician strata, the Stolonoidea are rare and often poorly preserved. They were benthic, encrusting graptoloids with irregularly branching stolons. Also, the stolons exhibit an irregular fusellar structure that disappears in the more expanded parts of the tubes. Some of the

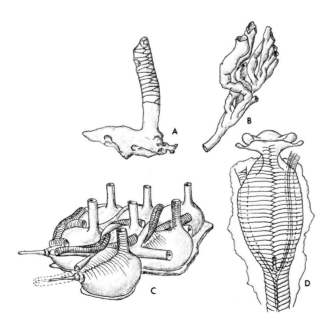

Figure 12.10   Representative of some minor graptolite groups. (A) *Idiotubus* (Ordovician), ×24. (B) *Stolonodendrum* (Ordovician), ×20. (C) Restoration of *Bithecocamara* (Ordovician), ×50. (D) Restoration of *Bulmanicrusta* (Ordovician), ×40. (After O. M. B. Bulman. In C. Teichert (Ed.), *Treatise on Invertebrate Paleontology*, Pt. V (revised), Graptolithina. Boulder, CO, and Lawrence, KS: Geological Society of America and University of Kansas Press, 1975.

open terminations of the stolothecae appear to be autothecae. *Stolonodendrum* (Fig. 12.10B) from the Ordovician of eastern Europe serves as an example of this group.

Camaroidea   Camaroids have an encrusting habit. Autothecae, bithecae, and black stolon systems are present. The autothecae are distinctive in that a slender tubular portion arises from a peculiarly inflated basal portion. Bithecae are narrow tubes that have a random distribution across the colony (Fig. 12.10C). The Early Ordovician genus *Bithecocamera* is representative.

Crustoidea   As implied by their name, crustoids are encrusting graptolites characterized by distinctive, intricately infolded autothecal apertures. These apertures arise from short apertural necks. Autothecae, simple tubular bithecae, and sinuous stolothecae are arranged in triads. Although fossils are rare and poorly preserved, they are adequate to indicate that the crustoids were structurally more similar to living rhabdopleurid pterobranchs than any members of the Graptolithina. An example is *Bulmanicrusta* (Fig. 12.10D) from the Early Ordovician of Europe.

GRAPTOLITES THROUGH TIME

The fossil record for graptolites begins in the Middle Cambrian with the appearance of the many-branched, fan-shaped dendroids (Fig. 12.11). These early sessile forms lived attached to objects on the seafloor, but soon gave rise to dendroids, which assumed a secondarily planktonic habit by living attached to seaweed and other floating debris. Near the end of the Early Ordovician, the floating dendroids gave rise to the truly planktonic graptoloids. The transitional forms were members of the dendroid family Anisograptidae.

As noted earlier, except for a few primitive forms, graptoloids differ from their dendroid ancestors in having monomorphic thecae. Early graptoloids had four or more branches, but in subsequent evolution the number of branches was reduced to two. Changes also occurred in the way thecae were added relative to the sicula. The result was the appearance of pendent, horizontal, reclined, and eventually scandent growth patterns. By the Late Ordovician, reclined, two-stipe forms like *Dicellograptus* had appeared, along with graptoloids having biserial thecae, as seen in *Climacograptus*. Somewhat later, near the beginning of the Silurian, uniserial forms evolved by loss of thecae on one side. Often called monograptids, these uniserial graptoloids typified faunas of the Silurian and Early Devonian.

The dendroids, which populated seas from Cambrian to Pennsylvanian time, were the earliest graptolites to appear and the last to become extinct. After the Silurian, however, fossils of dendroids become uncommon. Graptoloids became prevalent from the Early Ordovician to the Early Devonian. Dendroids were also present, and these bushy forms persisted into the Upper Carboniferous. The demise of many graptolite families during the Late Paleozoic may have been related to the expansion of plankton-feeding fishes.

A revision of the geologic range of graptolites that would extend their range to the present now seems likely. In samples taken from deep water off New Caledonia (about halfway between Brisbane and Fiji), marine biologist Noel Dilly found an organism with a remarkable resemblance to a graptolite. Dilly named the animal *Cephalodiscus graptolitoides*, and in a 1993 report stated: "The animal is probably classifiable as a graptolite and thus a living fossil representing a group thought to be

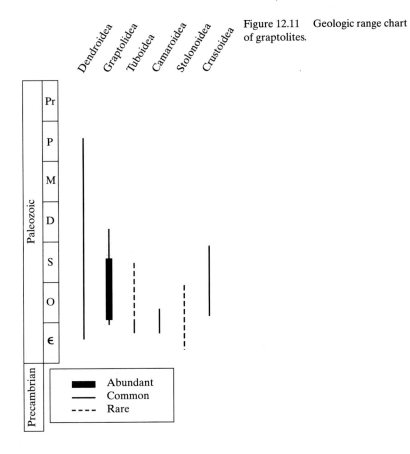

Figure 12.11   Geologic range chart of graptolites.

extinct for 300 million years." Further study and analysis of the criteria used to taxonomically differentiate between graptolites and pterobranchs will determine if *Cephalodiscus graptolitoides* is truly a living genus of graptolites, a pterobranch that closely resembles a graptolite, or whether graptolites are merely an extinct group of pterobranchs. In any case, having the good fortune to be able to observe a living counterpart to extinct graptolites has provided answers to many questions about how the organism constructed its skeletal components.

Since the original discovery of *Cephalodiscus graptolitoides*, Noel Dilly has found additional specimens while vacationing in Bermuda. Thus, these "living fossils" appear to have global distribution.

GRAPTOLITE HABITS AND HABITATS

One generalization that can be made about graptolite habits is that most dendroids were benthic and most graptoloids were planktonic. Minor graptolite orders such as the tuboids, camaroids, and crustoids included encrusting members. These were primarily inhabitants of shallow, nearshore areas. Many of the shrubby dendroids probably also populated the seafloors of shelf areas. Recognition of the habitats of graptolites led to their early use in making paleogeographic interpretations. As early as

1897, Charles Lapworth used graptolite occurrences to infer that British graptolite-bearing rock sequences were deposited in embayments that were open on the west to an influx of plankton from the Atlantic (actually the Proto-Atlantic) Ocean. (Lapworth also employed graptolites to establish the Ordovician System, thus ending the famous dispute over placement of the Upper Cambrian and Lower Silurian in England.) In the 1930s, paleontologists noted a general similarity of organisms found in graptolitic shales to those living today in the Sargassum Meadows of the central Atlantic. The contribution of carbon from phytoplankton in such environments may account for the frequency with which graptolites occur in dark shales.

Graptoloids exhibit many features that confirm their status as zooplankton. As is common in zooplankton, they are widely distributed. The exoskeleton is light yet pliable and sturdy enough to withstand turbulence. Some graptoloids developed disks and vanelike structures to assist in flotation. It is even likely that some contained pockets of gas to improve buoyancy. For the same reason, others evolved loosely woven meshlike structures, spines, or were joined to other rhabdosomes in a radial pattern that resisted sinking. Spiral forms like *Monograptus* may have been able to spiral upward under the influence of currents. In general, such spiral motion would also have enhanced feeding. Finally, if the zooids of graptolites functioned somewhat like their living pterobranch counterparts, they may have used the action of cirri (filaments) on tentacles to lift the colony in the water column.

Studies of the distribution of graptoloids indicate an increase in diversity above the seaward margins of shelves and shelf slopes. This also occurs in many other forms of planktonic organisms, for in such areas upwelling currents bring nutrients to the surface, fostering the growth of a variety of plants and animals throughout the marine food chain.

REVIEW QUESTIONS

1. What features possessed by hemichordates indicate their close evolutionary relationship to chordates?
2. What features of graptolites suggest they are closely related to living pterobranchs?
3. Describe the wall structure of a graptolite theca.
4. Describe the stipes in scandent, reclined, and pendent rhabdosomes.
5. Describe the budding sequence in dendroid graptolites.
6. What were synrhabdosomes? What was the function of the central bodies in these graptolites?
7. What generalizations can be made about the relationship between number of stipes and geologic age of graptoloid graptolites?

SUPPLEMENTAL READINGS AND REFERENCES

Berry, W. B. N. 1977. Graptolite biostratigraphy. In E. Kauffman & J. E. Hazel, (Eds.), *Concepts and Methods of Biostratigraphy.* Stroudsburg, PA: Dowden Hutchinson, and Rice.

Dilly, P. N. 1993. *Cephalodiscus graptolitoides* sp. nov., a probable extant graptolite. *Journal of Zoology*, 229:69–78.

Kozlowski, R. 1966. On the structure and relationships of graptolites. *Journal of Paleontology,* 40 (3):489–501.

Reudemann, R. 1947. Graptolites of North America. *Geological Society of America* Memoir 19, Boulder, CO: Geological Society of America.

Rigby, S. 1913. Graptolites come to life. *Nature,* 362:209–210.

Teichert, C. (Ed.). 1970. *Treatise on Invertebrate Paleontology,* Pt. V, Graptolithina (revised). Lawrence, KS, and Boulder, CO: University of Kansas Press and Geological Society of America.

Urbanek, A. 1986. The enigma of graptolite ancestry: Lesson from a phylogenetic debate. In A. Hoffman & M. H. Nitecki, (Eds.), *Problematic Fossil Taxa.* New York: Oxford University Press.

CHAPTER 13

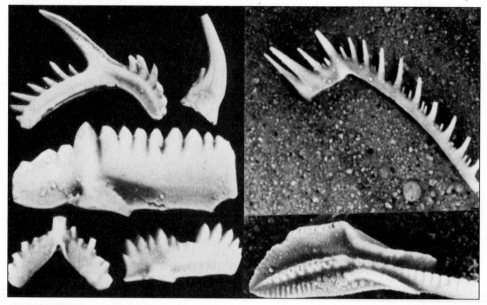

Conodont elements, x20. (courtesy of Karl M. Chauff)

Conodonts

Nearly all recent students of conodonts have classed them as teeth of primitive fishes, and we concur in that classification.

E. B. Branson and M. G. Mehl, 1933

The evidence of soft-part anatomy, together with features of element histology, show that conodonts are vertebrates.

R. J. Aldridge, D.E.G. Briggs, M. P. Smith,
E.N.K. Clarkson, & D. L. Clark, 1993

Conodonts are toothlike or jawlike structures composed of a tiny crystallites of durable fluorapatite. They occur worldwide in marine rocks of Paleozoic and Triassic age. Conodonts were discovered in 1856 by Christian Heinrich Pander. In the belief that he had found microscopic fish teeth, Pander called his fossils *Conodonten* (conodonts in English), from the Greek meaning "cone teeth". Whereas many paleontologists supported Pander's contention that conodonts were components of some sort of primitive vertebrate, others disagreed. The result has been nearly continuous debate over the probable function and zoological affinities of these tiny, intriguing fossils. Their importance in biostratigraphic zonation, however, was not at all diminished by the controversy. Recent discoveries based on fossils of the living animal appear to favor Pander's original contention that the conodont-bearing animal was a vertebrate. Nevertheless, conodonts are included in this volume of invertebrate fossils because they are present among marine invertebrate fossil assemblages worldwide, and they are exceptionally useful in biostratigraphy.

CONODONT ELEMENT MORPHOLOGY AND ULTRASTRUCTURE

The individual conodont fossil is usually less than 1 mm in size, and is referred to as a conodont **element**. Because conodont elements are resistant to chemical attack, dilute organic acids can often be used to free them from their rock matrices. Conodont elements obtained in this way, however, are unavoidably dissociated from natural groupings they may have had in the rock. To determine which elements occur naturally together (that is, emplaced in the rock at the time of death of the parent organism and not dispersed) it is necessary to find and examine conodonts in place along the bedding surfaces of strata. When, with good fortune and diligent search, one finds such specimens, it can be seen that particular conodont elements occur in groupings that are presumed to be the natural assemblage as it existed in the original conodont organism. Such natural groupings are termed **associations**.

The three basic forms of conodont elements are coniform, ramiform, and pectiniform. Each one of these includes numerous subvarietal forms. **Coniform** elements (Fig. 13.1A) usually take the form of a curved cone or **cusp** (the upper or oral end), and an expanded **base**, which bears an aboral attachment cavity. The large cusps may be followed by one or more smaller cusps. Coniform elements are found in rocks ranging in age from Late Cambrian to Devonian.

Ramiform elements (Fig. 13.1B) characteristically have a main cusp as well as smaller additional cusps or denticles along their entire length. They often have a barlike or bladelike form, and in some there are laterally expanded processes. Ramiform elements range from the Ordovician to the Triassic. The bladelike ramiforms are widely used in the correlation of strata ranging in age from Silurian to Triassic.

Pectiniform conodont elements probably evolved from bar or blade types by developing lateral or shelflike platforms. Nodes, pits, and ridges ornament the surface of the platform. At the posterior end, a denticulate plate or **carina** rises perpendicularly to the surface of the platform (Fig. 13.1C). A basal attachment cavity may or may not be present on the lower surface of the element. Platform conodonts range from the Ordovician to the Triassic. Because of their rapid evolution, they produced numerous

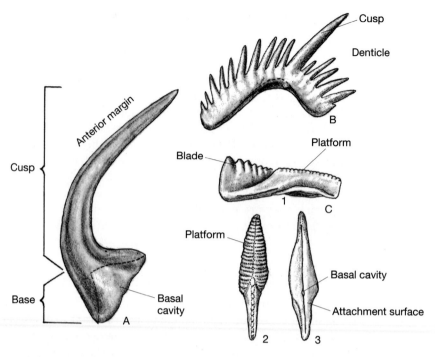

Figure 13.1   Coniform, ramiform, and pectiniform conodonts. (A) Coniforms, as seen in *Drepanodus*, ×45; (B) ramiform, as seen in *Prioniodina*, ×30; (C) pectiniform, as seen in *Taphrognatuus*, ×35. Lateral, oral, and aboral views are shown in 1–3. (After D. L. Clark et al. In R. A. Robison (Ed.), *Treatise on Invertebrate Paleontology*, Pt. W, Miscellanea. New York and Lawrence, KS: Geological Society of America and University of Kansas Press, 1981.)

species having short geologic ranges, and these are particularly valuable for biostratigraphic zonation.

Conodont elements are constructed of an organic matrix in which crystalites of carbonate fluorapatite have been secreted as thin parallel **lamellae**. The lamellae are added successively around an initial growth center. In many conodont elements the basal plate has a somewhat higher content of organic matter, and this may indicate the base on which the element was attached to soft tissue. A nonlamellar, nearly opaque, light-colored material called **white matter** is also seen in thin sections of conodonts. Histologic studies of white matter in Carboniferous conodont elements from Scotland indicate it is composed of cellular bone. Enamel homologues and probable calcified cartilage have also been reported in these conodonts.

CONODONT ASSEMBLAGES

Prior to about 1960, the taxonomy of conodonts was necessarily an artificial system in which each discrete element was treated as a separate taxonomic entity. Artificial "form genera" were used in naming the elements. This system of naming conodonts

Figure 13.2   The feeding apparatus of the conodont *Idiognathus* (×25). (Sketched from a photograph of a scale model prepared by Mark A. Purnell of the University of Leicester.)

was adequate for many biostratigraphic studies. It was not, however, biologically valid. The discovery of complete conodont apparatuses (Fig. 13.2) composed of natural groupings of elements indicated that quite different form genera were the components of a single conodont animal. For this reason, it has become the aim of conodont specialists to name genera and species on the basis of the observed or inferred conodont apparatus. When the complete apparatus with elements in place cannot be found, the natural assemblage of elements can sometimes be reconstructed from the constant ratios of different elements in the samples being examined.

THE MAJOR CONODONT GROUPS

Following a proposal by W. C. Sweet (1988), conodonts can be grouped into two classes, the Cavidonti and Conodonti. The former group is of lesser importance. It includes elements that typically have thin walls, smooth surfaces, and coniform shape. Cavidonts range from Late Cambrian into the Devonian.

Members of the Class Conodonti are far more abundant and biostratigraphically important than the cavidonts. About 140 biostratigraphic zones have been established on the basis of Conodonti. The class originated in the Late Cambrian and persisted into the Late Triassic. Individual elements of Conodonti are generally striated and have relatively short bases. Coniform, ramiform, and pectiniform elements all occur in this large class of conodonts. Among the more familiar forms are *Oistodus* (Fig. 13.3C), *Prioniodina* (Fig. 13.3D), *Ozarkodina* (Fig. 13.3E), *Hibbardella* (Fig. 13.3F), *Hindeodella* (Fig. 13.3G), *Palmatolepis* (Fig. 13.3H), *Idiognathodus* (Fig. 13.3I), and *Polygnathus* (Fig. 13.3J). Of the above, *Oistodus* occurs in Early Ordovidian sections. *Prioniodina*, *Hibbardella*, *Ozarkodina*, and *Hindeodella* range from the Middle Ordovician to the Middle Triassic. Species of *Idiognathodus* serve as Early to Late Carboniferous guide fossils.

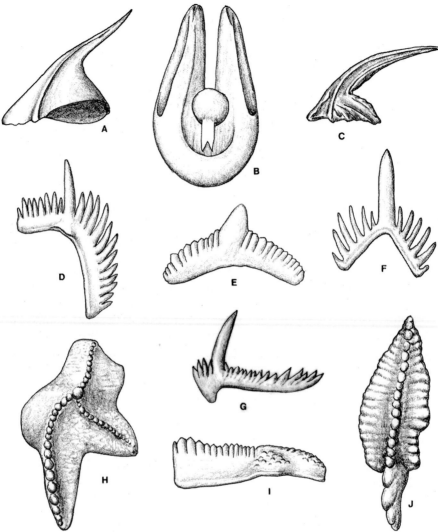

Figure 13.3   Representative conodont elements, about ×35. (A) *Furnishina*; (B) *Westgaardodina*;
(C) *Oistodus*; (D) *Prioniodina*; (E) *Ozarkodina*; (F) *Hibbardella*; (G) *Hindeodella*;
(H) *Palmatolepis*; (I) *Idiognathodus*; (J) *Polygnathus*. (After D. L. Clark et al. In R. A. Robison
(Ed.), *Treatise on Invertebrate Paleontology*, Pt. W, Miscellanea. New York and Lawrence, KS:
Geological Society of America and University of Kansas Press, 1981.)

THE CONODONT ANIMAL

It has been nearly a century and a half since conodont elements were first observed,
yet questions remain concerning their function and origin. At one time or another, it
has been proposed that conodonts belonged to such diverse groups as annelids,
arrow worms (Chaetognatha), fishes, snails, cephalopods, and arthropods. Recently,
however, the choices were narrowed by the discovery of fossils of the conodont ani-

3 mm

Figure 13.4   Restoration of the conodont animal as interpreted from fossil remains found in the Lower Carboniferous Granton shrimp beds of Scotland. (After R. J. Aldridge, E. E. J. Briggs, M. P. Smith, E. N. K. Clarkson, & N. D. L. Clark, *Phil. Trans. R. Soc. London* B, 340:405–421.)

mal. The most significant of these extraordinary fossils were obtained from Lower Carboniferous strata exposed near Edinburgh, Scotland. The fossils occur in rocks called the Granton shrimp beds because of the eumalcostracan shrimp remains contained there.

Examination of fossils of the conodont animal reveal an elongate creature with a short head at one end and a ray-supported tail at the other (Fig. 13.4). Specimens range in length from about 15 to 55 mm. Two lobate structures on the head were the probable location of large movable eyes. Extending dorsolaterally along both sides of the laterally compressed trunk and tail are serially arranged **V**-shaped structures representing blocks of muscle fibers. The structures, termed **myomeres**, are closely similar to those seen in the living marine lancelet *Branchiostoma* (a marine cephalochordate). In life, sequential, wavelike contractions of the myomeres provided the animal with forward propulsion. Some of the Granton specimens also exhibit two closely spaced parallel lines that are thought to represent the margins of a longitudinal supportive structure called a **notochord**. Like the myomeres, the presence of a notochord links the conodonts to the chordates. A notochord in the conodont animal would prevent the body from shortening during contraction of the myomeres, and thereby help to provide the undulatory motion of the body needed in swimming. In fossils of the conodont animal, the notochord tapers to a point at both ends and terminates anteriorly near the feeding apparatus, which is composed of conodont elements *in situ*. In most of the Granton specimens, the feeding apparatus consists of comblike ramiform elements that form a sort of basket suitable for capturing prey. Behind the ramiform elements are pectiniform elements that may have functioned in processing food. A dark, obscure axial line can be discerned above the putative notochord in two of the Granton specimens. Perhaps additional discoveries will validate the suggestion that the dark line represents the trace of the animal's dorsal nerve cord.

The presence of myomeres (a feature present in cephalochordates and vertebrates), a notochord (shared by tunicates, cephalochordates, and vertebrates), large eyes with extrinsic eye muscles (revealed in the remains of a large Ordovician conodont animal *Promissum pulchrum* from South Africa), a possible dorsal nerve cord, and probable cellular bone (indicated in histological studies of white matter) may be taken as evidence supporting the hypothesis of vertebrate affinity for the conodonts. As-yet-undiscovered specimens are certain to reveal more about this group, once considered the most enigmatic of common fossils.

CONODONT HABITS AND HABITATS

Inferring the habits of a group of fossils known primarily from hard parts that are often the scattered components of an undiscovered animal can be a rather risky business. We do know that many conodont species achieved wide geographic distribution, and some may have been truly cosmopolitan. The association of conodonts with marine invertebrates such as corals, bryozoa, brachiopods, and cephalopods indicates they were ocean dwellers. As is the case with many fossil groups, they are generally more abundant in marine sediments of lower paleolatitudes. That some species were tolerant of brackish water is indicated by the fossils present in the Granton shrimp bed. Fossils from that site also indicate that the conodont animal was a swimmer, and therefore unlikely to be affected by conditions on the ocean floor. Support for this interpretation is seen in the common occurrence of conodont elements in black shales. The anoxic environment in which black shales characteristically accumulate is lethal to benthic creatures. In most instances, the fossils found in such rocks are of planktonic or nektonic organisms that apparently were killed by hydrogen sulfide-laden anoxic waters that presumably were present above the areas where black clays were accumulating on the seafloor. The host rocks of conodont elements, as well as for the invertebrates associated with the elements, suggest that particular conodont groups inhabited nearshore environments, whereas others preferred deeper offshore realms. There is some evidence that larger, more robust conodont elements were more prevalent in higher energy nearshore environments. *Icriodus, Polygnathus,* and *Cavusgnathus* are examples of such sturdier elements.

In addition, the faunas of the nearshore environments are less diverse than deeper-water faunas, show fewer discrete denticles, and are often more varied in morphology. Condodont elements from offshore areas tend to be more varied and more abundant than those in nearshore environments. In deeper-water sedimentary rocks, conodont elements are often found in association with cephalopods and in Cambrian rocks with fossils of pseudoplanktonic agnostid trilobites. Elements associated with deep-water dark shales and limestones are usually relatively delicate, and the total assemblage of elements less diverse. Palmatolepis and Ancyrodella are examples of conodont elements frequently seen in deeper lithofacies.

With regard to the food-gathering methods of conodonts, detailed electron microscopy examination of conodonts has revealed wear patterns that would suggest the elements were teeth. Surfaces are pitted and chipped in patterns produced in teeth by crushing food, and parallel striations indicate that certain elements were used for shearing. Taken along with their ability to swim and probable good vision, the toothlike wear patterns indicate that many conodonts were active predators.

CONODONTS THROUGH TIME

Conodonts of a questionable nature appear in the latest Proterozoic rocks of Siberia. From that time until the Late Cambrian, however, they are rarely encountered. The sparse remains that have been reported consist of tiny, weakly phosphatized elements like *Furnishina* and *Westergaardodina* (see Fig. 13.3A,B). Conodont populations

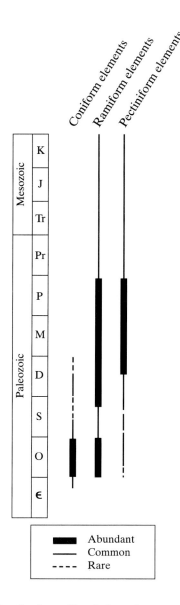

Figure 13.5   Geologic ranges of coniform, ramiform, and pectiniform conodont elements.

Coniform elements
Ramiform elements
Pectiniform elements

Mesozoic

K

J

Tr

Pr

Paleozoic

P

M

D

S

O

Є

▬▬▬   Abundant
———   Common
- - - -   Rare

increased by the Late Cambrian when members of the Conodonti (mostly simple coniform Conodonti) made their appearance (Fig. 13.5). By Early Ordovician, conodont elements having two or more kinds of elements (**multimembrate elements**) became prevalent. Ramiform and coniform elements are the most abundant in these Lower Ordovician rocks. Diversity and abundance increased markedly during the Middle Ordovician. when conodonts reached their evolutionary peak. Over 100 genera are known from the Ordovician.

Success during the Ordovician, however, did not carry over into the Silurian. During that period, diversity decreased to the point where only 11 biostratigraphic zones could be established (in contrast to 22 for the Ordovician). The situation improves in the Devonian when Conodonti experience a resurgence. Multimembrate ramiform with pectiniform elements predominate, not only in the Devonian but also in subsequent geologic periods. Genera such as *Ancyrognathus*, *Palmatolepis*, and *Ancyrodella* are important Devonian guide fossils. During the Mississippian, species of *Siphonodella* and *Gnathodus* became abundant and today are widely used in biostratigraphic correlations.

Following the Mississippian period, conodonts decline, reaching a low point in the Permian. Conodonts present in strata at the Permian–Triassic boundary are generally small forms that either lack or have reduced basal cavities. By the end of the Triassic, conodonts had vanished completely from the fossil record.

COLOR IN CONODONT ELEMENTS

Although casual examination of conodont elements indicates most have a yellow or brown color, more precise observations reveal a considerable variation in this property among conodonts from different localities. As demonstrated by Anita Harris and her colleagues, color variation in conodont elements is often associated with temperature increases resulting from burial and tectonic forces that affected the host rock. Color alteration in response to heating has been demonstrated in the laboratory, where conodont elements are heated to a maximum temperature of 600°C. As the temperature is gradually increased, conodont elements placed in the oven progressively change color. From the experimental observations, it is possible to establish a conodont alteration index, or CAI, that can be assigned to each color beginning with pale yellow (designated CAI-1), through shades of light brown and black (CAI-5), to changes from black to gray, and finally from white to clear (CAI-8). Each change in color is caused by an irreversible alteration of the organic matter within the conodont element.

The conodont color index provides geologists with information about the thermal history of the rock, and, by inference, knowledge of the compressional and deformational history as well. When plotted on maps, the indices reveal patterns of regional metamorphism in older mountain ranges like the Appalachians, Caledonians, and Urals. They provide clues to the locations of former hot spots above which tectonic plates moved. Petroleum geologists use conodont alteration indices to recognize rock sequences that are unlikely to contain petroleum or natural gas because of the high temperatures to which the rocks were once subjected. In the exploration for ores, conodont color can sometimes be used to assess the thermal effects of ore-bearing fluids. Thus, investigations into the color of conodont elements have added a new area of value to fossils long recognized for their importance in biostratigraphy.

REVIEW QUESTIONS

1. Why are conodonts considered to have chordate affinities?
2. What problems are associated with valid taxonomic identification of conodont elements?

3. Why are conodont elements particularly susceptible to the Lazarus Effect discussed in Chapter 2?

4. Describe coniform, ramiform, and pectiniform conodont elements. In which of these groups do the following "form genera" belong:
   a. *Prioniodina*
   b. *Polygnathus*
   c. *Drepanodus*
   d. *Taphrognathus*

5. Is the value of conodont "form genera" as stratigraphic markers affected by uncertainty about their phylogenetic relationships? Explain.

6. What evidence might substantiate the hypothesis that conodont animals were adapted to a wide range of depositional environments?

SUPPLEMEMENTAL READINGS AND REFERENCES

Aldridge, R. J. & Briggs, D. E. G. 1986. Conodonts. In A. Hoffman & M. H. Nitecki (Eds.), *Problematic Fossil Taxa.* New York: Oxford University Press.

Aldridge, R. J., Briggs, D. E. G., Smith, M. P., Clarkson, E. N. K. & Clark, N. D. L. 1993. The anatomy of conodonts. *Phil. Trans. R. Soc. London.* 340:405–421.

Austin, R. L. (Ed.). 1987. *Conodonts: Investigative Techniques and Applications.* Chichester: Ellis Horwood Ltd. for the British Micropaleontological Society.

Brasier, M. D. 1980. *Micropaleontology.* London: George Allen and Unwin.

Clark, D. L. *et al.* 1981. Conodonta. In R. A. Robison (Ed.), *Treatise on Invertebrate Paleontology,* Pt. W, Supplement 2. Boulder, CO, and Lawrence, KS: Geological Society of America and University of Kansas Press.

Dzik, J. 1986. Chordate affinities of the conodonts. In A. Hoffman & M. H. Nitecki (Eds.), *Problematic Fossil Taxa.* New York: Oxford University Press.

Epstein, A. G., Epstein, J. B. & Harris, L. D. 1977. Conodont color alteration—an index to organic metamorphism. *United States Geological Survey Professional Paper,* No. 935.

Gabbott, E. E., Aldridge, R. J. & Theron, J. N. 1995. A giant conodont with preserved muscle tissue from the Upper Ordovician of South Africa. *Nature* 374 (6525):800–803.

Purnell, M. A. 1995. Microwear on conodont elements and macrophagy in the first vertebrates. *Nature* 374 (6525):798–800.

Purnell, M. A., Aldridge, R. J., Donoghue, P. C. J. & Gabbott, S. E. 1995. Conodonts and the first vertebrates. *Endeavor* 19(1):20–27.

Sweet, W. C. 1988. *The Conodonta: Morphology, Taxonomy, and Evolutionary History of a Long-Extinct Animal Phylum.* Oxford: Oxford University Press.

Index

356